MAGI
SHELF No.
TRI
KV-371-229
1688

OXFORD MEDICAL PUBLICATIONS

Psychopharmacology of the
limbic system

300734398.

BRITISH ASSOCIATION
FOR PSYCHOPHARMACOLOGY MONOGRAPHS

1. Psychopharmacology of affective disorders
   edited by E. S. Paykel and A. Coppen

2. Psychopharmacology of anticonvulsants
   edited by Merton Sandler

3. Psychopharmacology of old age
   edited by David Wheatley

4. Psychopharmacology and sexual disorders
   edited by David Wheatley

5. Psychopharmacology of the limbic system
   edited by Michael R. Trimble and E. Zarifian

6. Pyschopharmacology: recent advances and
   future prospects
   edited by Susan D. Iversen

7. Psychopharmacology and food
   edited by Merton Sandler and Trevor Silverstone

8. Psychopharmacology and drug treatment of schizophrenia
   edited by P. B. Bradley and S. R. Hirsch

9. New brain imaging techniques and psychopharmacology
   edited by Michael R. Trimble

Psychopharmacology of the limbic system

BRITISH ASSOCIATION
FOR PSYCHOPHARMACOLOGY
MONOGRAPH
No. 5

EDITED BY

MICHAEL R. TRIMBLE
Consultant Physician in Psychological Medicine
The National Hospitals for Nervous Diseases, London
and Senior Lecturer in Behavioural Neurology, Institute of Neurology, London

AND

E. ZARIFIAN
Professor of Psychiatry, Hospital St. Jacques,
Clermont Ferrand, France

MAGDALEN COLLEGE LIBRARY

Oxford New York Toronto
OXFORD UNIVERSITY PRESS
1985

Oxford University Press, Walton Street, Oxford OX2 6DP

Oxford New York Toronto
Delhi Bombay Calcutta Madras Karachi
Kuala Lumpur Singapore Hong Kong Tokyo
Nairobi Dar es Salaam Cape Town
Melbourne Auckland

and associated companies in
Beirut Berlin Ibadan Nicosia

Oxford is a trademark of Oxford University Press

© The various contributors listed on pp. ix−xiv, 1984

First published 1984
Reprinted in paperback 1985

All rights reserved. No part of this publication may be reproduced,
stored in a retrieval system, or transmitted, in any form or by any means,
electronic, mechanical, photocopying, recording, or otherwise, without
the prior permission of Oxford University Press

This book is sold subject to the condition that it shall not, by way
of trade or otherwise, be lent, re-sold, hired out or otherwise circulated
without the publisher's prior consent in any form of binding or cover
other than that in which it is published and without a similar condition
including this condition being imposed on the subsequent purchaser

British Library Cataloguing in Publication Data
Psychopharmacology of the limbic system−
(British Association for Psychopharmacology monograph; 5)
1. Limbic system   2. Psychopharmacology
I. Trimble, Michael R.   II. Zarifian, E.
59901'8   QP383.2
ISBN 0-19-261425-8
ISBN 0-19-261575-0 (Pbk.)

Library of Congress Cataloging in Publication Data
Main entry under title:
Psychopharmocology of the limbic system.
(British Association for Psychopharmocology monograph; no. 5)
(Oxford medical publications)
Bibliography: p.
Includes index.
1. Neuropsychopharmacology.   2. Limbic system−Effect of drugs on.
I. Trimble, Michael R.   II. Zarifian, E.   III. Series.
IV. Series: Oxford medical publications.
[DNLM: 1. Limbic system−Drug effects−Congresses.
2. Emotions−Drug effects−Congresses.   3. Behavior−
Drug effects−Congresses.   4. Psychotropic drugs−
Pharmacodynamics−Congresses. W1 BR34D no. 5/WL 314 P974]
RM315.P757 1984   615'.78   84-921
ISBN 0-19-261425-8
ISBN 0-19-261575-0 (Pbk.)

Typeset by Cotswold Typesetting, Ltd., Cheltenham
Printed in Great Britain by St. Edmundsbury Press, Bury St. Edmunds.

Preface

It is fitting that the meeting on the limbic system on which this book is based was held in Paris. As pointed out by several authors in this book, the limbic lobe was first clearly defined by Paul Broca (1824–80). He was surgeon to the Bicêtre and a noted anthropologist. In the late-1870s he published descriptions of '*le grand lobe limbique*', noting in particular the importance of this area from comparative anatomical studies.

Initially, the limbic lobe held little interest for clinicians. Work in animal models suggested it had a mainly olfactory function – hence the term rhinencephalon. However, the rapid advance in neurophysiological and behavioural techniques in the mid-part of this century, and the delineation of the limbic system by Papez, led to a renewed search for the neurological control of emotional behaviour.

The discovery of successful psychopharmacological agents for the treatment of psychiatric illness, such as the major tranquillizers and antidepressants in the 1950s, stimulated a search for their possible modes of action. The postulated neurotransmitter theories which evolved led to a vast amount of neurochemical exploration of the relationship between the brain and behaviour. This received a major boost when specific neurochemical techniques allowed delineation of a number of neurotransmitter pathways, in particular the monoamine systems in the brain. From the viewpoint of neuroanatomy, a different kind of brain began to take shape. In the old days structures were related because they sat next to each other when viewed by the naked eye on the pathologist's bench. White matter separated different nuclei and connections between various parts of the brain at a distance from each other could not clearly be envisaged. With the discovery of neurochemical systems, however, links between various important areas of the brain became clarified but also led to a clearer delineation of the limbic system and its connections.

Although some would argue that such a 'system' does not exist in the brain, it is nevertheless intriguing that the structures initially discussed by Broca now form part of what seems to be a collection of nuclei and tracts that are intimately related to behavioural expression in animals and man and which, in addition, utilize a number of key neurotransmitters during their function. These neurotransmitters, which also have been shown to be involved in behaviour, are the very ones which psychopharmacology has concentrated on in terms of modes of action of psychotherapeutic drugs. Thus in the limbic system there are high concentrations of monoamines, such as noradrenaline, dopamine, and serotonin, and, more recently, evidence has accumulated showing a high concentration of peptides in similar brain structures.

In addition, neurophysiological evidence has accumulated showing that some areas of the limbic system, for example the hippocampus and amygdala, have very special properties which may be intimately related to their possible key role in modulating past with present experiences. Further, stimulation of these areas by experimental techniques such as kindling leads to progression of abnormal electrical activity, preferentially within certain areas of the brain, namely the limbic system itself.

It would thus seem that evidence is growing that the limbic system, which of course forms part of a brain that acts holistically in the regulation of behaviour, may properly be defined as a system and, of particular interest for psycho-pharmacologists, it may be the main area within the brain where many psycho-pharmacological agents act. In view of this, it was considered germane to produce a book on the psychopharmacology of the limbic system, discussing some of the ideas mentioned above, but also producing an up-to-date review of present knowledge about various classes of pyschotropic drugs, and their interaction with the limbic system.

The editors are extremely grateful to all the contributors for their manuscripts and hope that the volume will represent another small landmark in the delineation of the relationship between the brain and behaviour – and the contributions of psychopharmacology to that delineation.

London   M. R. T.
October 1983   E. Z.

Contents

Contributors

A. BARACESE-HAMILTON, B.Sc.,
Research Assistant,
Division of Medicine,
Royal Postgraduate Medical School,
London,
UK.

J. C. BLANCHARD,
Docteur Vétérinaire,
Rhône-Poulenc Santé,
Centre de Recherches de Vitry,
Vitry sur Seine,
France.

S. R. BLOOM, MA, D.Sc., MD, FRCP,
Professor Endocrinology,
Division of Medicine,
Royal Postgraduate Medical School,
London,
UK.

T. J. CROW, Ph.D., FRCP, FRCPsych.,
Division of Psychiatry,
Clinical Research Centre,
Northwick Park Hospital,
Harrow,
UK.

ROBERT DANTZER, DVM, Dr.ès.Sci.,
Neurobiologie des Comportements INRA,
Bordeaux,
France.

I. N. FERRIER, B.Sc., MRCP, MRCPsych.,
Wellcome Clinical Fellow and Honorary Senior Registrar in Psychiatry,
Divisional of Psychiatry,
Clinical Research Centre,
Northwick Park Hospital,
Harrow, and
Divisions of Medicine and Histopathology,
Royal Postgraduate Medical School,
London,
UK.

C. GARRET,
Licencié en sciences,
Rhône-Poulenc Santé,
Centre de Recherches de Vitry,
Vitry sur Seine,
France.

MICHEL HAMON, Ph.D.,
INSERM U 114,
Collège de France,
Paris,
France.

J. HERBERT, MB, Ch.B., Ph.D.,
Department of Anatomy,
University of Cambridge,
UK.

SUSAN D. IVERSEN, MA, Ph.D.,
Neuroscience Research Centre,
Merck Sharp and Dohme Ltd.,
Hertford Road, Hoddesdon,
Herts EN11 3BU.

MAURICE JALFRE, MD,
Centre de Recherche Delalande,
Rueil-Malmaison,
France.

E. C. JOHNSTONE, MRCP, MRCPsych., MD,
Clinical Scientist and Consultant Psychiatrist: Division of Psychiatry,
Clinical Research Centre,
Harrow; and
Divisions of Medicine and Histopathology,
Royal Postgraduate Medical School,
London,
UK.

L. JULOU,
Docteur Vétérinaire,
Head of Biology Department,
Rhône-Poulenc Santé,
Centre de Recherche de Vitry,
Vitry sur Seine,
France.

ANN E. KELLEY, Ph.D.,
Laboratoire de Neurobiologie des Comportements,
Université de Bordeaux II,
Bordeaux,
France.

C. J. LEBRUN;
INSERM, U 176,
Bordeaux,
France.

Y. LEE, B.Sc.,
Research Assistant,
Division of Medicine,
Royal Postgraduate Medical School,
London,
UK.

KENNETH G. LLOYD, Ph.D.,
Neuropharmacology Group,
L.E.R.S.-Synthélabo,
Bagneux,
France.

G. McGREGOR, BA, M.Sc.,
Research Officer,
Division of Medicine,
Royal Postgraduate Medical School,
London,
UK.

BRIAN MELDRUM, MB, B.Chir., Ph.D.,
Senior Lecturer,
Department of Neurology,
Institute of Psychiatry,
London,
UK.

ANDRÉ OBLIN, Ph.D.,
Neurochemistry Group,
L.E.R.S.-Synthélabo,
Bagneux,
France.

D. O'SHAUGHNESSY, Ph.D.,
Research Student,
Division of Medicine,
Royal Postgraduate Medical School,
London,
UK.

D. G. C. OWENS, MRCP, MRCPsych.,
Consultant Psychiatrist,
Division of Psychiatry,
Clinical Research Centre,
Northwick Park Hospital,
Harrow; and
Division of Medicine and Histopathology,
Royal Postgraduate Medical School,
London,
UK.

J. M. POLAK, MD, MRC.Path.,
Department of Histochemistry,
Hammersmith Hospital,
London,
UK.

ROGER D. PORSOLT, Ph.D.,
Centre de Recherche Delalande,
Rueil-Malmaison,
France.

ROBERT M. POST, MD,
Acting Chief,
Biological Psychiatry Branch,
National Institute of Mental Health,
Bethesda,
Maryland,
USA.

D. A. POULAIN,
INSERM, U 176,
Bordeaux,
France.

J. N. P. RAWLINS, MA, D.Phil.,
Henry Head Fellow of the Royal Society and Senior Research Fellow,
University College,
Oxford,
UK.

E. H. REYNOLDS, MD, FRCP,
Consultant Neurologist,
Maudsley and Kings College Hospitals,
London,
UK.

G. W. ROBERTS, Ph.D.,
Division of Psychiatry,
Clinical Research Centre,
Northwick Park Hospital,
Harrow,
UK.

BERNARD SCATTON, Ph.D.,
Head,
Neurochemistry Group,
L.E.R.S.-Synthélabo,
Bagneux,
France.

PHILIPPE SOUBRIÉ, Ph.D.,
INSERM,
Collège de France,
Paris,
France.

LOUIS STINUS, Ph.D.,
Laboratoire de Neurobiologie des Comportements,
Université de Bordeaux II,
Bordeaux,
France.

J. M. STUTZMANN,
Docteur en Neurophysiologie,
Rhône-Poulenc Santé,
Centre de Recherche de Vitry,
Vitry sur Seine,
France.

D. T. THEODOSIS,
INSERM, U 176,
Bordeaux,
France.

MARIE-HÉLÈNE THIÉBOT, Ph.D.,
Département de Pharmacologie,
Faculté de Médecine Pitié-Salpétrière,
Paris,
France.

MICHAEL R. TRIMBLE (editor), MRCP, FRC.Psych.,
Consultant Physician in Psychological Medicine and
Senior Lecturer in Behavioural Neurology,
National Hospital,
London,
UK.

THOMAS W. UHDE, MD,
Biological Psychiatry Branch,
National Institute of Mental Health,
Bethesda,
Maryland,
USA.

J. D. VINCENT,
INSERM, U 176,
Bordeaux,
France.

MARIE-THÉRÈSE WILLIGENS,
Neuropharmacology Group,
L.E.R.S.-Synthélabo,
Bagneux,
France.

MARIA WINNOCK, BA,
Laboratoire de Neurobiologie des Comportements,
Université de Bordeaux II,
Bordeaux,
France.

PAUL WORMS, Ph.D.,
Neurobiology Group,
Centre de Recherches Clin-Midy,
Montpellier,
France.

EDOUARD ZARIFIAN (Editor),
Hopital St. Jacques,
BP 69,
63000-Clermont Ferrand,
France.

BRANIMIR ZIVKOVIC, Ph.D.,
Head,
Neuropharmacology Group,
L.E.R.S.-Synthélabo,
Bagneux,
France.

1

Recent advances in the anatomy and chemistry of the limbic system

SUSAN D. IVERSEN

More than a century ago, Broca (1878) introduced the name '*grand lobe limbique*' to denote the somewhat distinct region of cortex nearest the margin (limbus) of the cortical mantle. The term referred to the hippocampus and gyrus fornicatus (gyrus cinguli, retrosplenial cortex, and parahippocampal gyrus). The interconnections between the hippocampus and the gyrus fornicatus and the efferent projection of this system via fornix to hypothalamus impressed Papez (1937) and he proposed this anatomical circuit as a neural substrate of emotion.

The term limbic system was introduced somewhat later by MacLean who was also concerned with the mechanisms of emotion. MacLean (1952) pointed out that amygdala, like the hippocampus, was strongly associated anatomically with the hypothalamus. Thus his term limbic system (or visceral brain) referred to amygdala and hippocampal formation plus gyrus fornicatus. The reasons for classifying these structures do not lie in their anatomical origin since the hippocampus and gyrus fornicatus are primitive cortex while the amygdala is unmistakenly subcortical grey matter. It is their direct association with the hypothalamus which unites this heterogeneous group of basal and medial telencephalic structures. Having defined the anatomical core of the limbic system, the full extent of it remains a matter of conjecture. If we can demonstrate strong and maybe even reciprocal connections with distant sites both at the telencephalic and at the lower levels of the central nervous system (CNS), shall we define those sites as limbic? If so, then the limbic system will be seen to extend from frontal cortex to the brainstem and indeed to impinge on brain structures which have not been considered limbic by any stretch of the imagination. For example, as we shall see later, the amygdala and the hippocampus project to the nucleus accumbens and the anterior part of the head of the striatum (caudate/putamen), a major component of the extrapyramidal motor system. Shall we therefore now class striatum as limbic? As Nauta (1982) points out the nervous system abounds with examples of neural structures that form part of one 'system' when viewed from one point of view and are part of quite a different one when considered in a different way. As he comments, 'Such ambiguities seem distressing only because of our tendency to expect a degree of separation of neural mechanisms that appears to be an exception rather than the rule in the more highly developed central nervous system.' Perhaps, therefore, it is more satisfactory to restrict the term limbic system as MacLean did, while emphasizing, both in anatomical and functional terms, the important influence of these limbic structures on widespread circuitries of brain.

There are many new findings concerning the efferent and afferent connections of the limbic system which are worth reviewing. Two factors account for the explosion of work and interest in limbic system. First, the introduction of new anatomical techniques. In this connection, the autoradiographic fibre-tracing method involving the anterograde transport of radioactively labelled amino acids is noteworthy. This technique largely eliminates the problems of damage to fibres of passage inherent in older fibre-degeneration studies based on lesions. Equally important is the method of retrograde labelling of cell bodies with horseradish–peroxidase (HRP), although it must be said that unwanted infusion of HRP into fibres of passage at the terminal injection site has meant that this technique has its limitations. The second factor has been the discovery that brain contains a number of pathways which contain specific neurotransmitters. These can be stained with specific histochemical methods which depend on fluorescent, immunological or radiographic reactions. The limbic system is not unique in having chemically coded afferent and efferent neural pathways, but the specification of the neurotransmitter links in the limbic circuitry has provided us with further insight into the organization of the system, facts which must be taken into account in trying to understand its functions.

DESCENDING COMPONENTS OF THE LIMBIC SYSTEM

The limbic structures of the forebrain are reciprocally connected with an uninterrupted continuum of subcortical grey matter that begins with the septum, continues from there caudalward over the pre-optic region and hypothalamus, and extends over a paramedian zone of the mesencephalon that reaches caudally to the hindbrain. The mesencephalic component of this continuum includes the ventral tegmental area (VTA), the dorsal raphe nucleus, the median raphe nucleus, the interpeduncular nucleus, and the dorsal and ventral tegmental nuclei of Gudden. Projections from the limbic system to the limbic midbrain area follow three principal routes: (a) the medial forebrain bundle (mfb); (b) the stria medullaris and fasciculus retroflexus involving *en route* the habenula nuclei; and (c) the mammillo–tegmental tract. The paramedian zone of the mesencephalon also receives ascending projections of the mfb and thus reciprocal connections exist between the limbic system and the midbrain.

The medial forebrain bundle

The hippocampal formation has two distinct routes of entry to the mfb.

    1. Via precommissural fornix to all parts of the septum;
    2. Via the massive postcommissural fornix to the mammillary bodies and hence to the mammillo–thalamic tract.

The amygdala gains access to the descending mfb by way of two fibre systems, the stria terminalis and the ventral amygdalo–fugal pathway. The stria terminalis in turn distributes fibres to its bed nucleus and to medial pre-optic areas and ventromedial hypothalamic cell groups, both of which contribute substantially to mfb. The ventral amygdalo–fugal pathway, a fibre complex containing, as bed-nucleus, the substantia innominata, provides reciprocal connection between the amygdala and the septo–pre-optico–hypothalamic region. The central amygdaloid nucleus has particularly strong projections in this pathway, some of which project as far as the nucleus of the solitary tract and the dorsal motor nucleus of the vagus – a route providing a direct influence of the limbic system on visceral centres of the hindbrain. The extent of the descending projections of the mfb is illustrated in Fig. 1.1.

Stria medullaris and fasciculus retroflexus

These fibre bundles provide the alternative route by which the limbic system influences the midbrain limbic sites. In contrast to the mfb, it is the habenula

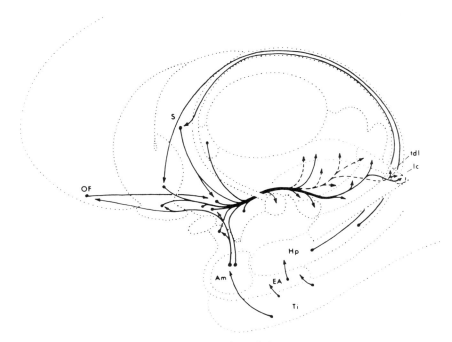

Fig. 1.1. Diagrammatic representation of the descending components of the medial forebrain bundle. Broken lines represent the bundle's lateral division, distributed to the nigral complex, the central regions of the mesencephalic reticular formation, the peripeduncular nucleus, and the parabrachial region. Not indicated are those lateral-division fibres that arise from the hypothalamus and central amygdaloid nucleus and continue caudalward to the nucleus of the solitary tract, the dorsal motor nucleus of the vagus, and the lateral horn of the spinal grey matter. Also included in the diagram are projections from the antero-inferior temporal cortex (Ti) to amygdala (AM) and – indirectly, by way of the entorhinal area (EA) – to hippocampus. Identification of structures: OF, orbitofrontal cortex; lc, locus coerleus; s, septum; t d l, nucleus tegmenti dorsalis lateralis. (Reproduced from Nauta (1982).)

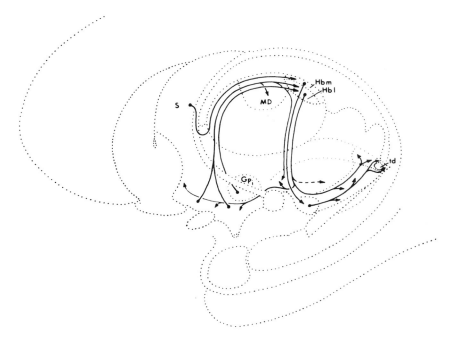

Fig. 1.2. Conduction pathways involving the habenula complex, as described in the text. Broken lines indicate projections from the lateral habenula nucleus (Hbl) to structures lateral to the parmedian 'limbic midbrain area' in particular substantia nigra, pars compacta, and tegmental reticular formation. (Reproduced from Nauta (1982).)

nuclei rather than the pre-optic/hypothalamic nuclei, which lie astride the pathway (Fig. 1.2). Like the mfb, the stria medullaris arises principally from the medial and basal forebrain structures to which the limbic structures project so heavily. But it also acquires input from the globus pallidus, an extra-pyramidal structure. Thus the habenula nuclei, like their target sites in the ventral tegmental area (VTA) of the mesencephalon, receive both limbic and motor projections. The basal forebrain sites form a projection to the lateral habenula nucleus, whereas a projection from the supracommissural septum innervates the medial habenula nucleus.

The habenula nuclei have recently become a focus of considerable anatomical interest. The medial habenula, since it receives input from the dorsal septum, is likely to convey hippocampal output and projects, via the interpeduncular nucleus (IPN), to the raphe nuclei. The lateral habenula receives input from the basal forebrain and the globus pallidus and distributes fibres widely at the level of the ventral tegmental area and substantia nigra before projecting on the raphe nuclei of the mesencephalic reticular formation. These two pathways through the habenula appear to remain distinct in their descending trajectories and only converge in the region of the raphe nuclei. A prominent group of substance-P-containing neurones are located in dorsal part of the medial habenula (Hokfelt *et al.* 1975) and constitute part of the pathway projecting to the IPN and raphe nuclei (Neckers *et al.* 1979).

AFFERENT PROJECTIONS TO THE LIMBIC SYSTEM

Classical anatomical studies revealed that the mfb contained ascending projections from mid- and hindbrain to the limbic system. But there have been few studies of these projections with the new tracing techniques, largely because their origins are obscure.

The last decade has seen a major breakthrough with the discovery that at least some of the ascending projections to limbic system contain specific chemical neurotransmitters. With the increasing availability of histochemical methods with which to stain neurotransmitters *in situ*, it has been possible to identify several highly organized chemical inputs to limbic system.

Ascending dopamine projections from the ventral tegmental area of the mesencephalon

With a histofluorescence technique, it is possible to reveal and distinguish the monoaminergic transmitters dopamine (DA), noradrenaline (NA), and serotonin (5HT).

Dopamine neurones located bilaterally in the VTA provide ascending projections to the amygdala, hippocampus, and, in greatest density, to certain structures intimately related to the limbic structures. These include the tuberculum olfactorium, the nucleus accumbens, and the septum. The DA neurones of the VTA are designated the A10 group and they also innervate the anterior part of the striatum, which, as we shall see, is now known to receive direct projections from the amygdala and hippocampus. Ungerstedt (1971) coined the term *Meso-limbic system* for this set of topographically organized DA projections (Fallon and Moore 1978). The A9 DA neurones of the zona compacta of the substantia nigra (SN) innervate the full extent of the striatum; this pathway has been designated the *Nigro–striatal* system. Those A9 neurones in the medial SN juxtaposed to the A10 neurone group (Fig. 1.3) innervate the anterior medial striatum whereas medial and lateral SN neurones innervate the mid, lateral, and posterior striatal regions. Thus, there is a region of anterior medial striatum which clearly receives input from both A9 *and* A10 DA neurones.

More recently a third DA projection from the mesencephalon to forebrain has been described. A group of A10 DA neurones located in the anterior dorsal part of the VTA juxtaposed to the interpeduncular nucleus provides an innervation of frontal cortex (which is defined as the projection field of the nucleus medialis dorsalis of thalamus (Fig. 1.4(a)). In the rat this cortex is located on the medial wall of the frontal pole (Kretteck and Price 1977*a*) and on the suprarhinal fissure. In other species the location of frontal cortex is different and certainly in primates increasingly extensive. Other areas of cortex which are anatomically linked with the limbic system and which are concerned with the integration of complex cortical information also receive a DA input from the

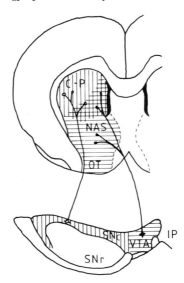

Fig. 1.3. Organization of the projection from the *nigral complex*: A9 (vertical stripes); A10 (horizontal stripes); overlap territory of A9 and A10 (cross-hatched) to the *striatal complex*. C–P, caudate/putamen; NAS, nucleus accumbens; OT, olfactory tubercle; VTA, ventral tegmental area, cell group A10; SNc, substantia nigra cell group A9, zona compacta; SNr, substantia nigra zone reticulata. (Modified from Domesick (1981*b*).)

VTA. These include the anterior cingulate cortex and the entorhinal cortex. It is notable in the rhesus monkey that DA concentrations in cortex are highest in the frontal and temporal lobe association cortex, but low in posterior sensory cortex (Bjorklund *et al.* 1978). As we shall see, not only is there convergence of input from nucleus medialis dorsalis and from DA neurones to frontal cortex (Divac *et al.* 1978) but the amygdala (Fig. 1.5(b)) has also input to these same areas (Kretteck and Price 1977*b*).

Phillipson and Pycock (1982) have extended further our conception of the ascending DA system with the description of organized projections from the VTA to both medial and lateral habenula nuclei. The origin of this pathway in the VTA appears to be similar to that of the meso-cortical innervation to frontal cortex, although the significance of this observation has yet to be evaluated. Thus, one can conceive of the dopamine projections influencing the limbic system at three levels, cortical, subcortical, and midbrain levels. The independence of these two DA terminal fields in the habenula complex (Fig. 1.4) mirrors the independence of the limbic efferent projections coursing through the medial and lateral habenula nuclei. It is worth noting that a group of neurones rich in the neuropeptide, Substance P (SP), is found in medial habenula in precisely the position of the DA and NA innervation. The significance of this anatomical and chemical overlap remains to be evaluated. The SP neurones have been shown to project via the fasciculus retroflexus to the interpeduncular nucleus and hence to raphe nuclei, again in strikingly similar

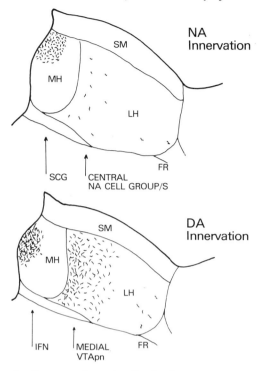

Fig. 1.4. A composite figure to show the distribution and most likely sources of the catecholamine innervation of habenula as deduced from a combination of published anatomical histochemical and biochemical data. It is not clear whether the territory in MH innervated by NA fibres corresponds exactly to that innervated by DA fibres. Abbreviations: SCG, superior cervical ganglion; FR, fasciculus retroflexus; SM, stria medullaris; IFN, interfascicular nucleus; VTA pn, paranigral neurones in VTA. (Reproduced from Phillipson and Pycock (1982).)

configuration to the limbic efferent projections themselves. Indeed, it is possible that some of these limbic efferent pathways defined with tracing techniques do indeed contain specific neurotransmitters.

The ascending projections from the raphe nuclei

Using amino-acid tracing techniques (Azmitia and Segal 1978), projections from the raphe nuclei to the hypothalamus, pre-optic area, and septum were noted. By applying the monoaminergic histofluorescence method, it has been demonstrated that a substantial part of this projection involves serotonin-containing neurones of the raphe. The ascending projections of the raphe, like those of the VTA, terminate in both striatum and limbic structures. Steinbush (1981) has provided the most detailed information on the serotonin projections and the extent of their terminal fields in the forebrain. 5HT innervation appears to be somewhat more extensive than DA in the limbic system, involving the amygdala and hippocampus and the length of the gyrus fornicatus including entorhinal cortex.

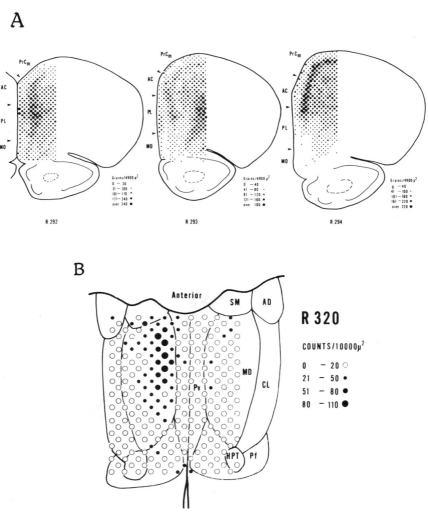

Fig. 1.5. (a) Graphic representation of grain counts obtained with the computer–microscope system over the medial frontal cortex which illustrates the differential distribution of label in different parts of the mediodorsal projection field following injections into different parts of the mediodorsal nucleus. The size of the dots on the sections is proportional to the grain density, with the open circles representing background grain densities. The distribution of DA innervation to medial frontal cortex from the A10 neurones is strikingly similar to the medalis dorsalis projection illustrated. (Reproduced from Kretteck and Price (1977*a*).) (b) diagrammatic representation of grain counts over the mediodorsal thalamic nucleus in horizontal section from the rat. This illustrates that following an injection into the basolateral amygdaloid nucleus (R320) grains are concentrated over the medial portion of the nucleus. (Reproduced from Kretteck and Price (1977*b*).)

The ascending noradrenaline projections from brainstem

Noradrenaline (NA) neurones of the pons and medulla form two major ascending pathways, the dorsal and ventral NA pathways which innervate cortex/limbic system and hypothalamus, respectively (Ungerstedt 1971). Anatomical and electrophysiological findings suggest that the NA, like the other ascending catecholamine pathways, releases its neurotransmitters to modulate CNS activity, rather than acting as the means for information transmission at specific synapses.

In studying the morphology of the NA terminals in cortex, Descarries *et al.* (1977) noted that NA axons of the locus-ceruleus neurones coursing through cortex formed large numbers of varicosites containing neurotransmitter vesicles. However, although the morphology of the axon terminals suggested that they release neurotransmitter, only at a very small percentage of sites were characteristic junctions with postsynaptic specialization observed (Beaudet and Descarries 1978). Descarries *et al.* (1977) therefore proposed that NA is released from large numbers of varicosites along the cortical trajectory of the NA axons, and thus 'NA afferents might exert a diffuse, desynchronized and tonic influence on vast neuronal assemblies, and thus modulate integrative and/or specific cortical function'.

Electrophysiological studies of the action of NA in hippocampus (also inner-vated by locus ceruleus) also suggest a modulatory role for NA. Segal and Bloom (1976) studied the effect of iontophoretically-applied NA on hippo-campal pyramidal neurones activated during a conditioning procecure when food presentation was signalled by a tone. NA enhanced the response of the neurones suggesting that its release normally enhanced the signal-to-noise ratio in hippocampal neurone responses. Subsequently, in the behaving animal, Foote *et al.* (1980) found a correlation between the activity in NA locus-coeruleus neurones and the arousal induced by sensory stimuli. They conclude that activity in these NA neurones 'biases target neurones to respond with enhanced signal to noise ratios to subsequent sensory stimuli'.

Cholinergic projections within the limbic system

Using acetylcholine esterase (ACh-degrading enzyme) staining to identify functionally significant acetylcholine (ACh)-containing pathways, it can be demonstrated that in addition to the prominent septo–hippocampal cholinergic input, ACh-containing neurones (Fig. 1.6) also exist in the basal forebrain and project strongly to the limbic system and cortex (Wenk *et al.* 1980). Cholinergic neurones in the septum project to the hippocampus (dentate gyrus) and contri-bute to the normal neurophysiological responses of the hippocampus and to the failure of fornix-sectioned rats to learn a spatial memory task. Grafts of foetal cholinergic septal neurones which reinnervate the hippocampus in fornix-sectioned rats restore behaviour on spatial tasks (Dunnett *et al.* 1982). Nagai *et al.* (1982) (Fig. 1.7) have demonstrated projections from basal forebrain to the amygdala. Cholinergic neurones in this same region of the basal forebrain also project to cortex, particularly to fronto-temporal cortex. Particular attention has been focused on these neurones since it has been reported that degeneration

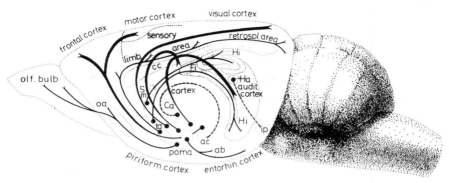

Fig. 1.6. Schematic drawing summarizing the direct monosynaptic cholinergic pathways from magnocellular nuclei of the basal forebrain of the rat to the cortex. Neocortical fields receive their cholinergic innervation from scattered cell groups situated in the substantia innominata (dark neurones not designated lying outside and between the nuclei of origin innervating paleo- and archicortical fields). The olfactory bulb and entorhinal cortex are innervated by cholinergic neurones of the nuc. preopticus magnocellularis (poma), the hippocampal formation, and the limbic cortex from cholinergic cells of the medial septum (sm) and Broca's diagonal tract nucleus (td). No quantitative data are yet available for the pathway designated by the broken line. For references see text. (Reproduced from Wenk *et al.* (1980).)

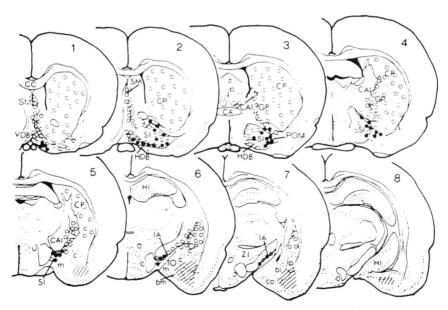

Fig. 1.7. The use of retrograde labelling and AChE staining to demonstrate Ach containing neurones of the basal forebrain projecting to the amygdala. The solid circles indicate cells labelled by Primuline following injection into the amygdala which also show AChE staining, while the open circles indicate those without such Primuline labelling. The shaded portion of the figure shows extent of primuline around the injection site. (Reproduced from Nagai *et al.* (1982).)

of this forebrain neurotransmitter pathway occurs in some patients dying with Alzheimer's pre-senile dementia. Levels of AChE and the synthetic enzyme choline acetyltransferase are reduced in frontal and temporal cortex (Perry *et al.* 1977) and in other studies it has been observed that the neurones of the basal forebrain show degeneration (Whitehouse *et al.* 1981).

Non-chemical afferent projections to limbic system

It has been appreciated gradually that not all the neurones in the afferent limbic projections contain monoaminergic transmitters or ACh, both of which can be identified easily. Some of these afferent systems may well contain chemical transmitters of other classes or indeed may not utilize chemical transmitters at all. The first step in identifying these pathways lies in anatomical-tracing studies and then in electrophysiological studies. Having defined the functions of such a pathway and its location, the search for its potential transmitter is easier. There are now a number of such studies on the limbic system. For example, Femano *et al.* (1979) have described a potent excitatory influence of substantia inominata (SI) on basolateral amygdala. The SI conveys brainstem visceral and somatic information to the amygdala and makes a substantial contribution to this pathway. Nagai *et al.* (1982) propose that ACh-containing neurones on the basal forebrain are involved in this effect. Turning to the prominent DA pathways of the forebrain, non-dopaminergic neurones of the VTA have been described by Thierry *et al.* (1980), which are thought to provide a further input to the limbic and striatal sites innervated by the A10 DA neurones.

PROJECTION OF THE AMYGDALA AND HIPPOCAMPUS OUTSIDE THE CLASSICAL LIMBIC CIRCUITS

In addition to its afferent and efferent connections, the limbic system also projects to frankly non-limbic areas of brain. Three recent studies emphasize the importance of such projections.

Amygdala thalamo–cortical connections

Kretteck and Price (1977*a, b*) have published a detailed description of the projections of the amygdala confirming and extending many earlier findings. One projection previously in dispute was clarified. The amygdala projects to the neocortex, frontal cortex in particular, both via the medalis dorsalis of thalamus (Fig. 1.5(a), (b)) and directly. It is notable therefore, that the A10 dopamine neurones, the amygdala, and the medalis dorsalis nucleus of thalamus all project to the same sector of frontal cortex (Divac *et al.* 1978). In the same series of studies by Kretteck and Price, it was reported that the basolateral amygdala projects to the association cortex of the temporal lobe. This projection has been verified and its reciprocal nature stressed in more recent retrograde (Turner *et al.* 1980) and anterograde (Aggleton *et al.* 1980) tracing studies. It is interesting that opiate receptors exist on these neurones and appear to become more and more densely represented as this temporal cortex projection focuses on anterior temporal lobe and amygdala (Lewis *et al.* 1981). A further dimension of the amygdalo–cortical interaction is afforded by the

ACh neurones of the basal forebrain, which arises from the bed nucleus of the stria terminalis and innervate cortex widely.

Amygdalo–striatal projections

As we have seen earlier, the limbic structures have efferent projections which impinge on non-limbic circuitries at lower levels. However, it is now clear that cross-talk of this kind occurs at the level of the forebrain. It has been demonstrated that the basolateral amygdala projects both to the nucleus accumbens, a limbic recipient of dopaminergic input, and to a large sector of the striatum (Kelley *et al.* 1982), the major focus of the nigro–striatal dopamine projection. Indeed, only that sector of striatum receiving cortical input from the sensory–motor cortex is devoid of amygdala input (Fig. 1.8).

DA release in the nucleus accumbens is able to modify the response of the neurones to input from the amygdala (Yim and Mogenson 1982). It was demonstrated that electrical stimulation of the basolateral amygdala produced strong excitatory responses in the medial part of nucleus accumbens and that iontophoretically-applied DA attenuated this excitatory response (Fig. 1.9). Presumably, DA is released mainly when the A10/nucleus accumbens pathway is activated and, in agreement with this interpretation, stimulation of VTA was found also to attenuate the excitatory response of the nucleus accumbens to

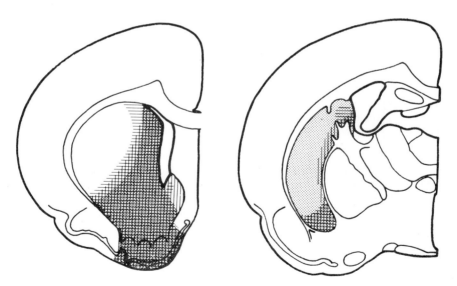

Fig. 1.8. Semidiagrammatic representation of three projections to the striatum: from the ventral tegmental area (vertical lines), the prefrontal cortex (horizontal lines), and the amygdala (stipple pattern). The projection from the ventral tegmental area has been adapted from Beckstead *et al.* (1979); that from the prefrontal cortex from Beckstead (1979). No attempt has been made in this figure to indicate irregularities ('clustering') in the distribution patterns, prominent especially in the prefrontal and amygdalar projections. (Reproduced from Kelley *et al.* (1982).)

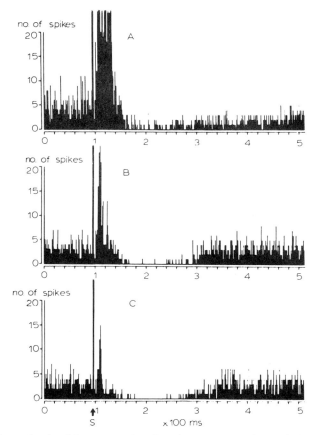

Fig. 1.9. Post-stimulus histograms showing the response of a unit in the nucleus accumbens to amygdala stimulation and its attenuation by iontophoretically-applied dopamine. Histograms were compiled from 300 sweeps at 1.5 Hz. Arrow marks stimulus artefact. (A) Response to single pulse stimulation of the amygdala at 600 μA, 0.15 ms. In this example, a single stimulation of the amygdala elicited a burst of 2–3 spikes, thus generating an excitation with a long spread in the histogram. (B) Dopamine was applied iontophoretically at 5 nA continuously during the sequence of stimulation. The spontaneous activity of the cell was depressed slightly but the elicited excitatory response from amygdala stimulation was markedly attenuated. (C) Ionto-phoretical application of dopamine at 10 nA further attenuated by the elicited excitatory response but the spontaneous activity was affected to a much lesser extent. (Reproduced from Yim and Mogenson (1982).)

amygdala stimulation in the same way as is achieved by direct iontophoretic application of DA.

The projection of the amygdala to the striatum provides a further and alternative output pathway for limbic structures. The nucleus accumbens, which receives a particularly dense amygdala input, has direct projections not only to hypothalamus, but also to the substantial nigra, a major integrative structure of the extrapyramidal motor system (Nauta *et al.* 1978). This connection provides what one might consider a 'feed-forward' projection of limbic structures on to the motor system (Domesick 1981*a*). Thus a second level of limbic/striatal crosstalk exists. Nauta and Domesick (1978) have emphasized the importance of this projection

. . . If it is correct to assume that at least some major function of the limbic system expresses itself in the domain of affect and motivation, while the corpus striatum is engaged more in particular in skeleto-muscular mechanisms, then it could be postulated that connections such as discussed here may represent 'the interface' between motivation and movement.

Hippocampal–striatal projections

Equally important is a recent report that the hippocampus also projects to the nucleus accumbens (Kelley and Domesick 1982) but the innervation is sparse when compared with the amygdala projection and localized to the medial part of this nucleus, an area which incidentally receives a rich noradrenaline input.

CONCLUSION

In this brief summary an effort has been made to pinpoint advances in our knowledge of the anatomy and chemistry of the limbic system achieved within the last five years. Many of these observations may prove to be relevant to our understanding of disease states in man which result in disturbances of emotional balance and cognition.

REFERENCES

Aggleton, J. P., Burton, M. J., and Passingham, R. E. (1980). Cortical and subcortical afferents to the amygdala of the rhesus monkey (Mucaca mulatta). *Brain Res.* **190**, 347–68.

Azmitia, E. C. and Segal, M. (1978). An autoradiographic analysis of the differential ascending projections of the dorsal and median raphe nuclei in the rat. *J. comp. Neurol.* **179**, 641–68.

Beaudet, A. and Descarries, L. (1978). The monoamine innervation of rat cerebral cortex: synaptic and nonsynaptic axon terminals. *Neuroscience* **3**, 851–60.

Beckstead, R. M. (1979). An autoradiographic examination of corticortical and subcortical projections of the medio-dorsal projection (prefrontral) cortex in the rat. *J. comp. Neurol.* **184**, 43–62.

——, Domesick, V. B., and Nauta, W. J. H. (1979). Efferent connections of the substantia nigra and ventral tegmental area in the rat. *Brain Res.* **175**, 191–217.

Bjorklund, A., Divac, I., and Lindvall, O. (1978). Regional distribution of catecholamines in monkey cerebral cortex, evidence for a dopaminergic innervation of the primate prefrontal cortex. *Neurosci. Lett.* **7**, 115–19.

Broca, P. (1878). Anatomie compare des circonvolutions cerebrales. Le grand lobe limbique et la scissure limbique dans la series des mammiferes. *Rev. d'Anthrop., Ser. 2* **1**, 285–498.

Descarries, L., Watkins, K. C., and Lapierre, Y. (1977). Noradrenergic axon terminals in the cerebral cortex of rat. III. Topometric ultrastructural analysis. *Brain Res.* **133**, 197–222.

Divac, I., Bjorklund, A., Lindvall, O., and Passingham, R. E. (1978). Converging projections from the mediodorsal thalamic nucleus and mesencephalic dopaminergic neurons in the neocortex in three species. *J. comp. Neurol.* **180**, 59–72.

Domesick, V. B. (1981*a*). The anatomical basis for feedback and feedforward in the striatonigral system. In *Apomorphine and other dopaminomimetics* (ed. G. L. Gena and G. U. Corsini), Vol. 1, pp. 27–39. Raven Press, New York.

—— (1981*b*). Further observations on the anatomy of nucleus accumbens and caudato-putamen in the rat. Similarities and contrasts. In *The nucleus accumbens* (ed. J. de France and R. Chronister), pp. 7–30. Haer Institute, New Brunswick.

Dunnett, S. B., Low, W. C., Iversen, S. D., Stenevi, U., and Bjorklund, A. (1982). Septal transplants restore maze learning in rats with fornix-fimbria lesions. *Brain Res.* **251**, 335–48.

Fallon, J. H. and Moore, R. Y. (1978). Catecholamine innervation of the basal forebrain IV. Topography of the dopamine projection to the basal forebrain and neostriatum. *J. comp. Neurol.* **180**, 545–80.

Femano, P. A., Edinger, H. M., and Siegel, A. (1979). Evidence of a potent excitatory influence from substantia innominata on basolateral amygdaloid units: a comparison with insula-temporal cortex and lateral olfactory tract stimulation. *Brain Res.* **177**, 361–6.

Foote, S. L., Aston-Jones, G., and Bloom, F. E. (1980). Impulse activity of locus coeruleus neurones in awake rats and monkeys is a function of sensory stimulation and arousal. *Proc. Nat. Acad. Sci.* **77**, 3033–7.

Hokfelt, T., Kellerth, J. O., Nilsson, G., and Pernow, B. (1975). Substance P localisation in the central nervous system and in some primary sensory neurons. *Science* **190**, 889–90.

Kelley, A. E. and Domesick, V. B. (1982). The distribution of the projection from the hippocampal formation to the nucleus accumbens in the rat. An anterograde and retrograde horseradish peroxidase study. *Neuroscience* **7**, 2321–35.

——, ——, and Nauta, W. J. H. (1982). The amygdalo–striatal projection in the rat – an anatomical study by anterograde and retrograde tracing methods. *Neuroscience* **7**, 615–30.

Kretteck, J. E. and Price, J. L. (1977*a*). The cortical projections of the mediodorsal nucleus and adjacent thalamic nuclei in the rat. *J. comp. Neurol.* **171**, 157–92.

—— and —— (1977*b*). Projections from the amygdaloid complex to the cerebral cortex and thalamus in the rat and cat. *J. comp. Neurol.* **172**, 687–722.

Lewis, M. E., Mishkin, M., Bragin, E., Brown, R. M., Pert, C. B., and Pert, A. (1981): Opiate receptor gradients in monkey cerebral cortex: correspondence with sensory processing hierarchies. *Science* **211**, 1166–9.

MacLean, P. D. (1952). Some psychiatric implications of physiological studies on frontotemporal portion of limbic system (visceral brain). *EEG clin. Neurophysiol.* **4**, 407–18.

Nagai, T., Kimura, H., Maeda, T., McGeer, P. L., Peng, F., and McGeer, E. G. (1982). Cholinergic projections from the basal forebrain of rat to the amygdala. *J. Neurosci.* **2**, 513–20.

Nauta, W. J. H. (1982). Neural associations of the limbic system. In *The neural basis of behaviour* (ed. A. Beckman), pp. 175–206. Spectrum Publ. Inc., New York.

—— and Domesick, V. B. (1978). Crossroads of limbic and striatal circuitry: hypo-thalamic–nigral connections. In *Limbic mechanisms* (ed. K. E. Livingston and O. Hornykeiwicz), pp. 75–93. Plenum Publishing Corp., New York.

——, Smith, G. P., Faull, R. L. M., and Domesick, V. B. (1978). Efferent connections and nigral afferents of the nucleus accumbens septi in the rat. *Neuroscience* **3**, 385–401.

Neckers, L. M., Schwartz, J. P., Wyatt, R. J., and Speciale, S. G. (1979). Substance P afferents from the habenula innervate the dorsal raphe nucleus. *Exp. Brain Res.* **37**, 619–23.

Papez, J. W. (1937). A proposed mechanism of emotion. *Arch. Neurol. Psychiat., Chicago* **38**, 725–43.

Perry, E. K., Perry, R. H., Blessed, G., and Tomlinson, B. E. (1977). Neurotransmitter enzyme abnormalities in senile dementia. *J. neurol. Sci.* **34**, 247–65.

Phillipson, O. T. and Pycock, C. J. (1982). Dopamine neurones of the ventral tegmentum project to both medial and lateral habenula. *Exp. Brain Res.* **45**, 89–94.

Segal, M. and Bloom, F. E. (1976). The action of norepinephrine in the rat hippocampus. IV. The effects of locus coeruleus stimulation on the evoked hippocampal unit activity. *Brain Res.* **107**, 513–25.

Steinbush, H. W. M. (1981). Distribution of serotonin-immunoreactivity in the central nervous system of the rat-cell bodies and terminals. *Neuroscience* **6**, 557–618.

Thierry, A. M., Deniau, J. M., Herve, D., and Chevalier, G. (1980). Electrophysiological evidence for non-dopaminergic mesocortical and mesolimbic neurons in the rat. *Brain Res.* **201**, 14.

Turner, B. H., Mishkin, M., and Knapp, M. (1980). Organisation of the amygdalopetal projections from modality-specific cortical association areas in the monkey. *J. comp. Neurol.* **191**, 515–43.

Ungerstedt, U. (1971). Stereotaxic mapping of the monoamine pathways in the rat brain. *Acta physiol. scand.* **367**, 1–48.

Wenk, H., Bigt, V., and Meyer, U. (1980). Cholinergic projections from magnocellular nuclei of the basal forebrain to cortical areas in rats. *Brain Res. Rev.* **2**, 295–316.

Whitehouse, P. J., Price, D. L., Clark, A. N., Coyle, J. T., and Delong, M. R. (1981). Alzheimer's disease: evidence for selective loss of cholinergic neurones in the nucleus basalis. *Ann. Neurol.* **10**, 122–6.

Yim, C. Y. and Mogenson, G. J. (1982). Response of nucleus accumbens neurones to amygdala stimulation and its modification by dopamine. *Brain Res.* **239**, 401–15.

2

Some neurophysiological properties of the septo–hippocampal system

J. N. P. RAWLINS

This chapter was originally intended to cover the neurophysiology of the limbic system. However, given the immensity of such a task (which seems to require the better part of a book of its own, e.g. Isaacson 1982), my aims must be very much more limited. I shall therefore concentrate on presenting data concerning the neurophysiology of a very limited portion of the limbic system, but first I shall try to place my chosen subarea into the wider context with which the chapter should have been concerned.

The limbic system is a title for a somewhat heterogeneous group of central nervous regions, which bear very little structural resemblance to one another and appear to have a considerable diversity of function. Originally, the limbic 'lobe' referred to the border area which includes the central nervous tissue surrounding the brainstem and lies beneath the six-layered neocortex. Most definitions of the limbic *lobe* would include the cingulate gyrus and induseum griseum; the septum; the hippocampus and dentate gyrus; the subiculum, presubiculum, parasubiculum, and entorhinal cortex; the prepiriform cortex; the amygdaloid nuclei; and the olfactory tubercle. Some definitions would also include such structures as the posterior orbital cortex in the frontal lobe and a number of areas of transitional cortex. The limbic *system* can be taken to include all of the above regions, but often includes further structures – for example, the hypothalamus, which has extensive connections with many of the structures in the limbic lobe (Isaacson 1982). Inclusion is sometimes based on anatomical grounds, and sometimes on functional ones.

This chapter will focus on the limbic structures closely associated with the hippocampus, which is to be taken here to mean both Ammon's horn and the dentate gyrus. My basic aim is to review a variety of electrophysiological studies which have been conducted at the gross, the semi-micro, and the single unit level, and relate these findings to some of the behavioural, neurochemical, and pharmacological factors which appear to modify hippocampal activity. This requires some description of the spontaneous activity seen in the hippocampus in freely-moving animals, an analysis of the anatomical basis for this activity, and some consideration of drug effects upon the normal patterns of activity observed. Next, I shall present data concerning the long-term changes in hippocampal excitability which can be induced by a variety of means, but most typically by tetanic activation of some of the afferent and intrinsic pathways. I

MAGDALEN COLLEGE LIBRARY

shall then consider the influence of the ascending monoamine pathways upon hippocampal electrical activity, and summarize some of the data on putative neurotransmitters in the hippocampus. Finally, I shall propose a new role for hippocampal theta to explain some recent observations concerning the behavioural and physiological properties of the septo–hippocampal projection, and outline a new theory of hippocampal function. But first, I should introduce the structure around which the chapter will be built and explain what makes it a suitable focus of attention.

The hippocampus is connected with most other limbic structures via its outputs to the lateral septum and the subiculum and has itself been the subject of an extraordinarily large number of studies, using a wide variety of quite different techniques. This has been for a number of reasons, but these can often be related to the properties of a simple archicortex. The clear, laminar structure lends itself particularly well to extracellular recording of population responses (Andersen *et al.* 1971*a*), in which the cell layers, which are highly homogeneous within each subfield, can be treated as 'open fields', so simplifying interpretation of the results. Furthermore, if hippocampal slices are maintained *in vitro*, the lamellar structure of the hippocampus (Andersen *et al.* 1971*b*; Rawlins and Green 1977) provides a trisynaptic loop of functional cortical tissue (Skrede and Westgaard 1971). This preparation allows mechanically stable recording, provides a near-ideal system for topical iontophoretic application of physiologically active compounds, and permits more general manipulations of the bathing medium in which the slice is maintained. It should however be remembered that experiments with slices, although having some obvious technical advantages, are almost inevitably being conducted on tissue which is to a greater or lesser extent damaged and whose degree of oxygenation and capacity for metabolite exchange are related to the thickness of the slice.

The hippocampus is therefore structurally well suited to stimulation and recording studies. The discovery of long-term potentiation of hippocampal responses to afferent stimulation, following tetanization of the appropriate afferent pathway (Bliss and Lømo 1973), resulted in still greater attention being paid to hippocampal electrophysiology, since the phenomenon provides a striking model for the kinds of changes which might underlie learning and memory.

In addition to stimulation-induced responses, the spontaneous electrical activity seen in the hippocampus of lower mammals has also been of interest, particularly the slow (3–7 Hz in rabbits, cats, and dogs; 6–12 Hz in rats), large-amplitude, and almost sinusoidal activity often referred to as the hippocampal theta rhythm (Jung and Kornmuller 1938; Green and Arduini 1954). The clarity of the theta rhythm and perhaps its ease of recording combined with its clear behavioural correlates at least in some species (Vanderwolf 1969) have led to much attention being paid to its means of generation (in terms both of extrinsic and intrinsic circuitry) and to its possible function. The ease with which hippocampal theta can be recorded is likely to be a further consequence of the open, laminar nature of the hippocampus which ensures that, in a given cell subfield, the sources and sinks for extracellular current flow will all be located at the same laminar level when the cells in the subfield are synchronously activated. In contrast, no theta rhythm can be detected in EEG records taken from the septum, although single units in the medial septal nucleus and diagonal band nucleus, like units in the hippocampus, often fire in bursts which are phase-locked to the theta rhythm. This is probably because the septum does not share an open, laminar structure, but is organized rather more in concentric layers.

Studies of the hippocampus are further encouraged by two factors. First, investigators are considerably aided by our extensive knowledge of hippo-campal anatomy starting with the pioneering studies of Ramòn y Cajal (1911/1955) and Lorente de Nò (1934); both neurophysiological studies and lesion studies benefit greatly from this knowledge. Second, the location of the hippocampus, with one major source of afferent information ascending from the septum and the other major source descending via the perforant path from the entorhinal cortex, has made it a most tempting target for speculation about its function. As long ago as 1937, Papez suggested that the hippocampus was part of a circuit, including a number of other limbic structures, which was concerned with the experience of emotions. Gray (1982) has proposed that the hippocampus is an essential component of a behavioural inhibition system responsible for organizing the behavioural response to anxiogenic stimuli. Both these theories depend to some extent on the location of the hippocampus in the limbic system, away from the motor pathways and the primary sensory systems but able to receive polymodal information from a variety of neocortical areas via the entorhinal cortex (Van Hoesen *et al.* 1972, 1975; Van Hoesen and Pandya 1975). Such a position should allow the hippocampus to abstract information of all kinds, which would be essential, given the range of stimuli which might acquire emotional properties, without requiring direct participa-tion in the processing of such stimuli or direct responsibility for organizing the motor output.

THE HIPPOCAMPAL THETA RHYTHM

Generators

The hippocampal theta rhythm was first reported in rabbits, but it can also be recorded in a number of other lower mammals. It is readily recorded from the alvear surface of the dorsal hippocampus (Bland *et al.* 1975; Green and Rawlins 1979) and even from the neocortex (Stumpf 1965), but experiments with roving electrodes have demonstrated that this is volume-conducted activity, there being two generators somewhat deeper in the hippocampus. One is located in Ramòn y Cajal's *regio superior*, while the other is in the dentate gyrus; the two being separated by a null zone in the anaesthetized and the free-moving rabbit (Green *et al.* 1960; Bland *et al.* 1975; Winson 1976*b*) and in the urethane-anaesthetized and the curarized rat (Winson 1976*a*), though not in the free-moving rat (Winson 1974). That curare effects theta topography is interesting in view of the evidence suggesting that theta may depend upon the integrity of a cholinergic projection; Bland *et al.* (1974) identified both muscarinic and nicotinic synapses on hippocampal pyramid cells.

While theta has been recorded in *regio inferior*, it does not appear to have an amplitude any greater than that recorded in the neocortex. The phase changes which occur over a very short distance when a recording electrode is moved through *stratum radiatum* of CA1 occur only gradually when moving an electrode through CA3, and never end in a complete phase reversal (Bland and Whishaw 1976). Equally, although theta can be recorded from the subiculum (Gray 1970; James *et al.* 1977), roving electrodes traversing the subiculum also detect only gradual phase changes (Bland and Whishaw 1976). These observations led Bland and Whishaw to suggest that theta activity is not generated in either CA3 or the subiculum, in the sense that the active sources and sinks for extracellular current flow are not located there; theta recorded from these regions is taken to represent volume-conducted activity from the generators located in CA1 and the dentate gyrus (cf. Green and Rawlins 1979).

Recently, however, at least one, and possibly two, generators of theta activity have been identified in the medial entorhinal cortex by using roving electrodes to construct laminar profiles in freely moving rats (Mitchell and Ranck 1980). In the deeper cortical layers, the theta rhythm is approximately in phase with the CA1 theta rhythm, while in the more superficial layers it is approximately in phase with the dentate gyrus theta rhythm, the two being about 180° out of phase with one another. It seems possible that theta activity may be generated by cells located both in layer III and in layer II (using the terminology of Ramòn y Cajal 1911/1955), since the phase reversal occurs rapidly around the border between layers III and II, or perhaps in the deep half of layer II. However, the data would also be consistent with the possibility that theta is generated only by layer-III cells, as the authors point out. The implications of these findings will be discussed later.

Two further reports have suggested that there may also be generation of theta in the cingulate cortex, overlying the medial border of the dorsal hippocampus in the urethane-anaesthetized rat (Feensta and Holsheimer 1979; Holsheimer 1982). However more usually investigators have observed neo-cortical de-synchrony associated with hippocampal theta (Green and Arduini 1954), or have believed theta recorded simultaneously in the neo-cortex and hippo-campus to be volume-conducted from the latter, as discussed above.

Correlates of theta activity

Observations of behaviour in the free-moving rat have demonstrated that theta activity is normally present when the rat is engaging in any of one group of behaviours, 'type-1 behaviour' – including locomotion, rearing, manipulation of objects with the forelimbs, postural shifts, or isolated movements of the head or one limb; and is absent when the rat engages in any of another group, 'type-2 behaviour' – including alert immobility in any posture, licking, chewing, grooming, or scratching (Vanderwolf *et al.* 1975). When an animal engages in behaviour from both these groups simultaneously (e.g. licking, while at the same time shifting posture), then theta activity is seen.

While type-1 and type-2 behaviours in the normal rat have clearly different relations to theta, the relationship breaks down when recording from rabbits or cats, which can show clear theta activity while remaining motionless (e.g. Winson

1972). Equally, in the urethane-anaesthetized rat, clear theta can be recorded, though its frequency is lower (4–6 Hz) than in walking rats (7–8 Hz). Following up these behavioural observations has led to the suggestion that there are two, pharmacologically separable pathways responsible for theta generation (see review in Vanderwolf and Robinson 1981).

Drug effects on spontaneous theta

Given the close relationship between behaviour and theta activity, it is clear that any treatment, be it a drug or a tail pinch, which changes the rat's pattern of movement will, at the same time, produce changes in theta activity. Since large hippocampal lesions do not prevent the occurrence of activities normally associated either with the presence of theta or with its absence, it is clear that theta does not in any critical sense cause movement to occur. Therefore, to gain really convincing evidence of a direct pharmacological effect on theta generation, the normal relationship between behaviour and theta must be changed. This should be borne in mind whenever considering treatment effects on theta activity.

Cholinergic agonists and antagonists have clear effects on theta activity. Experiments using urethane-anaesthetized rabbits have shown that systemic administration of physostigmine, while unaccompanied by behavioural activation, induces reliable theta activity which can be abolished by making appropriate septal lesions (Stumpf 1965). Theta is also seen in the urethane-anaesthetized rabbit following appropriate brainstem stimulation, or sometimes following sensory stimulation, or even spontaneously; the theta seen under urethane is always blocked by administering atropine or scopolamine (Vanderwolf and Robinson 1981). Vanderwolf refers to this as 'atropine-sensitive' theta. Similarly, the theta activity seen in anaesthetized rats can be abolished by atropine (Vanderwolf *et al.* 1975), so this is also regarded as being atropine-sensitive.

If unanaesthetized rabbits are used, it appears that the theta seen in this species when the animal is immobile vanishes if atropine is administered, but that the type-1 theta seen when the animal is shifting posture or hopping around persists almost unchanged. Giving atropine to an unanaesthetized rabbit makes the correlation between theta and behaviour resemble that seen in the normal rat (Kramis *et al.* 1975). If atropine or scopolamine is given to the unanaesthetized rat, even in high doses (up to 150 mg kg^{-1}, or 10 mg kg^{-1}, respectively), the theta seen during type-1 behaviour is not abolished. If such an animal is then given an anaesthetic, theta is no longer seen during the stage where struggling has ceased, but some movement persists. This led to the suggestion that atropine-resistant theta is sensitive to anaesthetics (Vanderwolf *et al.* 1975). Thus theta associated with type-1 and type-2 behaviour in rats relies on pharmacologically separable systems.

Unit activity during theta generation

A number of studies have demonstrated the existence, in the hippocampus, of single units which fire in a fixed relation to concurrent theta activity. These units have been observed in paralysed (Green *et al.* 1960) and free-moving animals (O'Keefe and Dostrovsky 1971; Ranck 1973), and can be found both in CA1 and in the dentate gyrus in urethane-anaesthetized rabbits (Bland *et al.* 1980). Multi-unit recording has demonstrated that cells in layer III of the entorhinal cortex also fire in phase with the theta rhythm in the free-moving rat (Mitchell and Ranck 1980). There are also a number of reports that cells in the medial septal nucleus and the diagonal band nucleus fire in phase with hippocampal theta activity (Petsche *et al.* 1962; Ranck 1973; Apostol and Creutzfeldt 1974; Wilson *et al.* 1976). The possible role of these septal cells in providing an external pace maker for the theta rhythm will be considered at greater length in a subsequent section.

While there is general agreement that hippocampal cells do sometimes fire in phase with the theta rhythm, there is some doubt about which cells are actually entrained in this way. In studies conducted using free-moving rats, 'theta cells' – that is, cells which increase their rate of firing above their normal rate if, and only if, there is a regular theta rhythm in the hippocampus (Ranck 1973) – were found most commonly in *stratum oriens* of CA1; in *stratum lucidum, stratum radiatum,* and *stratum moleculare* of CA3; and in the hilus of the dentate gyrus (Fox and Ranck 1975). There were thus relatively few theta cells in the pyramidal-cell layers of CA1 or CA3, and in the granule-cell layer in the dentate gyrus. However, it should be noted that the 'complex spike cells' which constituted the other major cell type from which Ranck recorded – defined as cells having a group of action potentials occurring within 1.5–6 ms interspike intervals, in which the size of each spike changes, usually decreasing – can also become entrained with theta (Ranck, in Elliott and Whelan 1978, pp. 313–15). Complex spike cells appear to be located predominantly in the pyramidal-cell layers of the hippocampus proper and in the granule-cell layer of the dentate gyrus. It is therefore reasonable to suppose that theta cells are interneurones and complex spike cells are pyramid cells and granule cells.

Further evidence that complex spike cells do sometimes entrain to concurrent theta activity has been obtained in the urethane-anaesthetized rabbit. Bland *et al.* (1980) identified units by determining their responses to antidromic activation of the efferent mossy fibres from the dentate gyrus, or hippocampal efferent pathways in the fimbria or the alveus, and found the strongest correlation between unit firing and concurrent theta actuvity in dentate granule cells and CA1 pyramidal cells. Basket cells and CA3 pyramid cells showed much poorer correlations. There is thus agreement that both interneurones and neurones giving rise to efferent connections *can* participate in theta activity; what is uncertain is to what *extent* each class of cells does so. One possibility referred to by Bland *et al.* (1980) is that the different classes' contributions to theta depend upon the animals's behavioural and pharmacological state; this is based on the suggestion, made on separate behavioural and pharmacological grounds, that there may be two distinct types of theta activity (reviewed by Vanderwolf and Robinson 1981).

Intracellular recordings have been made both from CA1 cells (Fujita and Sato 1964) and dentate granule cells (Andersen and Schwartzkroin 1978). In both cases, regular membrane potential oscillations were observed which were in phase with the gross extracellular theta activity. These results suggest that the cause of theta waves is 'nearly synchronous rhythmical postsynaptic potentials in hippocampal formation neurons . . .' (Bland *et al.* 1980). How is the activity of these neurones synchronized?

Circuitry required for theta generation

Since the generators in the hippocampus and in the entorhinal cortex show identical frequency modulation and since the behavioural correlates of theta in the generators also appear identical (Mitchell and Ranck 1980), it is reasonable to suppose either that one of these structures drives or paces the activity in the other or that activity in both is paced by some further structure. Since sectioning the perforant path (via which the input from the entorhinal area reaches the dentate gyrus) does not abolish the dentate theta rhythm in urethane-anaesthetized rabbits, it seems that the entorhinal cortex does not provide an input critical for hippocampal theta (Andersen *et al.* 1979). However there have been reports that retrohippocampal lesions in the subiculum can abolish theta in the corresponding hippocampal lamella (Chronister *et al.* 1974). Andersen *et al.* (1979) did not see such effects and, given the ease of damaging the blood supply to the hippocampus (Nilges 1944) and thereby attenuating hippocampal activity (Boast *et al.* 1976) when making lesions around the distal border of CA1, it seems possible that Chronister *et al.* may have produced their result through an indirect means.

Stimulation studies intended to demonstrate the extrinsic circuitry necessary for the production of hippocampal theta activity suffer from one simple problem of interpretation. Since, as we have already seen, hippocampal theta in the rat reliably appears when the rat engages in appropriate behaviours (including running, jumping, or REM sleep) and disappears when he is engaging in other specified behaviours (including sitting still, licking, or slow-wave sleep) apparently regardless of *why* the rat is doing so (Vanderwolf 1969; Black and Young 1972), *any* stimulus which induces one of these behaviours will induce the state of hippocampal electrical activity which accompanies it. For this reason, it is important to know what the rat is doing, when recording hippocampal activity.

A number of studies have shown that stimulation at various midbrain sites can elicit hippocampal theta. Green and Arduini (1954) described a pathway originating in the mesencephalic reticular formation, traversing the hypothalamus to the septum and finally reaching the hippocampus via the fornix, upon whose integrity theta depended. High-frequency electrical stimulation at some points along this pathway elicits theta activity; midbrain reticular stimulation (e.g. Green *et al.* 1960; Stumpf 1965) or posterior hypothalamic stimulation (Bland and Vanderwolf 1972) both elicit theta activity whose initial frequency and amplitude is related to the stimulation voltage, though after a short while the frequency tends to drop to a steady level regardless of voltage. Breaking the continuity of this system, by making lesions or by injecting local anaesthetic, abolishes spontaneous theta and theta elicited by stimulation of the pathway caudal to the block (Green and Arduini 1954;

Stumpf 1965). Thus septal lesions, or total fornix–fimbria lesions, prevent the appearance of theta under any circumstances, except possibly in response to high-frequency perforant path stimulation (preliminary report in Andersen *et al.* 1979).

As well as being the final relay in this pathway to the hippocampus, the role of the septum seems to be of unique importance in theta generation; at this level of the pathway, high-frequency stimulation no longer elicits theta activity, but blocks it (Stumpf 1965; Ball and Gray 1971). However, single pulses delivered to the medial septum can elicit slow-wave responses in the hippocampus which look very much like theta and have the same frequency as the septal stimulation, as long as the stimulation is kept within the theta range (Stumpf 1965; Ball and Gray 1971). Thus, important changes seem to take place at this level of the pathway. Behaviourally, reticular or hypothalamic stimulation which elicits theta in the free-moving rat also elicits the kinds of behaviour usually associated with spontaneous theta. This septal stimulation which elicits 'theta' does not do so. Physiologically, high-frequency afferent information appears to be transformed into theta-frequency activity for transmission to the hippocampus. The existence of cells in the medial septal and diagonal band nuclei showing bursting activity in phase with hippocampal theta (Petsche *et al.* 1962; Apostol and Creutzfeldt 1974; Ranck 1976) adds weight to this suggestion. It has therefore been proposed that hippocampal theta is 'paced' by the rhythmic activity of these septal neurones.

It is known that the medial septal area (a term used here to refer to both the medial septal nucleus and the nucleus of the diagonal band) contributes a major projection to the hippocampus via the fornix and the fimbria (Raisman 1966; Meibach and Siegel 1977) and there is evidence to suggest that this projection may be cholinergic (Lewis and Shute 1967; Mellgren and Srebro 1973; Dudar 1975). However, there has been doubt about how well the distribution of fibres responsible for theta matches the distribution of fibres from the septum as mapped by histochemical or labelled transport studies. This raises the possibility that there may be two populations of cells in the medial septum, both projecting to the hippocampus, but possibly differing in their functions, the courses taken by their axons, and their transmitters. Myhrer (1975) demonstrated that lesions restricted to the dorso-medial portion of the fornix, leaving the fimbria intact, abolished hippocampal theta, even though most of the fibres from the medial septal area to the hippocampus travel in the fimbria (Meibach and Seigel 1977). Lynch *et al.* (1978, p. 12) have reported that they have observed, labelled by horseradish peroxidase injections into the hippocampus, medial septal cells which did not stain for acetylcholinesterase (AChE). Although the authors pointed out that this could have stemmed from a methodological problem, it could also be taken to provide evidence of non-AChE septal projections to the hippocampus.

A series of experiments conducted in our laboratory was designed to throw further light on these problems, while at the same time examining one theory of how the rhythmic, bursting activity of medial septal neurones is maintained. McLennan and Miller (1974, 1976) suggested that feedback from the hippocampus via the fimbria to the dorsolateral septum activates a frequency-gating mechanism to control medial septal activity via interneurones. This theory therefore predicted that interruption of the hippocampal feedback loop to the septum should disrupt hippocampal theta activity as effectively as do lesions of the medial septal area, or its efferents carrying the output from the pacemaker cells.

The projection from the medial septal area appears to be topographically organized, with the projection to the adseptal pole (a term used as a synonym for the more widely used 'septal pole' to designate that end of the hippocampus located nearest the septum) travelling in the medial part of the fornix and the projection to the temporal pole (the farthest end of the hippocampus) travelling in the most lateral part. In view of this organization, it seemed wise to repeat Myhrer's (1975) experiment, but using recording electrodes in the temporal region as well as in the adseptal region of the hippocampus. The experiments were carried out in free-moving rats, using chronically-implanted recording electrodes, which allowed recordings to be made both before and after the lesion for comparative purposes. The results demonstrated that while theta was abolished in the dorsal/adseptal hippocampus by dorsomedial fornix lesions, it remained intact in the ventral/temporal hippocampus (Rawlins *et al.* 1979). Staining of the brain for AChE showed that temporal hippocampal AChE staining was unaffected by the lesion, while adseptal staining was practically abolished.

Further experiments showed a precisely reversed pattern of results following fimbria lesions (adseptal theta and AChE staining unaffected; temporal theta and AChE staining abolished), while medial septal lesions abolished theta and AChE staining at all hippocampal recording sites. These data demonstrate that the presumed cholinergic septo–hippocampal pathway, and the septo–hippocampal pathway responsible for theta are at the very least clearly co-extensive. The additional observation that dorso-lateral septal lesions neither abolish hippocampal theta nor modify hippocampal AChE staining shows that, contrary to McLennan and Miller's (1974, 1976) suggestions, only damage to septo–hippocampal fibres abolishes theta; damage to hippocampo–septal ones does not. Given that Ranck (1976) saw no lateral septal cells firing in phase with theta, but saw many cells in the medial septal area firing in phase with theta, some even in the absence of hippocampal theta activity, it is hard to avoid the conclusion that hippocampo–septal feedback does not contain frequency-specific information necessary for the maintenance of hippocampal theta, although the information carried might *modulate* theta frequency. Experiments carried out in the urethane-anaesthetized rabbit by Andersen *et al.* (1979) generally confirmed some of our findings for dorsal hippocampal theta, although Andersen *et al.* were recording dentate theta, while we were recording activity originating in *regio superior*.

Because entorhinal theta is isomorphic with hippocampal theta, we conducted further experiments at the Johns Hopkins University to see whether entorhinal theta would be similarly affected by our lesions, and, if so, to deter-mine whether this depended on activation relayed via the hippocampus, or was a result of a direct input from the medial septal area. The methods used were essentially the same as those employed in the earlier study; recordings were made from free-moving rats, which were forced to walk in a treadmill. This method ensures that long trains of clear theta activity will be emitted. We were able to record simultaneously from the adseptal/dorsal hippocampus and the medial entorhinal cortex, in most cases before and after making lesions. AChE staining was again used to see if AChE loss corresponded with theta loss.

The results (Mitchell *et al.* 1982) demonstrated that medial septal lesions disrupt entorhinal theta activity, just as they do hippocampal theta (see Fig. 2.1). The extent of theta loss in the medial entorhinal cortex matched very

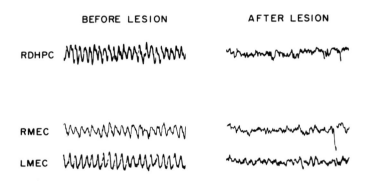

Fig. 2.1. EEG recordings taken simultaneously from the right dorsal hippocampus (RDHPC) and the right and left medial entorhinal cortex (RMEC and LMEC), before and seven weeks after a medial septal lesion. Calibration bars: 500 μV and 1 second. (From Mitchell *et al.* (1982).)

closely the extent of loss in the homolateral dorsal hippocampus. There was a marked depletion of AChE staining throughout the hippocampus and the entorhinal cortex. This left open the possibility that entorhinal theta is paced directly by medial septal activity, which would be consistent with the observed loss of AChE staining, but did not exclude the possibility that the placing is relayed via the hippocampus, and entorhinal theta was affected only because hippocampal theta was being directly affected by the lesion. Since the only monosynaptic hippocampal projection to the entorhinal cortex is derived from the CA3 pyramid cells in the temporal one-third of the hippocampus (Hjorth-Simonsen 1971), we investigated the effects of unilateral fimbria lesions upon medial entorhinal theta activity. These lesions have already been shown to abolish temporal hippocampal theta (Rawlins *et al.* 1979), and so should prevent any rhythmic impulses from being transmitted from this area to the entorhinal cortex. This might produce a dissociation between dorsal hippocampal theta, which would be left intact, and entorhinal theta, which would be abolished. However, entorhinal theta was not affected by these lesions, indicating that it is not driven monosynaptically from the hippocampus and that any direct projection from the medial septum to the medial entorhinal cortex does not run in the fimbria. AChE staining in the entorhinal cortex was affected ipsilateral to the lesion, but only in the ventral lateral entorhinal cortex and not in the medial entorhinal cortex, where we were recording.

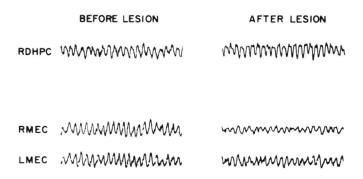

Fig. 2.2. EEG recordings taken simultaneously from the right dorsal hippocampus (RDHPC) and the right and left medial entorhinal cortex (RMEC and LMEC), before and one day after a right unilateral dorsal fornix lesion. Calibration bars: 500 µV and 1 second. (From Mitchell *et al.* (1982).)

These results suggested that the entorhinal theta activity may be paced directly by a medial septal input. In a final experiment, we attempted to cut the projection from the medial septum to the entorhinal cortex, while leaving the projection to the hippocampus intact. We therefore made a lesion in the dorsal fornix, caudal to the point at which fibres destined for the dorsal hippocampus would have diverged from the fibre bundle (Andersen *et al.* 1979; Rawlins *et al.* 1979). The results (Mitchell *et al.* 1982) showed that it is indeed possible to induce theta loss in the medial entorhinal cortex while leaving hippocampal theta intact (see Fig. 2.2). We further observed that large lesions of this kind in the dorsal fornix left hippocampal AChE staining completely unaffected, but produced a loss of AChE staining in the medial entorhinal cortex only. Moreover, data from two rats with poorly placed lesions showed substantial loss of dorsal hippocampal theta, with minimal loss of entorhinal theta; this finding is inconsistent with the possibility that entorhinal theta is paced by a polysynaptic input from the dorsal hippocampus. It therefore seems likely that entorhinal theta, like hippocampal theta, is driven by pacemaker cells located in the medial septal area. Furthermore, the increased evidence for a relation between the topographies of theta loss and of AChE loss makes it still more probable that the final pathway for controlling theta is cholinergic.

Whether the septo–entorhinal theta pathway consists of collaterals from the same cells of origin as the septo–hippocampal pathway is unclear. The high correlation we observed between the extent of theta loss in hippocampus, and the loss in entorhinal cortex following medial septal lesions suggests that the same medial septal cells may project to both areas. The alternative possibility is that the cells of origin are thoroughly intermixed in the medial septal area. Recent experiments with double labelling have suggested that the two projections may arise from collateral branches from the same cell (C. Kohler, personal communication). In either case, the topography of the projections to the hippocampus and entorhinal area through the fornix and fimbria look very similar, with the fibres to the most dorsal and medial portions of both structures running largely through the dorsomedial fornix, and the fibres to the most ventral and lateral areas apparently running largely through the fimbria. In essence, this means that the fibres to the different parts of each structure simply take the shortest route (see Fig. 2.3).

Medial septal lesions also appear to abolish theta recorded in the ventral tegmental area, in close proximity to the A10 dopaminergic cell group (Le Moal and Cardo 1975). Given that there also seems to be a medial septal projection to the cingulate cortex (Swanson and Cowan 1979) in which theta generation has also been reported (Holsheimer 1982), it is tempting to suppose that theta is paced in a similar way in any region in which it is generated, though this has yet to be demonstrated. These results have a number of implications.

First, because the generation of extra-hippocampal theta, at least in the entorhinal cortex, appears to depend on exactly the same afferent information as hippocampal theta, any reference to special features of hippocampal intrinsic circuitry as being critical for generating theta activity may be redundant unless there is parallel circuitry in the entorhinal cortex. This problem is especially important, since the hippocampal pyramid cells are so well equipped with inhibitory feedback circuits, probably mediated by the

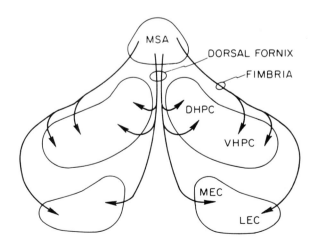

Fig. 2.3. A schematic diagram showing the organization of the cholinergic theta producing pathways from the medial septal area to the dorsal and ventral hippocampus (DHPC and VHPC) and to the medial and lateral entorhinal cortex (MEC and LEC). We have no theta records from the LEC, but fimbria lesions abolish cholinesterase staining there. (From Mitchell *et al.* (1982).)

hippocampal basket cells which are themselves activated by collaterals from the axons from the pyramids (Kandel *et al.* 1961; Andersen *et al.* 1964). Inhibitory circuits of this kind could clearly contribute to the rhythmic patterns of activation underlying theta activity by producing inhibition followed by rebound excitation: the activity of the basket cells is first reduced, when the pyramids are inhibited, and then enhanced as the pyramids are released from the basket-cell-induced inhibition, and so the activation level of each population oscillates. Because the basket-cell inhibition of pyramidal cells is so effective, it has been suggested that the septum may in fact drive theta through activating the basket cells, each of which would inhibit and then release large numbers of pyramid cells (Lynch *et al.* 1978, pp. 16–18). This suggestion would fit well with the distribution of theta cells reported by Fox and Ranck (1975). It would be interesting to know whether physiologically analogous circuitry exists in the entorhinal cortex.

Second, the influence of the medial septum can be seen to extend up into precisely that area of transitional cortex which provides the hippocampus with its polymodal afferent information. The cells in entorhinal cortex from which the projections to the hippocampus via the perforant path arise – the layer-III pyramid cells, projecting to CA1 pyramid cells, and the layer-II star cells, projecting to dentate granule cells (Steward and Scoville 1976) – are also the cells constituting the laminae most likely to be generating theta activity (Mitchell and Ranck 1980). The activity of medial septal neurones can thus influence the passage of entorhinal cortical information through the hippocampus in two ways. First, the size of the population spike elicited in the dentate granule-cell layer by entorhinal stimulation has been shown to vary systematically with the concurrent phase of the theta rhythm (Rudell *et al.* 1980). Thus, the medial septum can pace or 'gate' the passage of information through the hippocampus. Second, by modifying activity in the entorhinal cortex itself, the medial septum can alter the information before it is even transmitted to the hippocampus. This means that any view of the hippocampus as being an interface between ascending brainstem information (exemplified as theta, paced by the septum) and neocortically processed, polymodal information descending via the entorhinal cortex, needs to be modified somewhat. The initial 'meeting of minds' seems to take place in the entorhinal cortex itself.

Finally, it is not clear whether pharmacological treatments which modify hippocampal theta also modify theta generated elsewhere in the same way. As we have seen, there may be separate pathways, at some level probably caudal to the septum (see below), for generating hippocampal theta at different frequencies, or under different pharmacological or behavioural conditions. No research has so far determined whether what is true for the hippocampus is also true for the entorhinal cortex. Data of this kind might help to cast light on the locus of action of the drugs studied.

Fitting these data into the other available data on pathways responsible for theta generation is difficult. First, there is the evidence that the medial septal input to the hippocampus may be cholinergic, which might suggest that it subserves only type-2 theta. However, medial septal lesions abolish *all* theta;

there is to date no known lesion which abolishes only type-1 theta. Further, it is still doubtful whether any lesion which does not modify medial septal activity can abolish theta of either kind. It therefore seems probable that any pharmacological separation between the two sorts of theta must reflect an action caudal to the septum; otherwise, a completely effective cholinergic blockade might be expected to abolish all theta, like a medial septal lesion. Equally, anaesthetic effects most probably act caudal to the septum, unless perhaps they act intra-hippocampally, or upon a noncholinergic septo–hippocampal projection; it is possible, as discussed earlier, that there is a noncholinergic projection from the medial septum to the hippocampus, but this has not been convincingly demonstrated. Thus the experimental data we have considered have not provided an obvious anatomical or physiological substrate for Vanderwolf's two types of theta; our data suggest that at this level of the theta-producing system we are dealing with a final common pathway required for theta generation.

Long-term potentiation

While the hippocampal response to entorhinal stimulation can be modulated by concurrent theta activity, it should be noted that the previous history of afferent stimulation of some kinds can also modify the response to a repetition of the stimulation. This apparent plasticity of response, while existing elsewhere, is particularly well marked in the hippocampus. More importantly, studies of the time course of these plastic changes seen in the hippocampus, which were initially conducted over minutes or hours, have subsequently been extended to days or weeks, encouraging the belief that such phenomena may serve as a model for learning and memory.

Bliss and Lømo (1973) observed that, following trains of stimulation of the perforant path at 10–20 Hz for 10–15 s, or 100 Hz for 3–4 s, the size of the dentate granule-cell population response to a single afferent volley was enhanced along a number of parameters. The latency of the population spike was decreased, the size of the population excitatory postsynaptic potential (EPSP) was increased, and the amplitude of the population spike itself was increased to an extent which could not entirely be accounted for by the increased EPSP. During the stimulation trains at 10–20 Hz, there was massive potentiation of the population spike, reflecting frequency potentiation. However during stimulation at 100 Hz, there was suppression of the population spike. Frequency potentiation is therefore not a necessary precursor of all long-term potentiation.

The intervals between final tetanization of the perforant path and subsequent assessments of the response to a single perforant-path volley ranged from 30 minutes to 10 hours – about the maximum feasible length for experiments of this kind conducted in anaesthetized subjects. At this longer interval, the amplitude of the population spike was still greater than its initial value, though it had fallen somewhat from its maximum value.

Subsequent experiments in chronically implanted unanaesthetized subjects have greatly increased the durations over which long-term potentiation (LTP) can be observed. Bliss and Gardner-Medwin (1973) demonstrated potentiation up to, but not exceeding, three days after a single tetanization of the perforant path. However, after a series of tetanizing trains at different stimulus intensities, one subject still showed apparent potentiation 16 weeks later. Similar maintenance of potentiation for up to two months has been reported in the free-moving rat (Douglas and Goddard 1975). When assessing the duration of LTP over these sorts of time-span, the stability of the chronic preparation becomes critically important. Ideally, a control input on another non-potentiated afferent pathway should be used to ensure that the recording electrode is correctly located. However, it is also important to ensure that the stimulating electrode is still correctly located and is exciting the same number of afferent fibres as before. This can sometimes be done by recording a non-potentiated collateral response from the same stimulation site (Andersen *et al.* 1973), or by recording the size of the presynaptic afferent volley elicited by the stimulation in the target cell field, but this is itself very difficult when stimulating at any distance from the electrode used to record the volley (Andersen 1978). Thus some caution is required when assessing the significance of very long-lasting but small changes in the absence of exhaustive control procedures.

Although changes which last this long are certainly durable enough to provide a basis for long-term memory storage, it should be emphasized at this stage that there is no doubt that general long-term memory storage does *not* depend upon an intact hippocampus. Animals with the most extensive hippo-campal lesions can clearly form and use long-term memories. Thus LTP in the hippocampus may provide a good model for long-term memory storage, but does not represent its physical substrate.

While there is no question that LTP is a powerful and readily demonstrable effect, there is still controversy, both over how it can be induced and over the nature of the changes which underlie it. Although this controversy is not yet resolved, it is worth considering some of the evidence supporting the various points of view, since any understanding of drug action, or other treatment effect, upon this kind of potentiation is going to depend on our understanding of the basis for the phenomenon itself.

There are two particularly interesting problems to consider. One is whether the changes underlying long-term potentiation stem from pre- or postsynaptic modifications. The other is whether LTP can be *heterosynaptic*: that is, whether the changes seen are confined to the particular pathway which was activated. This could be observed as the induction of potentiation in a non-stimulated pathway, following tetanus-induced LTP of a different pathway, or as LTP following simultaneous or near-simultaneous activation of more than one pathway, when it does not follow stimulation of either pathway on its own. This is of particular interest to models of learning and memory, since associa-tive memory represents a link between two or more previously unrelated items. Co-operativity between different inputs has some of the features one might expect to see in an associative memory network (McNaughton *et al.* 1978; Levy and Steward 1979).

There seems to be no doubt that stimulation of a single afferent pathway is sufficient for LTP to occur; and the potentiation is usually restricted to the afferent pathway which as been tetanized (e.g. Andersen 1978). In Andersen's experiments, conducted in the *in vitro* hippocampal slice, it was possible to record both the amplitude of the presynaptic afferent volley and the amplitude of the population spike. Long-term potentiation can then be seen particularly clearly: for a given size of presynaptic volley, the amplitude of the population spike elicited was increased, following tetanization of the afferent pathway. In these experiments, the response to stimulating another, independent, afferent path was also recorded. This response was unaffected by the potentiation of the tetanized pathway, providing clear evidence that the changes produced by tetanizing were restricted to the pathway thus activated.

However, there are examples of apparent interactions between pathways. Under different circumstances, LTP can be enhanced or retarded by approximately simultaneous stimulation of more than one afferent pathway. LTP was reported by McNaughton *et al.* (1978) following simultaneous high-frequency activation of the lateral and the medial perforant path (whose synapses are on the outer third and middle third of the dendrites of the dentate granule cells, respectively) whereas stimulation of either pathway alone at the same intensity produced no lasting changes. While this is particularly remarkable, as the synapses are quite clearly in separate locations, McNaughton *et al.* produced related evidence showing that stimulation of a given afferent pathway only led to LTP if above a certain intensity threshold. In itself this is unsurprising, but McNaughton *et al.* were actually able to record a population EPSP in the dentate gyrus at levels of stimulation below that at which LTP could be induced, showing that their stimulation was effective in producing a release of neurotransmitter. Thus there is also apparently a necessary co-operativity amongst the fibres of a single afferent pathway. Co-operation in producing enhancement has also been reported by Levy and Steward (1979), who stimulated the ipsilateral and contralateral perforant paths.

Interactions preventing LTP have also been reported. While a single pulse activation of the crossed commissural pathway initially excites the dentate granule cells, it then inhibits granule-cell responses to perforant-path stimulation. The granule-cell response to tetanization can be blocked in this way, without concurrently preventing the induction of LTP (McNaughton *et al.* 1978). However, trains of 400-Hz stimulation of the commissural pathway *can* block LTP produced by near-simultaneous tetanization of the perforant path (Douglas *et al.* 1982). Both forms of commissural stimulation can inhibit the granule cells' response to perforant-path stimulation, so this inhibition does not seem to be the critical factor. Further, the inhibition produced is not presynaptic inhibition of the perforant path terminals, so neurotransmitter release should be normal during perforant-path tetanization. Thus timing of activation, identity of activation, and frequency of activation are all important in determining the outcome of interactive stimulation.

These kinds of consideration become most important when trying to derive the functional properties of neural networks so as to describe how entire systems work. To add to the complexity, though stimulating CA3 does not induce LTP in the hippocampal projection to the dorso-lateral septum nor in the collateral commissural projection to the contralateral hippocampus, it does produce LTP in a third collateral projection to the ipsilateral CA1. Thus the capacity for potentiation can differ amongst the different axons from a single cell (Andersen *et al.* 1973; Skrede and Malthe-Sørensson 1981). It therefore seems impossible safely to generalize by inferring from what is known about one pathway to the properties of another pathway, even a very closely related one. Accordingly, while heterosynaptic potentiation is not seen in the afferent input to CA1 running in *stratum radiatum* when the afferents running in *stratum oriens* have been tetanized and potentiated (Andersen 1978), heterosynaptic LTP of this kind has been reported in CA3. Tetanizing the mossy-fibre input running in the hilus of the dentate gyrus could produce enhanced CA3 pyramid-cell responses to antidromic, but synaptically mediated, stimulation of Schaffer collaterals in CA1 *stratum radiatum* (Misgeld *et al.* 1979).

The data concerning heterosynaptic LTP are also clearly relevant to the question of whether LTP is based on pre- or postsynaptic changes. It is tempting to suppose that an indirectly-induced change seen in the response to an input which has not undergone any experimental manipulation to produce LTP must reflect some change in the target cell itself. This is less clear where heterosynaptic stimulation is used, since in such cases there is more obvious opportunity for direct interactions between the axon terminals excited.

It is clear that an overall increase in membrane excitability of the target cell cannot underlie LTP, since such an increase would result in enhanced responses to all afferent stimulation and this is not necessarily seen (Andersen 1978). Andersen tested this possibility directly in the experiments we have just considered, by studying the cell responses to constant depolarizing pulses. No general change in response was observed even though LTP had been induced. On the other hand, no general change in the projection cell can account for all the findings, since it is possible to observe LTP in the projection field of an axon without its being present in the field of another collateral from the same cell. It is necessary to look for more restricted changes.

While LTP does not appear to depend on the target cells' firing, it is known that high-frequency stimulation under conditions of low extracellular Ca^{2+} (which should prevent exocytosis and hence block neurotransmitter release) does not produce LTP; equally, treatments thought to prevent binding of glutamate or aspartate prevent LTP at the perforant-path synapse (Dunwiddie *et al.* 1978). This suggests that both release of neurotransmitter and successful binding are prerequisites for LTP to take place. These findings are suggestive of postsynaptic effects, though not totally incompatible with presynaptic ones. Postsynaptic models also gain support from the morphological changes seen in the dendritic spines of dentate granule cells, following tetanization of the perforant path: the mean spine diameters were measurably increased and the increase seemed restricted to spines in the perforant-path terminal field (Van Harreveld and Fifkova 1975). These changes are seen following electrophysiologically monitored LTP, but not following stimulation which does not produce LTP; it has been suggested that cytoplasmic actin contracts, as a consequence of stimulation-induced Ca^{2+} release, to shorten and widen the spine stalk, and thus change the electrical properties of the spine (Anderson and Fifkova 1982; Fifkova and Delay 1982).

Presynaptic influences are most strongly suggested by recent work on hippocampal slices which had been preloaded with tritiated D-aspartate (Skrede and Malthe-Sørensson 1981), in which the release of the labelled amino acid was measured. Slices cut in different planes were employed: one plane contained the CA3-to-lateral-septum projection, while the other contained the CA3-to-CA1 projection. The former pathway does not show potentiation, while the latter does. Stimulation of either pathway produced calcium-dependent D-aspartate release but, after tetanization in the longitudinal septal slice, the release fell back to the control level measured in a non-stimulated slice, while, after tetanization in the transverse CA1 slice, the baseline level of release remained elevated above control values. The authors therefore suggest that tetanization leads to an increased transmitter release from the presynaptic elements, which would mean that an afferent volley of a fixed amplitude should elicit a more powerful EPSP. Comlementary results, using an *in vivo* preparation, also suggest that the LTP following perforant-path tetanization is associated with increased glutamate release (Dolphin *et al.* 1982). Interestingly, in this experiment, some rats did not show clear LTP following the perforant-path tetanization. These rats did not show an increase in glutamate release; only the group which did show LTP following tetanization showed a prolonged and significant increase in glutamate release compared to non-tetanized controls. These results indicate very strongly that tetanization can result in lasting increases in transmitter release.

Direct testing of a postsynaptic model based on an increased sensitivity for glutamate has produced at least some negative findings. Apparent LTP can be induced in the absence of any stimulation in the hippocampal slice, simply by doubling the concentration of Ca^{2+} in the bathing medium for 10 minutes (Turner *et al.* 1982). This procedure elicited a clear increase in the CAI response to *stratum radiatum* stimulation, which outlasted the transient increase in calcium levels by at least 2½ hours. The procedure produced no consistent changes in the amplitude of the afferent volley, and did not produce an increased response to iontophoretically applied glutamate. Unlike Andersen's (1978) experiment, in which LTP was induced by stimulation, the CA1 response to *stratum oriens* stimulation was also found to be enhanced by transient increases in calcium levels, indicating, not surprisingly, a more general consequence of the treatment compared to stimulation of a specific afferent pathway. While these data are clearly compatible with a presynaptic model and while the specific postsynaptic model tested was not confirmed, it does not seem that there is yet any evidence which demonstrates an exclusive role for either postulated process.

Functional studies of long-term potentiation

The first question which must be answered in considering a functional role for LTP is whether the conditions required to induce it ever exist in the normal animal. Here the data look encouraging at first, since single units in the entorhinal cortex of free-moving rats do show brief bursts of appropriate, high-frequency activity (e.g. Segal 1973). However, it will be recalled that co-operativity amongst input fibres seems a requirement for LTP; whether the entorhinal cortex is likely to generate a big enough synchronous afferent volley is uncertain. There is some evidence that dentate granule cells *are* synchronously activated during the acquisition of conditioning to a tone and that different evoked responses are elicited following a procedure which differentially conditions two tones; it further seems that the differentially variable component of this auditorily evoked potential is mediated by the perforant path (Deadwyler *et al.* 1979).

There appears to be a dynamic fluctuation of the perforant-path-mediated evoked response: the amplitude of this response to a tone on any given trial, during mixed sequences positively or negatively correlated with food availability, seems to be determined by the identities of the tones presented in the three preceding trials. Thus a sequence of three negatively correlated tone trials results in a different response to the tone on the next trial (regardless of whether this trial is a positively or a negatively correlated one) from a sequence of three positively correlated trials. Different mixed sequences of positively and negatively correlated trials also produce predictably different evoked potential amplitudes (West *et al.* 1982). Thus the entorhinal projection to the dentate gyrus can produce a synchronous afferent volley and, very interestingly, the amplitude of the evoked response varies as a function of recent experience, as though the system reflected the contents of some information input buffer. However, these observations were only made under behaviourally and sensorily rather closely controlled circumstances, (Deadwyler *et al.* 1981), which would confer a very restricted behavioural function on mechanisms dependent on synchronous activation.

Another approach has been to look at LTP in animals which are believed to show behavioural signs of hippocampal dysfunction. Various tasks are highly sensitive indicators of a malfunctioning hippocampal system; if LTP represents some critical hippocampal function, animals which fail to show LTP for some reason should perform these sensitive tasks badly. There have been at least two interesting observations made: the first concerns senescent rats; the second, adult rats which were malnourished in infancy. Both old age and infant malnutrition result, like hippocampal lesions (O'Keefe and Nadel 1978), in poor performance on complex spatial memory tasks. (Barnes 1979; Jordan *et al.* 1981). Both factors reduce the number of dentate granule cells or synapses on them (Smart *et al.* 1974; Geinesman and Bandareff 1976). Both factors result in an abnormally rapid decay of LTP. There is one difference between the two factors, however, in that senescent rats show normal LTP at short retention-testing intervals, while it is hard to establish LTP at all in the malnourished rats (Barnes 1979; Jordan and Clark, in preparation). These correlative studies encourage the belief that problems with LTP indicate important functional problems in the hippocampus. However, it must be remembered that related changes are very likely to have occurred at other central nervous system sites, and these changes may be at least as important in producing the behavioural effects observed as are the changes presumed to underlie the failure of hippocampal LTP. Thus LTP might indicate a more generally functional central nervous system; still, the data do fit together well so far.

The behavioural state of the animal also influences the transmission of responses to perforant-path stimulation. Winson and Abzug (1978) showed that the response in the free-moving rat to low-frequency perforant-path stimulation was greatest at various intra-hippocampal recording sites when the animal was in slow-wave sleep. Responses were lessened when the animal was in rapid eye movement sleep, or was awake and moving around – both behaviours accompanied by prominent theta activity in the rat – and were smallest when the animal was immobile and alert. It would be interesting to know if LTP is more easily induced in animals in slow-wave sleep.

There are, then, a number of ways in which behaviour can be related to changes in the electrophysiological properties of the hippocampus; however it cannot yet be said that a functional role for LTP-induced changes in impulse transmission has been established, though one can clearly expect further relevant data to be collected.

THE INFLUENCE OF ASCENDING MONOAMINE PATHWAYS ON HIPPOCAMPAL ACTIVITY

The hilus of the dentate gyrus is well innervated by both noradrenergic and serotonergic pathways. The noradrenergic pathway originates in the locus ceruleus, and travels via the dorsal ascending noradrenergic bundle, which innervates a good deal of the forebrain including not only the hippocampus, but also the neocortex, the medial and lateral septal nuclei, parts of the amygdala, and the hypothalamus (Livett 1973). The serotonergic pathway to the dorsal hippocampus originates in the median raphe, which also innervates the medial and dorsolateral septal areas, while fibres from the dorsal raphe project to the ventral hippocampus and the anteroventral part of the lateral septal area (Azmitia and Segal 1978). The role of ascending brainstem pathways in theta generation has recently been reviewed by Vertes (in press).

Stimulating the locus ceruleus does not appear to have the dramatic effects on spontaneous hippocampal activity that other brainstem stimulation can have. Robinson and Vanderwolf (1978), found that theta activity was elicited only at very high current intensities, when current spread was likely to be a problem. Assaf *et al.* (1979), studying the granule-cell response to perforant-path stimulation, found that locus ceruleus stimulation produced a decrease in spontaneous unit activity, but an increased response to the perforant-path stimulation. Raphe stimulation had the same kind of effect (Assaf and Miller 1978*a*). Lesions in the locus ceruleus do not abolish theta recorded in unanaesthetized rats engaging in locomotion (Kolb and Whishaw 1977). However, there is a recent report that LTP of the perforant path is reduced in animals depleted in hippocampal monoamines (Bliss, Goddard, and Riives, cited in Bliss and Dolphin 1982), which suggests that ascending monoamine pathways can play a role in hippocampal plasticity, though the mechanism remains unclear. A series of experiments by Segal showed that locus ceruleus stimulation produced possibly related phenomena (Segal and Bloom 1976; Segal 1977). In one of these experiments, the development of hippocampal unit responses to a tone, which was associated with food, was enhanced if locus ceruleus stimulation preceded the tone presentation. This might mean that increased noradrenergic activity increases hippocampal plasticity. Again, we may hope for further related data.

Raphe stimulation, as noted previously, appears to have some features in common with locus ceruleus stimulation. However, it also has some features which are quite different. Median raphe stimulation can induce theta, up to a maximum frequency in the free-moving rat of about 7.7 Hz (Graeff *et al.* 1980). This theta activity is associated with immobility, and can be blocked by scopolamine. However, since theta activity in the rat is so clearly correlated with behaviour and since theta, of about this frequency and with a similar sensitivity to scopolamine, can be seen in an animal which is immobile for other reasons, as Graeff *et al.* noted, it is not clear that the theta was directly elicited by the stimulation. It may have been a consequence of the behaviour elicited by the stimulation. This indirect view is perhaps somewhat encouraged by the observation that spontaneous hippocampal theta is not abolished by lesions of the serotonergic projections to the medial septal area (McNaughton *et al.* 1980).

Contrary results from median raphe stimulation have been reported by Vertes (1981), who saw stimulation-induced *de*-synchronization of the hippocampal EEG. This was probably not as a consequence of over-activation of a synchronizing system, since at low stimulation currents, there was either no effect at all on the hippocampal EEG, or de-synchronization. Vertes (in press) has suggested that Graeff *et al.*'s (1980) result stemmed from spread of the stimulation current to the medial pontine reticular formation, or to fibres running in the medial longitudinal fasciculus, from both of which sites theta can be elicited. The suggestion that the median raphe nucleus may be the source of a de-synchronizing input to the hippocampus (Vertes 1981, in press) gains support from the observation that stimulation here disrupts the bursting discharge of septal pacemaker cells (Assaf and Miller 1978*b*). Still more significantly because the normal relationship seen between behaviour and theta activity can be seen to be changed by a lesion, Maru *et al.* (1979) reported that median raphe lesions resulted in the appearance of theta during behavioural immobility in the rat.

Thus, both the noradrenergic and serotonergic inputs to the hippocampus appear to play a significant modulatory role, but do not appear to be essential for at least some of the normal hippocampal functions. However their ability to modulate hippocampal function has led to the suggestion that these pathways might be a major mediating factor both in the experience of anxiety and its alleviation by anxiolytic drugs (Gray 1982). In accordance with this conclusion, lesions in these ascending pathways appear to have some, but by no means all, the behavioural properties of large septal or hippocampal lesions (Gray 1982).

At present, other theories of hippocampal function have not generally assigned a role to these projections.

BEHAVIOURAL CORRELATES OF HIPPOCAMPAL UNIT ACTIVITY

Some data concerning hippocampal single-unit activity in the free-moving rat have already been presented. Thus, there are changes in unit activity during various conditioning procedures, though, as ever, it is important to know whether the stimuli used came to induce, for example, immobility (type-2 behaviour) or orientation of the head towards the stimulus (type-1 behaviour), either of which would produce characteristically different patterns of hippocampal activity. The existence of theta cells has already been referred to; such cells are most likely to fire when the animal is actively moving, and therefore showing theta activity.

The most obvious and frequently reported correlation shown by non-theta cells in the rat hippocampus, is with the animal's location. Complex spike cells recorded in the hippocampus of the awake rat have, if active at all, an extraordinarily high (80–90 per cent) probability of firing when the animal is in a particular place (O'Keefe and Nadel 1978). In normal rats, the cues which seem to define a 'place' as far as the cells are concerned appear to be distal cues, often located in the room in which the experimental apparatus is located, rather than being cues on the experimental apparatus itself (O'Keefe and Black 1978). More recently it has been suggested that, for 'place' cells to fire, the rat must be facing in a specific direction at the time. (McNaughton *et al.* 1982). One interesting feature of these place cells is that, while they will fire in one location in a given testing environment, if the animal is moved to another testing environment, there may be a location there, as well, in which the cell will fire selectively (Ranck, in Elliott and Whelan 1978, p. 310). The cells are not specific to a uniquely defined place.

The other situation in which complex-spike cell activity is very clearly increased is in slow-wave sleep. In this state, in which theta is absent from the hippocampal EEG, not only are 'place' cells already identified by their behavioural correlates likely to fire, regardless of where the rat sleeps, but other complex spike cells which have been 'silent' during behavioural testing are likely to be recorded on the same electrode (e.g. Ranck, in Elliott and Whelan 1978, pp. 309–10). A similar phenomenon can be observed under barbiturate anaesthesia: the number of complex spike cells which fire increased dramatically, compared to the number recorded by a given electrode in the awake rat (P. Best, personal communication). Thus both slow-wave sleep and high doses of barbiturates produce an apparent disinhibition of firing in a population which most probably consists of pyramid cells. This may be part of the cellular basis for the large, irregular activity seen in the hippocampal EEG during slow-wave sleep, though theta cells also apparently fire, in bursts, during the spikes which typically appear in the hippocampal EEG (O'Keefe and Nadel 1978).

Rapid-eye-movement sleep is associated with clear, large-amplitude theta activity in the hippocampus. As would therefore be predicted, theta cells are active in this state, while the complex-spike cells reduce their firing rate from the relatively high rates they show in slow-wave sleep. If overall levels of unit activity were to be taken as the best indicators of hippocampal function, we should probably say that its primary role is in sleep!

PUTATIVE NEUROTRANSMITTERS IN THE SEPTO-HIPPOCAMPAL SYSTEM

The orderly laminar organization of the hippocampus has clear advantages for attempts to relate the distribution of possible neurotransmitters to the known regions of termination of particular pathways. Selective micro-dissection is facilitated, since a given pathway will generally terminate within a given lamina; furthermore, assays of micro-dissected material can often be most usefully combined with selective denervation. Iontophoretic application of transmitter candidates at positions very close to their presumed sites of action is made simple in the *in vitro* hippocampal slice. It is often possible to bring more than one of these techniques to bear on particular problems. The excellent review by Storm-Mathisen (1978) includes a great deal of the relevant data; I shall only very briefly summarize parts of it.

Acetylcholine

The existence of a possible cholinergic input from the medial septal area to the hippocampus has already been referred to. The distribution of this pathway appears co-extensive with that of the septo–hippocampal 'theta' pathway, as indicated by lesion-induced changes in AChE staining and theta activity. The normal distribution of AChE in the hippocampus is clearly laminated and the pattern of lamination fits fairly well with the location of the terminal fields of the septo–hippocampal pathway as assessed by other means (Lynch *et al.* 1978). Just as AChE activity is markedly reduced by lesions in the septo–hippocampal pathway, so too is hippocampal choline acetyltransferase (Storm-Mathisen 1978). These effects are seen following medial, but not lateral, septal lesions. Interestingly, it has been reported that direct, intra-hippocampal injection of oxotremorine, a cholinergic agonist, can induce hippocampal theta activity (Ott *et al.* 1977).

However, while septal stimulation does induce acetylcholine release from the hippocampus and while the release can be prevented by sectioning the septo–hippocampal projection (Dudar 1975), a note of caution should be sounded before concluding that the only septo–hippocampal projection is definitely a cholinergic one. First, there is the evidence that atropine and scopolamine do not block type-1 theta. Second, there is evidence that hippocampal responses to septal stimulation also survive muscarinic and nicotinic blockade (Fantie, cited in Gray 1982, p. 89). Possibly some of the anticholinergics do not cross the blood–brain barrier. However, it is interesting to note that, while incomplete medial septal lesions, or procaine injections into the medial septal area, produce decreases in theta amplitude without obvious changes in theta frequency or probability of occurrence (e.g. Brust-Carmona *et al.* 1973; Rawlins *et al.* 1979; Mitchell *et al.* 1982), no parallel effects have been reported following systemic administration of anticholinergics.

It should also be noted that there is evidence in the monkey for a cholinesterase-rich projection to the hippocampus which does not originate in the medial septal area, but seems to be a projection from nucleus basalis of Meynert (Kitt *et al.* 1982). Possibly the existence of this pathway might explain the existence of both muscarinic and nicotinic receptors in the hippocampal formation.

GABA

Glutamate decarboxylase (GAD) appears to be concentrated in specific laminae in the hippocampus, and GAD activity remains normal after extensive lesions in the known hippocampal afferent pathways. The pattern of zones showing higher than average uptake of tritiated GABA maps well onto immuno-histochemically visualized areas of high GAD content (all reviewed by Storm-Mathisen 1978). It has been suggested that the basket cells may deliver GABA as their inhibitory transmitter (Curtis *et al.* 1970), and it has been shown that GAD-positive cells in the hippocampus include cells which appear like basket cells in their structure, their position, and their termination (Ribak *et al.* 1978).

In a recent experiment, Andersen *et al.* (1982) iontophoretically applied GABA on to various levels of the CA1 pyramid cells using the *in vitro* slice preparation. The pyramid cells were activated by orthodromic stimulation in *stratum radiatum* or *stratum oriens*, or by glutamate application. GABA application near the cell body reduced the population spike to an apical dendritic input, but was even more effective when applied at some points in the apical dendritic tree. However, when stimulating *stratum oriens* to activate the basal dendrites, somatic application of GABA still had a clear effect but GABA applied close to the apical dendrites was relatively ineffective. Ejection of GABA among the basal dendrites, however, *was* effective in reducing the population spike. These data suggest that there may indeed be a means of producing overall inhibition of pyramid cells via a GABA-mediated input (probably from basket cells) on to the cell soma, but that there may also be more locally effective inhibitory inputs mediated by some other interneurones which can selectively inhibit the synaptic activation of dendrites' particular afferent pathways.

There are thus grounds for believing that GABA may act as an inhibitory neurotransmitter in the hippocampus. In view of this conclusion, it is interesting to recall that barbiturates (which facilitate GABA-ergic transmission) apparently *disinhibit* pyramid-cell activity in the hippocampus.

Glutamate and aspartate

The excitatory effects of glutamate application in the hippocampus have already been alluded to, as has the calcium-dependent change in D-aspartate release, apparently associated with stimulation of the projection from CA3 to CA1, and the change in resting levels of release following tetanization of this pathway.

Tritiated glutamate uptake studies again reveal a laminar pattern of activity, which can be selectively disrupted by making lesions. Lesions in the perforant path produce a loss of uptake in the outer two-thirds of the dentate molecular layer, which is the terminal zone of this pathway; lesions in CA3/4 produced a loss of uptake in CA1 *stratum oriens* and *stratum radiatum*, and in the innermost part of the dentate molecular layer, which are the terminal zones for CA3 projection (all reviewed in Storm-Mathisen 1978). These acidic amino acids thus appear reasonable candidates as possible hippocampal transmitters.

Monoamine distributions in the hippocampus have already been considered and will not be discussed further in this section.

PEPTIDES AND HORMONES

It has been known for a number of years that labelled corticosterone has binding sites in the hippocampus (McEwan *et al.* 1969). Autoradiographic studies have shown that the binding takes place in the *stratum granulosum* of the dentate gyrus and in the *stratum pyramidale* of the hippocampus proper (Stumpf 1971). Manipulating corticosterone levels experimentally seems to modify some of the electrophysiological properties of the hippocampus. Its injection alters the threshold at certain frequencies for elicitation of apparent theta by septal stimulation (Valero *et al.* 1977) and its absence, as a consequence of adrenalectomy, appears to inhibit LTP (Dana *et al.* 1982). In the latter experiment, diurnal changes in plasma corticosterone levels appeared to correlate with changes in the magnitude of the LTP seen in normal rats.

There have also been a number of reports demonstrating that opiates and opioid peptides modify hippocampal pyramid-cell excitability when iontophoretically applied both *in vivo* and in the *in vitro* slice preparation (e.g. Zieglgänsberger *et al.* 1979; Henriksen *et al.* 1982). These investigators suggested that the opioids may inhibit the inhibitory interneurones, thus allowing a rebound increase in pyramidal-cell excitability. It has also been suggested that the increased pyramid-cell responsiveness to excitatory inputs produced by opiates may result from an alteration of the coupling between the dendritic EPSP and the spike-initiation mechanisms (Lynch *et al.* 1981). These studies also showed that those effects of the opioids could be antagonized by naloxone, suggesting that the effects are mediated by opiate receptors, though not all hippocampal effects of opioid peptide administration can be blocked in this way (Henriksen *et al.* 1982).

A recent immunohistochemical study has shown that there are enkephalin-containing fibres in the hippocampal formation of the guinea pig, which have a clear laminar organization (Tielen *et al.* 1982). The density of these fibres is greatest towards the temporal pole of the hippocampus. In a small number of experimental preparations, the authors also observed apparently enkephalin-containing cell bodies. There appear to be four groups of enkephalin-containing fibres, of which two correspond to the projections of the dentate granule-cell axons running transversely to the longitudinal axis of the hippocampus; one runs longitudinally along the hippocampal axis, and one runs along the same trajectory as the perforant path fibres from the entorhinal cortex. There is thus reasonable evidence that opioids do play some functional role in hippocampal activity.

CONCLUDING REMARKS

After surveying a fraction of the extensive literature concerning hippocampal system physiology, one is perhaps entitled to step back a little from the data and ask what it is all for. As suggested in the introduction, the position of the hippocampus, in a desirable central location in the limbic system, coupled with its intriguing electrophysiological properties, has made it the target for much speculation. Although its monosynaptic projections are limited in extent, the degree of interconnectivity of the region, and particularly the extensive projections from the subiculum, ensures that its potential influence is widespread, extending not only through the limbic system, but also back down into brainstem regions and up to the neocortex.

The possible physiological function of the hippocampal theta rhythm has also received attention. There is a natural tendency to propose that it in some way 'times' or 'steps' the passage of information through the system (O'Keefe and Nadel 1978; Gray 1982). This has some empirical physiological support, reviewed earlier, but it also leads to some difficulties, since such theories predict that a hippocampus in which theta has been destroyed (e.g. by a medial septal lesion) will be in many respects completely dysfunctional. However, lesion data do not support this view. Rats with medial septal lesions which have eliminated hippocampal theta throughout most of the hippocampal formation do not show the same behavioural changes as rats with large hippocampal lesions (e.g. compare Feldon and Gray 1979 and Rawlins *et al.* 1980), and can show apparent recovery on tasks at which rats with large hippocampal lesions never improve (Mitchell *et al.* 1982). More interestingly still, reimplanting septal tissue into rats with total fornix–fimbria lesions, and thus inducing a reappearance of AChE staining in the hippocampus, also induces a partial behavioural recovery (Dunnett *et al.* in press). Since these rats do not show a recovery of theta activity (Low *et al.* in press), the recovery suggests that perhaps the septal input has an important tonic influence on hippocampal function, and is not only important because of the phasic modulation it produces.

If then the theta rhythm is important as a tonic influence and is not purely a carrier of time-locked information, could this give some new clue as to its function? A novel possibility, untested so far at the neurophysiological level, is that the septal input somehow helps to maintain hippocampal storage of specific information from *other* sources. Testing such a hypothesis should be simple: one need only determine the time-course of decay of LTP in animals in which the theta rhythm has been experimentally abolished. Attributing an 'enabling' function of this kind to the theta rhythm would be consistent with a view that the septal input has a tonic function. But would it fit the data concerning the behavioural correlates of theta activity, which the specific information theories mentioned earlier (O'Keefe and Nadel 1978; Gray 1982) handle so well with respect to movement-related theta, though not, perhaps, with respect to REM sleep?

At a general descriptive level, the answer is 'yes'. If we assume that the greatest threats to storage are decay (as a function of time since activation) and interference (resulting from the input and storage of further items), then the greatest need to protect stored information from decay or over-writing would be during sleep (when extrinsic information is not being received and the normal waking pattern of hippocampal activity is radically changed) and during exploration or movement (when new stimuli are constantly being presented). This would account for the occurrence of bouts of REM sleep, in which large-amplitude theta is generated, interrupting the hippocampal large irregular EEG activity seen in slow-wave sleep and would also account for the relationship between movement and theta: the hippocampus could be seen as protecting itself and its contents. Furthermore, the apparent absence of theta in primates might be attributable to an increased cortical input to the hippo-campus, sufficient to provide enough tonic input without regular bouts of rhythmic activation being required. Thus theta would represent not new information flowing into the hippocampus but, effectively, an attempt to maintain the old, possibly in part by restricting the information being newly registered.

This kind of view, while suggesting a role for theta in assisting information storage, gives no clue as to what kind of information might in fact be stored in the hippocampus. A number of current theories of hippocampal function suggest that the hippocampus is concerned with handling information of particular kinds, which are seen as differing qualitatively from other classes of data. Examples of this kind of theory are the cognitive map hypothesis (O'Keefe and Nadel 1978), the recognition memory hypothesis (Gaffan 1972), the working memory hypothesis (Olton *et al.* 1979), and the behavioural inhibi-tion theory, in its current form proposed by Gray (1982). All these theories share the belief that the hippocampus is critically important in dealing with stimuli or information of some particular specifiable kind: they differ in their identification of the critical data class.

In contrast to this kind of approach, I propose that the hippocampus actually handles data of *all* kinds, regardless of the nature of the data, and regardless of the operationally defined nature of the task to be solved. I suggest that the hippocampus acts as an intermediate-term, high-capacity memory buffer, acting in parallel with a limited-capacity, short-term memory system located elsewhere. Incoming information of all kinds is fed into the hippo-campus and is maintained there over relatively long periods of time. This intermediate memory system would thus allow items to be associated with one another, even though they might be presented at quite separate times. On the other hand, an item which is closely followed by a second item will still be represented in the short-term memory store at the time the second item is presented. In this case the two items could be associated with each other, and the association subsequently stored by a conventional long-term memory system, without the hippocampus needing to be involved at all. Thus hippo-campal damage will only obviously affect the performance of tasks in which the short-term memory system is inadequate for the memory load: such tasks would be those in which temporally discontiguous items must be associated with one another, or tasks in which a large number of items need temporary storage, even if only for a short time. The most obvious difference between this view of hippocampal function and those already referred to lies in the prediction that tasks of *any* kind can be made either more or less sensitive to hippocampal dysfunction simply by manipulating the temporal relationships between items whose association is essential for the solution of the task. This classification cuts right across previous classifications in that it regards different task sensitivities as stemming from *quantitative* differences between tasks, rather than qualitative ones.

Behavioural experiments recently carried out in Oxford and designed to test this view have so far tended to support it (Rawlins *et al.* 1982). Whether it will eventually be able to incorporate the vast body of literature concerning the behavioural effects of hippocampal lesions is uncertain. However, it has so far been successful in predicting the outcomes of experiments designed specifically to test its predictions against the contrary predictions made by a number of other theories. This suggests that it represents at least an extra factor which must be considered in refining our theories of hippocampal function.

ACKNOWLEDGEMENTS

I should like to thank Dr Susan Mitchell for providing copies of the figures; I should also like to thank Ms Claire Nash and Mrs Lesley King for their hours of work on the various versions of this manuscript.

The author is the Royal Society Henry Head Research Fellow in Neurology.

REFERENCES

Andersen, P. (1978). Long-lasting facilitation of synaptic transmission. In *Functions of the septo–hippocampal system* (ed. K. Elliott and J. Whelan), Ciba Foundation Symposium 58 (new series), pp. 87–102. Elsevier, Amsterdam.

——, Bie, B. O., and Ganes, T. (1982). Distribution of GABA sensitive areas on hippocampal pyramid cells. *Exp. brain Res.* **45**, 357–63.

——, Bliss, T. V. P., and Skrede, K. K. (1971a). Unit analysis of hippocampal population spikes. *Exp. brain Res.* **13**, 208–21.

——, ——, and —— (1971b). Lamellar organisation of hippocampal excitatory pathways. *Exp. brain Res.* **13**, 222–38.

——, Eccles, J. C., and Løyning, Y. (1964). Location of postsynaptic inhibitory synapses on hippocampal pyramids. *J. Neurophysiol.* **27**, 592–607.

——, Myhrer, T., and Schwartzkroin, P. A. (1979). Septo–hippocampal pathway necessary for dentate theta production. *Brain Res.* **165**, 13–22.

—— and Schwartzkroin, P. A. (1978). In *Functions of the septo–hippocampal system* (ed. K. Elliot and J. Whelan), pp. 311–12. Elsevier, Amsterdam.

——, Teyler, T. J., and Wester, K. (1973). Long-lasting change of synaptic transmission in a specialised cortical pathway. *Acta physiol. scand. Suppl.* **396**, 34.

Anderson, C. L. and Fifkova, E. (1982). Morphological changes in the dentate molecular layer accompanying long-term potentiation. *Soc. Neurosci. Abs.* **8**, 279.

Apostol, G. and Creutzfeldt, O. D. (1974). Crosscorrelation between the activity of septal units and hippocampal EEG during arousal. *Brain Res.* **67**, 65–75.

Assaf, S. Y., Mason, S. T., and Miller, J. J. (1979). Noradrenergic modulation of neuronal transmission between the entorhinal cortex and the dentate gyrus of the rat. *J. Physiol., Lond.* **292**, 52P.

—— and Miller, J. J. (1978a). Neuronal transmission in the dentate gyrus: role of inhibitory mechanisms. *Brain Res.* **151**, 587–92.

—— and —— (1978*b*). The role of a raphe serotonin system in the control of septal unit activity and hippocampal desynchronisation. *Neuroscience* **3**, 539–50.

Azmitia, E. C. and Segal, M. (1978). An autoradiographic analysis of the differential ascending projections of the dorsal and median raphe nucleus in the rat. *J. comp. Neurol.* **179**, 641–8.

Ball, G. G. and Gray, J. A. (1971). Septal self-stimulation and hippocampal activity. *Physiol. Behav.* **6**, 547–9.

Barnes, C. A. (1979). Memory deficits associated with senescence: A neurophysiological and behavioral study in the rat. *J. comp. Physiol. Psychol.* **93**, 74–104.

Black, A. H. and Young, G. A. (1972). The electrical activity of the hippocampus and cortex in dogs operantly trained to move and to hold still. *J. comp. physiol. Psychol.* **79**, 128–41.

Bland, B. H., Andersen, P., and Ganes, T. (1975). Two generators of hippocampal activity in rabbits. *Brain Res.* **94**, 199–218.

——, ——, and Sveen, O. (1980). Automated analysis of rhythmicity of physiologically identified hippocampal formation neurons. *Exp. brain Res.* **38**, 205–19.

——, Kostopoulos, G. K., and Phillis, J. W. (1974). Acetylcholine sensitivity of hippocampal formation neurons. *Can. J. Physiol. Pharmacol.* **52**, 966–71.

—— and Vanderwolf, C. H. (1972). Diencephalic and hippocampal mechanisms of motor activity in the rat: effects of posterior hypothalamic stimulation on behavior and hippocampal slow wave activity. *Brain Res.* **43**, 67–88.

—— and Whishaw, I. Q. (1976). Generators and topography of hippocampal theta (RSA) in the acute and freely moving rabbit. *Brain Res.* **118**, 259–80.

Bliss, T. V. P. and Dolphin, A. C. (1982). What is the mechanism of long-term potentiation in the hippocampus? *Trends Neurosci.* **5**, 289–90.

—— and Gardner-Medwin, A. R. (1973). Long-lasting potentiation of synaptic transmission in the dentate area of the unanaesthetised rabbit following stimulation of the perforant path. *J. Physiol., Lond.* **232**, 357–74.

—— and Lømo, T. (1973). Long-lasting potentiation of synaptic transmission in the dentate area of the anaesthetised rabbit following stimulation of the perforant path. *J. Physiol., Lond.* **232**, 331–56.

Boast, C. A., Reid, S. A., Johnson, P., and Zornetzer, S. F. (1976). A caution to brain scientists: unsuspected hemorrhagic vascular damage resulting from mere electrode implantation. *Brain Res.* **103**, 527–34.

Brust-Carmona, H., Alvarez-Leefmans, F. J., and Arditti, L. (1973). Differential projections of septal nuclei to ventral and dorsal hippocampus in rabbits. *Exp. Neurol.* **40**, 553–66.

Chronister, R. B., Zornetzer, S. F., Bernstein, J. J., and White, L. E., Jr. (1974). Hippocampal theta rhythm: Intra-hippocampal formation contributions. *Brain Res.* **65**, 13–28.

Curtis, D. R., Felix, D., and McLennan, H. (1970). GABA and hippocampal inhibition. *Br. J. Pharmacol.* **40**, 881–3.

Dana, R. C., Gerren, R. A., Sternberg, D. B., Martinez, J. L., Hall, J., Stansbury, N. A., and Weinberger, N. M. (1982). Long-term potentiation is impaired by adrenalectomy and restored by corticosterone. *Soc. Neurosci. Abs.* **8**, 316.

Deadwyler, S. A., West, M., and Lynch, G. (1979). Activity of dentate granule cells during learning: differentiation of perforant path input. *Brain Res.* **161**, 29–43.

——, ——, and Robinson, J. H. (1981). Entorhinal and septal inputs differentially control sensory-evoked responses in the rat dentate gyrus. *Science* **211**, 1181–3.

Dolphin, A. C., Errington, M. L., and Bliss, T. V. P. (1982). Long-term potentiation of the perforant path *in vivo* is associated with increased glutamate release. *Nature, Lond.* **297**, 496–8.

Douglas, R. M. and Goddard, G. V. (1975). Long-term potentiation of the perforant path–granule cell synapse in the rat hippocampus. *Brain Res.* **86**, 205–15.

——, ——, and Riives, M. (1982). Inhibitory modulation of long-term potentiation: evidence for a post-synaptic locus of control. *Brain Res.* **240**, 259–72.

Dudar, J. D. (1975). The effect of septal nuclei stimulation on the release of acetylcholine from the rabbit hippocampus. *Brain Res* **83**, 123–33.

Dunnett, S. B., Low, W. C., Iversen, S. D., Stenevi, U., and Björklund, A. (1982). Septal transplants restore maze learning in rats with fornix–fimbria lesions. *Brain Res.* **251**, 335–48.

Dunwiddie, T., Madison, D., and Lynch, G. (1978). Synaptic transmission is required for initiation of long-term potentiation. *Brain Res.* **150**, 413–17.

Elliott, K. and Whelan, J. (eds.) (1978). *Functions of the septo–hippocampal system.* Ciba Foundation Symposium 58 (new series). Elsevier, Amsterdam.

Feensta, B. W. A. and Holsheimer, J. (1979). Dipole-like neuronal sources of theta rhythm in dorsal hippocampus, dentate gyrus and cingulate cortex of the urethane-anesthetized rat. *EEG clin. Neurophysiol.* **47**, 532–8.

Feldon, J. and Gray, J. A. (1979). Effects of medial and lateral septal lesions on the partial reinforcement extinction effect at one trial a day. *J. exp. Psychol.* **31**, 653–74.

Fifkova, E. and Delay, R. J. (1982). Cytoplasmic actin in dendritic spines as a possible mediator of synaptic plasticity. *Soc. Neurosci. Abs.* **8**, 279.

Fox, S. E. and Ranck, J. B., Jr. (1975). Localization and anatomical identification of theta and complex spike cells in dorsal hippocampal formation of rats. *Exp. Neurol.* **49**, 299–313.

Fujita, Y. and Sato, T. (1964). Intracellular records from hippocampal pyramid cells in rabbit during theta rhythm activity. *J. Neurophysiol.* **27**, 1011–25.

Gaffan, D. (1972). Loss of recognition memory in rats with lesions of the fornix. *Neuropsychologia* **10**, 327–41.

Geinesman, Y. and Bandareff, W. (1976). Decrease in the number of synapses in the senescent brain: a quantitative analysis of the dentate gyrus in the rat. *Mechanisms Ageing Develop.* **5**, 11–23.

Graeff, F. G., Quintero, S., and Gray, J. A. (1980). Median raphe stimulation, hippocampal theta rhythm and threat-induced behavioural inhibition. *Physiol. Behav.* **25**, 253–61.

Gray, J. A. (1970). Sodium amobarbital, the hippocampal theta rhythm and the partial reinforcement extinction effect. *Psychol. Rev.* **77**, 465–80.

—— (1982). *The neuropsychology of anxiety: an enquiry into the functions of the septo–hippocampal system.* Oxford University Press, Oxford.

Green, J. D. and Ardunini, A. (1954). Hippocampal electrical activity in arousal. *J. Neurophysiol.* **17**, 533–57.

——, Maxwell, D. S., Schindler, W. J., and Stumpf, Ch. (1960). Rabbit EEG 'Theta' ryhthm: its anatomical source and relation to activity in single neurones. *J. Neurophysiol.* **23**, 403–20.

Green, K. F. and Rawlins, J. N. P. (1979). Hippocampal theta in rats under urethane: generators and phase relations. *Electroenceph. clin. Neurophysiol.* **47**, 420–9.

Henriksen, S. J., Chouvet, G., and Bloom, F. E. (1982). Response of rat hippocampal neurons to electrophoretically and pneumatically applied dynorphin and other opioid peptides. *Soc. Neurosci. Abs.* **8**, 228.

Hjorth-Simonsen, A. (1971). Hippocampal efferents to the ipsilateral entorhinal area: an experimental study in the rat. *J. comp. Neurol.* **142**, 417–38.

Holsheimer, J. (1982). Generation of theta activity (RSA) in the cingulate cortex of the rat. *Exp. brain Res.* **47**, 309–12.

Isaacson, R. L. (1982). *The limbic system* (2nd edn.). Plenum Press, New York.

James, D. T. D., McNaughton, N., Rawlins, J. N. P., Feldon, J., and Gray, J. A. (1977). Septal driving of hippocampal theta rhythm as a function of frequency in the free-moving male rat. *Neuroscience* **2**, 1007–17.

Jordan, T. C., Cane, S. E., and Howells, K. F. (1981). Deficits in spatial memory performance induced by early undernutrition. *Dev. Psychobiol.* **14**, 317–25.

—— and Clarke, G. A. (1983). Early undernutrition impairs hippocampal long-term potentiation in adult rats. *Behav. Neurosci.* (In press.)

Jung, R. and Kornmuller, A. E. (1938). Eine Methodik der Ableitung localisierter Potentialschwankungen aus Subcorticalen hirngebieten. *Arch. Psychiat.* **109**, 1–30.

Kandel, E. R., Spencer, W. A., and Brinley, F. J. (1961). Electrophysiology of hippocampal neurons. I. Sequential invasion and synaptic organization. *J. Neurophysiol.* **24**, 225–42.

Kitt, C. A., Price, D. L., DeLong, M. R., Struble, R. G., Mitchell, S. J., and Hedreen, J. C. (1982). The nucleus basalis of Meynert: projections to the cortex, amygdala, and hippocampus. *Soc. Neurosci. Abs.* **8**, 212.

Kolb, B. and Whishaw, I. Q. (1977). Effects of brain lesions and atropine on hippocampal and cortical electroencephalograms in the rat. *Exp. Neurol.* **56**, 1–22.

Kramis, R. C., Vanderwolf, C. H., and Bland, B. H. (1975). Two types of hippocampal rhythmical slow activity in both the rabbit and the rat: relations to behaviour and effects of atropine, diethylether, urethane and pentobarbital. *Exp. Neurol.* **49**, 58–85.

Le Moal, M. and Cardo, B. (1975). Rhythmic slow wave activity recorded in the ventral mesencephalic tegmentum in the rat. *Electroenceph. clin. Neurophysiol.* **38**, 139–147.

Levy, W. B., and Steward, O. (1979). Synapses as associative memory elements in the hippocampal formation. *Brain Res.* **175**, 233–45.

Lewis, P. R. and Shute, C. C. D. (1967). The cholinergic limbic system: projections to hippocampal formation, medial cortex, nuclei of the ascending reticular system, and the subfornical organ and supra-optic crest. *Brain* **90**, 521–40.

Livett, B. G. (1973). Histochemical visualisation of peripheral and central adrenergic neurones. In Catecholamines (ed. L. L. Iversen). *Br. Med. Bull. Suppl.* **29**, 93–9.

Lorente, de Nò, R. (1934). Studies on the structure of the cerebral cortex. II. Continuation of the study of the ammonic system. *J. Psychol. Neurol.* **46**, 113–77.

Low, W. C., Lewis, P. R., Bunch, S. T., Dunnett, S. B., Thomas, S. R., Iversen, S. D., Björklund, A., and Stenevi, U. (1982). Function recovery following neural transplantation of embryonic septal nuclei in adult rats with septohippocampal lesions. *Nature, Lond.,* in press.

Lynch, G. S., Jensen, R. A., McGaugh, J. L., Davila, K., and Oliver, M. W. (1981). Effects of enkephalin, morphine and naloxone on the electrical activity of the *in vitro* hippocampal slice preparation. *Exp. Neurol.* **71**, 527–40.

——, Rose, G., and Gall, C. M. (1978). Anatomical and functional aspects of the septo-hippocampal projections. In *Functions of the septo–hippocampal system* (ed. K. Elliott and J. Whelan), Ciba Foundation Symposium 58 (new series), pp. 5–20. Elsevier, Amsterdam.

Maru, E., Takahashi, L. K., and Iwahara, S. (1979). Effects of median raphe nucleus lesions on hippocampal EEG in the freely moving rat. *Brain Res.* **163**, 223–34.

McEwan, B. S., Weiss, J. M., and Schwartz, L. S. (1969). Uptake of corticosterone by rat brain, and its concentration by certain limbic structures. *Brain Res.* **16**, 227–41.

McLennan, H. and Miller, J. J. (1974). The hippocampal control of neuronal discharges in the septum of the rat. *J. Physiol., Lond.* **237**, 607–24.

—— and —— (1976). Frequency-related inhibitory mechanisms controlling rhythmical activity in the septal area, *J. Physiol.* **254**, 827–41.

McNaughton, B. L., Barnes, C. A., and O'Keefe, J. (1982). Directional specificity of hippocampal 'place' cells on a radial 8-arm maze. *Soc. Neurosci. Abs.* **8**, 316.

——, Douglas, R. M., and Goddard, G. V. (1978). Synaptic enhancement in fascia dentata: cooperativity among coactive afferents. *Brain Res.* **157**, 277–93.

McNaughton, N., Azmitia, E. C., Williams, J. H., Buchan, A., and Gray, J. A. (1980). Septal elicitation of hippocampal theta rhythm after localized deafferentation of serotonergic fibres. *Brain Res.* **200**, 259–69.

Meibach, R. C. and Siegel, A. (1977). Efferent connections of the septal area in the rat: an analysis utilizing retrograde and anterograde transport methods. *Brain Res.* **119**, 1–20.

Mellgren, S. I. and Srebro, B. (1973). Changes in acetylcholinesterase and distribution of degenerating fibres in the hippocampal region after septal lesions in the rat. *Brain Res.* **52**, 19–36.

Misgeld, U., Sarvey, J. M., and Klee, M. R. (1979). Heterosynaptic postactivation potentiation in hippocampal CA3 neurons: long-term changes of the post-synaptic potentials. *Exp. brain Res.* **37**, 217–30.

Mitchell, S. J. and Ranck, J. B. Jr. (1980). Generation of theta rhythm in medial entorhinal cortex of freely-moving rats. *Brain Res.* **189**, 49–66.

——, Rawlins, J. N. P., Steward, O., and Olton, D. S. (1982). Medial septal lesions disrupt theta rhythm and cholinergic staining in medial entorhinal cortex and produce impaired radial arm maze behavior in rats. *J. Neurosci.* **2**, 292–302.

Myhrer, T. (1975). Locomotor, avoidance, and maze behaviour in rats with selective disruption of hippocampal output. *J. comp. Physiol. Psychol.* **89**, 759–77.

Nilges, R. G. (1944). The arteries of the mammalian cornu ammonis. *J. comp. Neurol.* **80**, 177–90.

O'Keefe, J. and Black, A. H. (1978). Single unit and lesion experiments on the sensory inputs to the hippocampal map. In *Functions of the septo–hippocampal system* (ed. K. Elliott and J. Whelan), Ciba Foundation Symposium 58 (new series), pp. 179–92. Elsevier, Amsterdam.

—— and Dostrovsky, J. (1971). The hippocampus as a spatial map. Preliminary evidence from unit activity in the freely-moving rat. *Brain Res.* **34**, 171–5.

—— and Nadel, L. (1978). *The hippocampus as a cognitive map.* Oxford University Press, Oxford.

Olton, D. S., Becker, J. T., and Handelman, G. E. (1979). Hippocampus, space and memory. *Behav. Brain Sci.* **2**, 315–65.

Ott, T., Malisch, R., and Krug, M. (1977). Pharmacological analysis of hippocampal 'theta rhythm'. *Acta Neurobiol. Exp.* **36**, 720–1.

Papez, J. W. (1937). A proposed mechanism of emotion. *Arch. Neurol. Psychiat.* **38**, 725–43.

Petsche, H., Stumpf, Ch., and Gogolak, G. (1962). The significance of the rabbit's septum as a relay station between the midbrain and hippocampus. I. The control of hippocampal arousal activity by the septum cells. *Electroenceph. Clin. Neurophysiol.* **14**, 202–11.

Raisman, G. (1966). The connexions of the septum. *Brain* **89**, 317–48.

Ramòn y Cajal, S. (1911/1955). *Studies on the cerebral cortex (limbic structures)* (Trans. by L. M. Kraft). Lloyd-Luke Ltd., London, and The Yearbook Publishers Inc., Chicago.

Ranck, J. B., Jr. (1973). Studies on single neurons in dorsal hippocampal formation and septum in unrestrained rats. *Exp. Neurol.* **41**, 461–555.

—— (1976). Behavioural correlates and firing repertoires of neurons in septal nuclei in unrestrained rats. In *The septal nuclei* (ed. J. De France), pp. 423–62. Plenum Press, New York.

Rawlins, J. N. P., Feldon, J., and Gray, J. A. (1979). Septo–hippocampal connections and the hippocampal theta rhythm. *Exp. brain Res.* **37**, 49–63.

——, ——, and —— (1980). The effects of hippocampectomy and of fimbria section upon the partial reinforcement extinction effect in rats. *Exp. Brain Res.* **38**, 273–83.

——, ——, and —— (1982). Behavioral effects of hippocampectomy depend on inter-event intervals. *Soc. Neurosci. Abs.* **8**, 22.

—— and Green, K. F. (1977). Lamellar organization in the rat hippocampus. *Exp. brain Res.* **28**, 335–44.

Ribak, C. E., Vaughn, J. E., and Saito, K. (1978). Immunocytochemical localization of glutamic acid decarboxylase in neuronal somata following colchicine inhibition of axonal transport. *Brain Res.* **140**, 315–32.

Robinson, T. E. and Vanderwolf, C. H. (1978). Electrical stimulation of the brain stem in freely moving rats. II. Effects on hippocampal and neocortical electrical activity and relations to behaviour. *Exp. Neurol.* **61**, 485–515.

Rudell, A. P., Fox, S. E., and Ranck, J. B. Jr. (1980). Hippocampal excitability phase-locked to the theta rhythm in walking rats. *Exp. Neurol.* **68**, 878–96.

Segal, M. (1973). Dissecting a short-term memory circuit in the rat brain. I. Changes in entorhinal unit activity and responsiveness of hippocampal units in the process of classical conditioning. *Brain Res.* **64**, 281–92.

—— (1977). The effects of brainstem priming stimulation on interhemispheric hippocampal responses in the awake rat. *Exp. brain Res.* **28**, 529–41.

—— and Bloom, F. E. (1976). The action of norepinephrine in the rat hippocampus. IV. The effects of locus coeruleus stimulation on evoked hippocampal unit activity. *Brain Res.* **107**, 513–25.

Skrede, K. K. and Malthe-Sørensson, D. (1981). Increased resting and evoked release of transmitter following repetitive electrical tetanisation in hippocampus: a biochemical correlate to long-lasting synaptic potentiation. *Brain Res.* **208**, 436–41.

—— and Westgaard, R. H. (1971). The transverse hippocampal slice: a well defined cortical structure maintained *in vitro*. *Brain Res.* **35**, 589–93.

Smart, J. L., Adlard, B. P. F., and Dobbing, J. (1974). Further studies of body growth and brain development in 'small for dates' rats. *Biol. Neonate* **25**. 135–50.

Steward, O. and Scoville, S. A. (1976). Cells of origin of entorhinal cortical afferents to the hippocampus and fascia dentata of the rat. *J. comp. Neurol.* **169**, 347–70.

Storm-Mathisen, J. (1978). Localization of putative transmitters in the hippocampal formation, with a note on the connections to septum and hypothalamus. In *Functions of the septo–hippocampal system* (ed. K. Elliott and J. Whelan), Ciba Foundation Symposium 58 (new series), pp. 49–79. Elsevier, Amsterdam.

Stumpf, Ch. (1965). Drug action on the electrical activity of the hippocampus. *Int. Rev. Neurobiol.* **8**, 77–138.

Stumpf, W. E. (1971). Autoradiographic techniques and the localisation of estrogen, androgen and glucocorticoid in the pituitary and brain. *Am. Zoologist* **11**, 725–39.

Swanson, L. W. and Cowan, W. M. (1979). The connections of the septal region in the rat. *J. comp. Neurol.* **186**, 621–56.

Tielen, A. M., Van Leeuwen, F. W., and Lopes da Silva, F. H. (1982). The localization of Leucine–Enkephalin immunoreactivity within the guinea pig hippocampus. *Exp. brain Res.* **48**, 288–95.

Turner, R. W., Bainbridge, K. G., and Miller, J. J. (1982). Calcium-induced long-term potentiation in the hippocampus. *Neuroscience* **7**, 1411–16.

Valero, L., Stewart, J., McNaughton, N., and Gray, J. A. (1977). Septal driving of the hippocampal theta rhythm as a function of frequency in the male rat: effects of adreno-pituitary hormones. *Neuroscience* **2**, 1029–32.

Vanderwolf, C. H. (1969). Hippocampal electrical activity and voluntary movement in the rat. *Electroenceph. clin. Neurophysiol.* **26**, 407–18.

——, Kramis, R., Gillespie, L. A., and Bland, B. H. (1975). Hippocampal rhythmical slow activity and neocortical low voltage fast activity: relations to behaviour. In *The hippocampus, Vol. 2. Neurophysiology and behaviour* (ed. R. L. Isaacson and K. H. Pribram), pp. 101–28. Plenum Press, New York.

—— and Robinson, T. E. (1981). Reticulo-cortical activity and behaviour: A critique of the arousal theory and a new synthesis. *Behav. Brain Sci.* **4**, 459–514.

Van Harreveld, A. and Fifkova, F. (1975). Swelling of dendritic spines after stimulation of the perforant path as a mechanism of post-tetanic potentiation. *Exp. Neurol.* **49**, 736–49.

Van Hoesen, G. W. and Pandya, D. A. (1975). Some connections of the entorhinal (area 28) and perirhinal (area 35) cortices of the rhesus monkey. I. Temporal lobe afferents. *Brain Res.* **95**, 1–24.

——, ——, and Butters, N. (1972). Cortical afferents to the entorhinal cortex of the rhesus monkey. *Science* **175**, 1471–3.

——, ——, and —— (1975). Some connections of the entorhinal (area 28) and perirhinal (area 35) cortices of the rhesus monkey. II. Frontal lobe afferents. *Brain Res.* **95**, 25–38.

Vertes, R. P. (1981). An analysis of ascending brain stem systems involved in hippocampal synchronisation and desynchronisation. *J. Neurophysiol.* **46**, 1140–59.

—— (1982). Brain stem generation of the hippocampal EEG. *Prog. Neurobiol.* **19,** 159–86.

West, M. O., Christian, E. P., and Deadwyler, S. A. (1982). Sensory evoked potentials in the rat dentate gyrus reflect processing of different types of information. *Soc. Neurosci. Abs.* **8,** 924.

Wilson, C. L., Motter, B. C., and Lindsley, D. B. (1976). Influences of hypothalamic stimulation upon septal and hippocampal electrical activity in the cat. *Brain Res.* **197,** 55–68.

Winson, J. (1972). Inter-species differences in the occurrence of theta. *Behav. Biol.* **7,** 479–87.

—— (1974). Patterns of hippocampal theta rhythm in the freely moving rat. *Electroenceph. clin. Neurophysiol.* **36,** 291–301.

—— (1976*a*). Hippocampal theta rhythm. I. Depth profiles in the curarized rat. *Brain Res.* **103,** 57–70.

—— (1976*b*). Hippocampal theta rhythm. II. Depth profiles in the freely moving rabbit. *Brain Res.* **103,** 71–80.

—— and Abzug, C. (1978). Neuronal transmission through hippocampal pathways dependent on behavior. *J. Neurophysiol.* **41,** 716–32.

Zieglgänsberger, W., French, E. D., Siggins, G. R., and Bloom, F. E. (1979). Opioid peptides may excite hippocampal pyramidal neurons by inhibiting adjacent inhibitory interneurons. *Science* **205,** 415–17.

3

Behaviour and the limbic system with particular reference to sexual and aggressive interactions

J. HERBERT

INTRODUCTION

We are in the midst of an avalanche of new information on the structure and composition of the nervous system, each issue of the journals bringing greater detail about the distribution and variety of monoamines, peptides, and amino acids, their possible role as chemical transmitters or messengers, and intriguing ideas of how these chemically distinct systems within the brain are arranged. At the same time, we are being presented with the increasingly various and specific drugs and other compounds that interfere with, modulate, or destroy these systems, making experimental manipulation of their activity more possible. We must ask whether the study of behaviour, the most complex and important way to explore the function of the brain, is meeting the challenge of this revolution in other branches of the neurosciences. I shall argue in this chapter that it is not and that this is because behavioural science is trying to do what has proved impossible elsewhere: answering new questions using old (if well tried) methods. This situation is not new; we must remember that detailed knowledge of monoaminergic systems and the substances necessary to destroy, activate, or block them have been with us for a decade or more. Yet we are still without any generally acceptable account of the function of any of these substances in the brain, though, of course, suppositions and countersuggestions have not been lacking (see below). We may even ask whether trying to equate a particular chemically defined system with a specific describable behavioural function is a sensible strategy. Nowhere is this dilemma more pointed than in the limbic system, for this part of the brain is where much of the new neuroanatomy and neurochemistry has been described. There are other areas (for example, the striatum) which, lying outside the conventional limbic system, are nevertheless reported to contain large quantities of the recently recognized chemotransmitters, but a special relationship between these regions and the limbic system is increasingly recognized, on both anatomical as well as functional grounds (see Chapter 1, this volume).

It is therefore appropriate to consider these problems in the context of the limbic system. I shall do so by special reference to the neural basis of sexual and aggressive behaviour, since these behaviours are quintessentially those attributed to this part of the brain and because they show, in their individual characteristics and their interactions, the difficulties of studying behaviour in the current neuroscientific ferment. I shall argue that the principal problem which currently afflicts the study of behaviour is that the historical simplification, by which behavioural variables were reduced to a minimum, though performing the invaluable function of allowing a complex subject to be rigorously explored experimentally and thereby yielding much fundamental and important information, has now become self-defeating. This is because the methods themselves, in seeking to reduce complexity, have at the same time removed from the area of study essential features related to the function of parts of the limbic system. Thus, eliminating from experimental conditions the necessity for animals to perform certain neural activities – particularly the selection of motivationally diverse stimuli and, consequently, decisions about which of a number of possible actions to take (Herbert 1968) – has meant that the role limbic areas play in these biologically important functions of the brain may also have been removed.

STUDYING SEXUAL BEHAVIOUR UNDER SIMPLIFIED CONDITIONS

The strengths and weaknesses of current methods of studying the limbic control of sexual behaviour can be assessed by considering the relatively simple method of pairing animals together for limited periods of time and observing their sexual interaction. This technique, still widely used, has all the attractions of simplicity, reproducibility, and quantifiability that characterize so many of the techniques currently used in behavioural neuroscience. The search for the neural substrate responsible for mating behaviour under these conditions was brightened by the finding that there were neural systems in the brain which accumulated and bound steroid hormones, an early and exemplary demonstration of a chemically specific system within the brain. Furthermore, the chemical nature of this system, unlike many more recently discovered ones, gave a clear hint about possible function.

Steroid-binding neurones are distributed in four main groups (though there are other, less conspicuous, ones): in the pre-optic area and anterior hypothalamus; in the ventro-medial/arcuate nucleus area of the hypothalamus; in the cortico-medial amygdala; and in the septum (McEwen 1981). It is still not entirely clear, on either anatomical or functional grounds, whether these four groups are separate systems; whether, for example, the septal binding neurones can really be separated from those in the underlying anterior hypothalamus other than by the fortuitous presence of the anterior commissure. Nevertheless, the expectation that at least part of this system would represent the neural site of steroid action is well founded. Electrolytic lesions in the anterior hypothalamus/pre-optic area prevent the hormonal activation of sexual behaviour of male rats (and other species) (Heimer and Larsson 1966), and, though less reliably, of females (Malsbury *et al.* 1977). The problem with such techniques is that, as is well known, they destroy all the cells in an area (including those not binding steroids) as well as the nerve fibres passing through it.

This is particularly significant in a region through which pass many connections between different parts of the limbic system – for example, from the amygdala and septum to the hypothalamus (Garris 1979). It seems likely that the newer cellular neurotoxins, such as ibotenic acid and n-methyl aspartate, will supersede earlier methods of producing localized lesions, though they have their own problems of specificity. For example, neurones in the pre-optic area are destroyed by injecting ibotenic acid, and the resultant lesions disrupt male sexual behaviour, reinforcing the view that a population of neurones in this part of the brain is critical (Hansen *et al.* 1982). This proposi-tion is also supported by a much more specific technique: implanting steroids directly into the brain. This type of chemical stimulation is particularly apt in this context, combining anatomical and chemical selectivity. Implants into the anterior hypothalamus of testosterone or its intracerebral metabolite oestradiol restore, to some degree at least, the behaviour of castrated male rats (Davidson 1966; Davis and Barfield 1979). Peripheral, hormonal-dependent mechanisms (such as the spines on the male's penis) are also necessary for sexual behaviour in the male to be fully expressed (Beach and Levinson 1950), reminding us that systemically-acting substances may have co-ordinated effects on both the limbic system and on peripheral structures.

But what of the functions of other steroid-binding neural systems? They may be involved in the regulation of pituitary function; for example, the arcuate nucleus is known to be involved in negative feedback, allowing plasma levels of steroids to regulate the amount of LH produced by the pituitary (review by Feder 1981*a*). The anterior hypothalamus may regulate positive-feedback effects of oestradiol and thus ovulation, though whether this includes part of the steroid-binding system is not entirely clear. But neither the septal nor cortico-medial amygdala can safely be allocated to this role, though endocrine changes have been demonstrated following manipulation of either. Septal lesions can alter sexual behaviour (as part of the septal 'hyper-reactivity' syndrome) (Brady and Nauta 1953) but whether the steroid-binding neurones here play a special or definable role in sexual activity, even in rodents, is unknown. The way that the cortico-medial amygdala is implicated in sexual behaviour will be discussed further below, but it is also difficult to fit its function into sexual behaviour as we understand it from 'pair test' experiments (Harris and Sachs 1975). Its large olfactory input suggests (but does not prove) that hormones may be able to operate on this important sensory modality (in rodents), while there is recent evidence that implants of the testosterone metabolite dihydrotestosterone (DHT) in this region may have a mildly stimu-lating effect on the castrated male rat's sexual behaviour (Baum *et al.* 1982).

'Males' and 'females' vs. 'masculine' and 'feminine'

When we consider the interpretation placed upon behaviour as it appears when studied under such conditions, further complications arise. Although 'male' and 'female' sexual behaviour have been discussed, in fact the term refers more to the pattern of the behaviour rather than the sex of the animal displaying it. 'Masculine' (mounting) or 'feminine' (lordosis) behaviour can be shown by both male and female rats under certain conditions, and the response to steroid manipulation differs according to the type of behaviour rather than the sex of the animal displaying it. For example, 'masculine' patterns of behaviour decline much more slowly after castration in both males or females than does lordosis. 'Heterotypical' behaviour is most easily induced by treating animals with the 'wrong' hormone combination and by pairing them with sexually active members of the same sex. We therefore have not one set of variables to consider, but three: sex of the recipients, sex of the partner, and the hormonal treatment of both, in determining the dependent function, the pattern of behaviour. Pairing tests show only that the threshold for a given pattern of behaviour can be varied as each of these conditions is itself varied.

It is now well established that treating female rats early in life with testosterone, or removing the testes from males, can alter this relationship, making heterotypical behaviour more easily inducible, and homotypical behaviour more difficult (Young 1961). What does this tell us about the function of the limbic system? The same hormonal treatments have been demonstrated to replicate or reduce sexually dimorphic patterns of organization in this part of the brain, including those in the synaptic input to the pre-optic area (Raisman and Field 1973), and the size of the medial pre-optic nucleus itself (Harlan *et al.* 1979). How these anatomical findings relate to sexual behaviour is still problematical, despite recent attempts to transfer sex differences in behaviour patterns by means of brain grafts and hence relate macroscopic structure to behavioural control systems (Arendash and Gorski 1982). Furthermore, although earlier workers described activation of the oestrous (lordosis) behaviour of female rats by intracerebral implantation of oestradiol in the anterior hypothalamus, more recently it has been claimed that the most sensitive site is, in fact, somewhat posterior: in the ventro-medial area which, if we recall, also contains steroid-binding neurones (Rubin and Barfield 1980). If there are indeed two sites in the brain, one responsible for the hormonal modulation of 'masculine' patterns and the other for 'feminine', we have to consider whether sexual differentiation of behaviour patterns by hormones (in rodents at least) involves a mechanism analogous to that operating on the internal genitalia in which two systems (Wolffian and Mullerian) exist side by side, each encouraged or discouraged by different endocrine conditions in early life. If this is so, 'bisexuality' implies the existence of two neural systems side by side. Alternatively, there may be a common site in the brain on which hormones act to activate either 'masculine' or 'feminine' behaviour depending both upon the antecedent endocrine experience of the animal and its current stimulus. This scheme thus compares more closely to the external genitalia, differentiating in either the 'male' (penile) or 'female' (vaginal) direction or, of course, arrested at some point in between (see also discussion by Feder 1981*b*). We cannot yet choose between these alternatives, even if we suspect that either of them represents a gross oversimplification, perhaps because the behavioural conditions we employ are usually so simple. Animals are made to respond to a given partner, not to choose from an array of possible partners, or even to decide to give preference to other kinds of response (for example, to eat or even to escape).

This experimental emphasis on behaviour patterns has not only limited our understanding of sexual behaviour in neural terms; it has also made extrapolation to clinical and human situations perilous, if tempting. Work on human behaviour has focused on the nature of the objects towards which sexuality is directed (e.g. hetero- vs. homosexuality) or upon the self-perception of gender identity (e.g. transsexualism). The inadequacy of some experimental models in the case of the former and their complete absence in the latter makes direct comparison of work on animals with humans doubtful and requiring much greater caution than is usually observed, a subject discussed more fully elsewhere (Herbert 1980). This is not to say that, were experimental procedures to be adapted, studies on the role of the limbic system of experimental animals might not yield information of great interest to students of human sexual behaviour. The behavioural methods of studying both must be made congruent if this is to happen.

Sensory input into the limbic system

The outline above explores the limitations of studying a limbic-controlled behaviour in a simplistic manner, and the consequent strange lack of information about brain–behaviour relations even in sexual activity, one of the better understood 'motivated' behaviours. The argument could be developed in other ways. The nature and neural processing of the stimuli eliciting sexual behaviour have received comparatively little attention experimentally and are not well understood. If we attempt to divide up properties of stimuli (or 'cues') in terms of the modalities by which the animal perceives them (this distinction may be more for convenience than anything else), we see that information concerned with sexual behaviour reaches the limbic system in significantly different ways (Herbert 1977). For example, olfactory input has comparatively direct access to the limbic system, bearing in mind the anatomical input from the olfactory bulb to the amygdala and temporal lobe and the way these structures themselves project to the rest of the limbic system. The question is how certain olfactory stimuli are classified as sexually exciting (rather than aggression-promoting, maternal-inducing, or predation-invoking) and how steroids or other neuroendocrine inputs modify this channelling of information. Nevertheless, electrophysiological and anatomical evidence clearly points to a general understanding of this phenomenon, even if the details remain obscure.

By contrast, visual information, also highly important, is processed very differently. In mammals, at least, visual processing is highly dependent upon the neocortex. The problem here is whether the cortex, having decoded and analysed the signal, then classifies it as 'sexual' and transmits this information (or some derivative of it) to those parts of the limbic areas that are directly responsible for sexual activity. Conversely, can the limbic system act recipro-cally upon the neocortex to alter this processing or the behavioural definition arrived at? Or is all sensory information received, in some form, by the limbic system and classified accordingly in this part of the brain?

USING MORE COMPLEX METHODS

Since I am primarily concerned here with behavioural techniques and concepts, I want to consider how modifying the circumstances under which behaviour is studied can greatly modify our ideas of how the limbic basis of sexual behaviour is organized, and the role that hormones play in it.

Most animals live in groups of some sort. If one studies endocrine manipulations in such groups, one begins to see that the neural mechanisms required for the expression of sexual behaviour are necessarily more complex than pairing experiments suggest. Groups of talapoin monkeys, for example, form a dominance hierarchy. This means that the animals can be arranged in an order (usually linear) which reflects the direction of aggressive interaction between the members. Of the males within such groups, only the most dominant show complete patterns of sexual interaction; subordinate males copulate very little, and even their sexual 'interest' in the females (for example sniffing or inspecting the female's genitalia) is sharply diminished (Yodyingyuad *et al.* 1982; Everitt *et al.* 1981). This differential distribution of sexual behaviour is dependent upon hormones to the extent that, if all the males are castrated, none of them shows much sexual interest; if all the females are ovariectomized, the dominant male displays sexual activity only if they are then treated with ovarian hormones. However, though the relation between sexual behaviour and the dominance hierarchy needs steroid hormones to reveal it, it is in no way altered by those hormones themselves. Thus, differential treatment of subordinate males (or females) with steroid hormones does not influence their position in this hierarchy directly (Keverne 1979). However, it should be noted that some species (e.g. rhesus monkeys) display the phenomenon of 'dependent' rank. If a comparatively low-ranking animal (e.g. a female) forms a consort relationship with a male during the time her ovarian hormones make her sexually attractive, she may acquire her consort's high rank, only to lose it once the consortship is over as her hormones wane (Zuckerman 1932). In this indirect way, steroid hormones can modulate a dominance hierarchy.

Two points important to this discussion follow from these observations. First, we need to enquire how the activating effects of hormones on the limbic system are constrained by the social context in which the animals live. Second, we must consider the role of aggressive interaction in the context of sexuality, and in limbic function.

Dominance and decision-making

The position of animals in the dominance hierarchy produces changes in steroid hormone levels but these cannot account for the suppression of sexuality in subordinate animals. Though testosterone levels are lower in subordinate than dominant males (Keverne 1979; Eberhart *et al.* 1980), they are nevertheless sufficient for sexual activity, since artificially maintaining castrated monkeys on comparable levels allows them to copulate adequately. Do immediate factors, for example the threat of attack by more dominant males, deter subordinates, therefore denying them access to a limited resource (in this case, sexually attractive females)? In a different context, these factors may act to deny subordinates access to food in direct competition. There is no doubt that dominant males may attack subordinates if the latter attempt to mount. However, even castrated dominants, themselves not very sexually active, may deter intact subordinates. Therefore there may be some other property of being subordinate which reduces such a male's sexual interest in females, even in the presence of adequate testosterone levels and (presumably) responsive hypothalamic neurones.

We can call the systems by which an animal's social structure modulates behavioural responses 'strategic' and 'tactical'. 'Strategy' refers to some mechanism which functions to the benefit of either a dominant animal (e.g. by removing the subordinate from sexual competitiveness) or the subordinate; for example, reducing the likelihood that he will make a 'wrong' decision, attempt to mate, and invite aggression('cost'). This might be considered as part of the 'coping' mechanism of subordination. 'Tactics' suggest that subordinate males may take moment-to-moment decisions about whether the 'benefit' of mating (in terms of immediate reward) is likely to outweight the 'cost' of being attacked – or of being rejected by females (who seem to find subordinate males sexually less attractive than dominant ones). The different use of the terms 'cost' and 'benefit' in either the context of strategies (or 'ultimate' factors) – the common usage in ethology – or as a tactical (or 'proximate') term must be noted, because the two categories will involve different neural mechanisms.

AGGRESSION AND THE LIMBIC SYSTEM

Since aggression has been invoked as both a determinant of the dominance hierarchy and a modulator of sexual activity, it is time to consider this category of behaviour and its relationship to the limbic system more closely. Even though different workers may argue about the definition of sexual behaviour, this is nothing compared to the unresolved debate about how to define and classify aggression. The most widely accepted definition, by Moyer (1968), is a curious mixture of causal, descriptive, and contextual terms, surviving only because it has proved difficult to produce a better one. It is important to note that most workers in this field consider aggressive behaviour in isolation, as a unitary phenomenon, even though in the life of most animals, aggressive behaviour occurs in the context of some other activity: feeding, mating, caring for the young, defending a resource either directly (e.g. by fighting over food) or indirectly (for example, preserving a territory by which these resources may be acquired). We may thus doubt whether aggression studied by itself, for example by pairing together male rodents, is likely to tell us everything we need to know about its function and, therefore, the activities of those parts of the limbic system implicated in aggressive behaviour.

These considerations have become important as we consider the functions of that part of the limbic system, particularly the amygdala, implicated in aggressive behaviour. Ever since the original description of the Kluver–Bucy syndrome (which included damage to other temporal structures besides the amygdala), it has been agreed that destroying all or part of the amygdala can lead to docility, lack of fear, and reduced ability to display aggression (Kluver and Bucy 1938). The form that this takes depends upon the conditions under which it is studied. For example, fear of new food may be reduced; so too is the response to a territorial intruder or passive avoidance behaviour studied under operant conditions, or taste aversion as one result of a previous unpleasant experience (Rolls and Rolls 1973; Grossman *et al.* 1975; Pellegrino 1968; Miczek *et al.* 1974). Electrical stimulation in the amygdala has generally produced converse effects: increased likelihood to attack an object, or (depending upon the exact method or site of stimulation) a defensive and fearful response in situations where this is otherwise unlikely to occur. Lesions of the ventro-medial hypothalamus, very close to those areas implicated in sexual activity and feeding, also cause increased aggressivity. There is good evidence that the amygdala and hypothalamus interact, so that lesioning one mitigates the effects of damage or stimulation to the other (reviewed by Siegel and Edinger 1982). The usual interpretation is that we are studying a neural system specifically involved with aggression, which somehow determines whether or not an animal will show this kind of behaviour – a highly questionable interpretation.

Aggression, like any other behaviour, results from the interaction between the stimuli (or cues) from outside the animal and its internal state, however that is determined. Marked changes in the stimuli which are effective can occur during an animal's life cycle. Males may attack other males, but only during the breeding season. Females may tolerate males outside the breeding season, actively solicit their company during it, but become highly aggressive to them once they have delivered a nest full of young (Rosenblatt *et al.* 1979). Those conditions, whatever they may be, which determine the principal motivational state of the animal also alter the context in which it shows aggression, and the two behaviours should be considered together and, more importantly, studied as part of a whole. A rather different picture emerges if one considers 'fear' or 'anxiety'. Such a response may be directed towards an object or individual (e.g. an aggressive and dominant member of the group) which the recipient has learned can be the source of dangerous or painful consequences. 'Defensive' aggression may occur in the same context as an alternative response to flight if and when the animal is actually attacked. Fear (anxiety) or defence in this context is thus the converse of aggressive behaviour. However, 'fear' or 'uncertainty' may also be engendered by lack of information. For example, an animal faced with a new or novel situation, or one in which it is unable to form any reliable or certain estimate of the nature of its predicament, may demonstrate fear or anxiety and seek to improve its knowledge by, perhaps, exploring its environment. We have therefore to postulate at least two kinds of neural activity in the limbic system related to 'fear' – one which determines the kind of stimulus that an animal may find fear-inducing, and another which gathers information within a behavioural context until a decision can be made about any likely threat. The behavioural responses to these two kinds of 'fear' (or 'anxiety') are demonstrably different.

The effects of lesions in the amygdala strongly suggest that these functions are highly impaired. The animal no longer behaves in a way that suggests that it is assessing its surroundings in terms of threat, and it therefore shows both diminished fear and aggressive responses, though this effect may be altered by the condition in which it is studied (Haber 1981). The problem of identifying that part of the limbic system which changes the bias or 'set' of an animal towards others according to its own internal physiological condition remains intriguing. Demonstrations that activity in the hypothalamus can determine sexual, feeding, or maternal activity suggest that it may, at the same time, confer on the amygdala some associated property concerned with determining the critical nature of aggression-inducing stimuli. Disturbances in the discrimination related to feeding and sexual behaviour have been described following damage to this part of the brain and form part of the original Kluver–Bucy syndrome. These considerations suggest that we shall only be able to tease out the complex function of the amygdala if we study correspondingly complex forms of behaviour; that is, if we allow the animals under study to display the variety of responses and decisions which are normally required for successful survival and reproduction. Furthermore, treating aggression in isolation as a separate category of behaviour is likely to be unrewarding. It is also important to determine whether this system relates in some way to the hippocampal–septal pathways which have been proposed, on a variety of grounds, to be concerned with generating 'anxiety' (Gray 1982).

If the amygdala is concerned with the neural processes whereby motivationally-important decisions are taken, then it must receive information about current sensory input. Recent anatomical evidence strongly supports this interpretation. Not only does the baso-lateral amygdala receive a plentiful input from the neocortex (either directly or indirectly) but this input comes mostly from the secondary sensory areas which, it is now known, deal with the processing or elaborations of information received from the primary ones (Turner 1973, 1981; Ottersen 1981). It is beginning to look as if the amygdala is an entrance to the limbic system for this kind of input, which can then act upon those other areas (e.g. septum, hypothalamus, and brainstem) which excite the relevant motivational response, in the light of the animal's estimate of 'cost' and 'benefit'.

We can postulate that such a motivational 'decision' has three parts

1. Defining the motivational category of the stimulus;
2. Assessing the value (or 'incentive') properties of the stimulus;
3. Assessing the likely consequences of responding to the stimulus in question.

It is probable that the amygdala is involved in all three, though this remains undemonstrated.

Arguing from a different standpoint, Mishkin and Aggleton (1981) have also proposed that the amygdala is an essential link between affective and cognitive processes which determine an animal's ability to learn an association between a complex stimulus and a behavioural response.

CATECHOLAMINERGIC SYSTEMS AND BEHAVIOUR

It is now time to turn to specified neural systems as an example of the way that behavioural science has attempted to relate function to chemically defined pathways. Studies on the catecholamines represent an earlier furore, corresponding in many ways to those currently under way concerning neuropeptides. There are increasingly excellent methods of selectively altering the function of catecholamines (i.e. drugs), as well as specific methods of removing them (neurotoxins) and for assessing changes in their activity (turnover, receptors, and metabolism). And yet, despite many attempts, we cannot confidently ascribe a function to any of the principal monoaminergic systems. Is this because we are attempting a mismatch between an anatomical and a functional entity – attempting, that is, an over-simplistic superimposition between two ideas? Or is it because our methods have failed us?

The disposition of the noradrenergic system is well known and will serve as our example (review by Robbins and Everitt 1982). It arises from groups of neurones located in the brainstem – one set (in the lateral reticular formation) connecting primarily with the limbic system, the other (largely from the locus ceruleus) with the cortex. A variety of functions has been ascribed to this system, a term which itself implies – perhaps incorrectly – that it operates as a whole. Operant experiments in which animals learn to bar-press for food have convinced some that cortical noradrenaline is concerned with learning, others that it is an essential component of 'attention' or 'anxiety' (Robbins and Everitt 1982). Yet others, using different behavioural methods which rely upon interaction between animals, come to an apparently different conclusion. For example, selective destruction of the ventral noradrenergic bundle, and hence input to the limbic system, impaired the usual sequence characterizing the sexual behaviour of the female rat. Such animals continued to display the behaviour (proceptivity) by which the male is incited to mount, but were unable to switch this response to the stereotyped immobilized reflex (lordosis) which enables him to intromit once proceptive behaviour has been successful.

It is difficult to ascribe such findings to alterations in 'attention' or 'learning' but rather as inability to act upon sensory information of a particular kind (Hansen *et al.* 1981). Another, and even more striking, example of a similar function for noradrenaline comes from experiments on mice. Female mice, recently mated, will resorb their pregnancy if exposed to the odours of a 'strange' male (that is, one other than their mate), an effect presumably related to the territorial behaviour of male mice under natural conditions. However, the removal of the noradrenergic input to her olfactory bulbs prevents the female's neuroendocrine system from recognizing 'her' male from others, so that, should he reappear, he now blocks his own pregnancy (Keverne and de la Riva 1982). These examples of neuroendocrine 'behaviour' show us that we need to consider noradrenaline in a rather different light, as a modulator of sensory input. Perhaps we should pay more attention to the increased signal-to-noise ratio demonstrated as a function of noradrenaline by neurophysiologists (Kanematsu and Heggelund 1982). It may be that particularly 'significant' sensory events, either themselves or the context in which they occur, activate noradrenaline. This results, in the short or longer term, in differences in response to some associated sensory input – in the two examples given, to tactile information from a mounting male, or to olfactory input from a recent mate. We begin to see how 'attention', 'learning', and 'print-now' mechanisms can be reconciled, and how we ought to take into account not only the function of noradrenaline itself but also its specific role in particular categories of behaviour and, hence, a particular sensory system. We also begin to see that studying behaviour – particularly a hormonally regulated one – on its own, without considering the endocrinologically mediated consequences of such a behaviour, may give us an incomplete story. This conclusion is further reinforced by reference again to the female rats deprived of their ventral noradrenergic bundle, who not only fail to display lordosis but, even if vaginally stimulated, do not enter the period of pseudo-pregnancy character-istic of the normal animal (Hansen *et al.* 1981). It is noteworthy that the act of mating sets up a kind of short-term neurodendocrine 'memory' in the rat, represented by twice daily surges of prolactin (which are needed for successful pregnancy or pseudo-pregnancy) (Neill 1980), and it is this process which is apparently dependent upon noradrenaline.

We have to try to understand the function of such a system in the context of the animal's behavioural repertoire. If such arguments apply to noradrenaline, then they apply with even greater force to dopamine, since dopamine-containing systems have equally been involved in many sorts of behaviour (review by Robbins and Everitt 1982). But we need to distinguish clearly between the function of such a system and that of the areas of the brain to which it projects.

PEPTIDERGIC SYSTEMS AND BEHAVIOUR

Almost in parenthesis, and as a preliminary argument only, we can consider some of the recent information on peptides. Some peptides, including the enkephalins, because of their widespread disposition within short-axoned neurones, are not readily imagined to have a single (behavioural) function, but a role which will depend upon the part of the brain in which they are found. Others (e.g. endorphins) seem to be located in restricted neurones which may branch throughout parts of the limbic system (Stengaard-Pedersen and Larsson 1981); such peptides may prove to have a different kind of role. The peripheral distribution of some of these peptides suggests that their function within the brain may somehow be linked with that outside it. Angiotensin given centrally causes animals to drink: peripherally it is released following haemorrhage, causing vascular constriction and fluid retention (Fitzsimons 1979). The two responses, one behavioural and the other physiological, contribute together toward adaptation to a threatening event. LHRH has been claimed to activate sexual behaviour (though the evidence is still not very strong) (Tennent *et al.* 1982). Peripherally the same peptide evokes LH release from the pituitary, and hence ovulation. Both behavioural and endocrine responses contribute towards fertility. Oxytocin is elevated by parturition and by suckling, and causes milk ejection and (in some species) uterine contractions. Centrally administered, it is claimed to induce maternal behaviour (Pedersen *et al.* 1982), a response adaptive in the context of both parturition and suckling. The clues are obvious, but there is much work before these (and some other) peptides can be invoked as co-ordinating behavioural and physiological adaptations. The three peptides considered here all have cell bodies localized in or near the hypothalamus, and these send branches to other parts of the limbic system and to the brainstem. The anatomical basis of this postulated co-ordinating role is easy to see, though how this might fit with other neuroendrocrine factors acting on the same behaviours is more difficult to discuss at the present time. The suspicion, however, is that there may be chemically identifiable limbic systems which correlate rather more closely with categories of behaviour (as these are usually defined) than is the case for the monoamines. There are many other peptides, however, for which such direct correlation with specific behaviour seems unlikely. The way that peptides and monoamines interact is an intriguing subject, and one which is rapidly developing, but is outside this discussion.

Extracerebral neuropeptides: prolactin

To complete this outline of the limbic system and its role in behaviour, we must consider the anterior pituitary, only recently recognized as a source of extra-cerebral neuropeptides. A number of relatively high-molecular-weight peptides are released from the anterior pituitary, of which prolactin (molecular weight c. 20 000) is of particular interest in this discussion. Studying both the function of the anterior pituitary itself and sexual behaviour in animals made hyperpro-lactinaemic by a variety of methods, clearly suggests that this hormone can have direct action upon the brain. Prolactin infused peripherally increases dopamine turnover in the ventro-medial hypothalmus, an effect replicated by prolactin injected directly into the third ventricle (Nicholson *et al.* 1980). More prolonged infusions of prolactin intraventricularly decrease the secretion of anterior pituitary prolactin, suggesting a negative feedback, and greatly reduce the ability of the gland to discharge prolactin in response to stimulatory substances (Herbert and Martensz 1982). There is clinical evidence, only partly replicated experimentally, that hyperprolactinaemia also alters the production of LHRH, and hence the release of gonadotrophins from the pituitary (Thorner 1977; McNeilly 1980). Both clinical and experimental results on sexual behaviour agree that, in the male at least, persistent hyperprolactinaemia reduces potency in the human male and impairs copulation in the male rat (Bailey and Herbert 1982). This effect occurs despite there being adequate amounts of testosterone in the blood, further suggesting that prolactin has a direct action upon the brain. What this action may be, and where it occurs, is still the subject of active enquiry, though recent reports of prolactin receptors in the hypothalamus (DiCarlo and Muccioli 1981) and the possibility that the metabolism of testosterone within the amygdala may be altered by hyperpro-lactinaemia (Bailey and Herbert, unpublished) clearly suggest that the limbic system may be a primary target for prolactin.

If this peptide is to be considered an important modulator of limbic-system function, then it must be able to enter the brain. Recent evidence shows that prolactin enters the brain slowly, in proportion to levels in the blood, so that, if serum levels are changed, this results in equivalent alterations in prolactin in the CSF. Since prolactin enters the CSF as easily from the peripheral venous system as from the animal's own pituitary, a privileged route of access to the limbic system from the pituitary along the portal vessels (by retrograde flow) seems unnecessary, in this context at least (Martensz and Herbert 1982).

Though information is less complete on some other anterior pituitary peptides, it seems that prolactin may be a special case (Dubey *et al.* 1983). LH, a peptide of similar molecular weight to prolactin, also enters the CSF in amounts related to those in the blood, but the CSF/plasma ratio is five or ten times lower than that of prolactin. By contrast, ACTH (with a molecular weight much smaller than the other two peptides) seems unable to enter the CSF from the vascular compartment; the same may also be true for β-endorphin, which is released from the anterior pituitary together with ACTH (Fukata *et al.* 1982). Why these peptides should differ so markedly in their ability to enter the brain, and hence alter neural function, is still unclear. One possibility is that peptides are rigidly excluded if there are intracerebral systems using the same substances as a neural transmitter, as seems to be the case for ACTH and β-endorphin. However, this argument is weakened by the possiblity, not yet fully proven, that there may be neural systems making LH and prolactin as well (Hofsetter *et al.* 1981).
neural systems making LH and prolactin as well (Hofsetter *et al.* 1981).

The important point relevant to this discussion is, however, that prolactin is a hormone which is highly sensitive to changes in an animal's environment. Changes in the time of day or the time of year result in marked fluctuations of prolactin levels in the blood. Internal stimuli (such as oestrogen) are also highly effective, though these are more obvious in rodents than in primates. Sensory input during mating releases prolactin (and hence results in pseudo-pregnancy in mice), and its suppression accounts for the phenomenon of pregnancy block described earlier in the chapter. Prolactin can be elevated in 'stressful' conditions, including those experienced by subordinate talapoin monkeys. All this indicates that this hormone is responsive to the behavioural as well as the physical environment. High levels of prolactin induced in such monkeys may be sufficient to prevent oestrogen-induced LH surges, similar to the condition described in hyperprolactinaemic women and hence reduce or eliminate fertility (Bowman *et al.* 1978). Whether socially induced hyperprolactinaemia also acts to depress the sexual behaviour of subordinate males or females, both of whom can show high prolactin levels, is less certain at the moment, though the possibility certainly exists. Whether prolactin interacts within the brain with other neural systems, including LHRH or steroid-binding neurones, and thus regulates sexual behaviour, or even whether its effect is limited to reproductive function and behaviour, remains unexplored. However, the role of this important neuropeptide in a behavioural context would not have been apparent had sexual activity been studied only in the simplified conditions described in the earlier part of this chapter. Even had the depressant effect of prolactin on sexual behaviour been detected, the relevance of such a finding would have remained obscure.

CONCLUSION

We must overhaul our methods of studying behaviour if we are to make functional sense of the new knowledge about the structure and organization of the limbic system. Behavioural studies have to be made as powerful analytically as the techniques in other branches of neuroscience to which they relate.

ACKNOWLEDGEMENTS

I am most grateful to my colleagues L. T. Dunn, B. J. Everitt, and E. B. Keverne for comments on this manuscript, and to Jane Pratt for her help in preparing it. The work from this laboratory is supported by an MRC Programme Grant.

REFERENCES

Arendash, G. W. and Gorski, R. A. (1982). Enhancement of sexual behaviour in female rats by neonatal transplantation of brain tissue from males. *Science* **217**, 1276–8.

Bailey, D. J. and Herbert, J. (1982). Impaired copulatory behaviour of male rats with hyperprolactinaemia induced by domperidone and pituitary grafts. *Neuroendocrinology* **35**, 186–93.

Baum, M. J., Tobet, S. A., Starr, M. S., and Bradshaw, W. G. (1982). Implantation of dihydrotestosterone propionate into the lateral septum or medial amygdala facilitates copulation in castrated male rats given oestradiol systemically. *Horm. Behav.* **16**, 208–23.

Beach, F. A. and Levinson, G. (1950). Effects of androgens on the glans penis and mating behaviour of castrated male rats. *J. exp. Zool.* **114**, 159–71.

Bowman, L. A., Dilley, S. R., and Keverne, E. B. (1978). Suppression of oestrogen-induced LH surges by social subordination in talapoin monkeys. *Nature* **275**, 56–8.

Brady, J. V. and Nauta, N. J. H. (1953). Subcortical mechanisms in emotional behaviour: affective changes following septal forebrain lesions in the albino rat. *J. comp. Physiol. Psychol.* **46**, 339–46.

Davidson, J. M. (1966). Activation of the male rat's sexual behaviour by intracerebral implants of androgen. *Endocrinology* **79**, 783–94.

Davis, P. G. and Barfield, R. J. (1979). Activation of masculine sexual behaviour by intracranial estradiol benzoate in male rats. *Neuroendocrinology* **28**, 217–27.

Di Carlo, R. and Muccioli, G. (1981). Changes in prolactin-binding sites in the rabbit hypothalamus induced by physiological and pharmacological variations of prolactin serum levels. *Brain Res.* **230**, 445–50.

Dubey, A. K., Herbert, J., Martensz, N. D., Beckford, U., and Jones, M. T. (1983). *Life Sciences* **32**, 1857–63.

Eberhart, J. A., Keverne, E. B., and Meller, R. E. (1980). Social influences on plasma testosterone levels in male talapoin monkeys. *Horm. Behav.* **14**, 247–66.

Everitt, B. J., Herbert, J., Keverne, E. B., Martensz, N. D., and Hansen, S. (1981). Hormones and sexual behaviour in rhesus and talapoin monkeys. In *Steroid hormone regulation of the brain* (ed. K. Fuxe), pp. 317–30. Pergamon, Oxford.

Feder, H. H. (1981a). Experimental analysis of hormone action on the hypothalamus, anterior pituitary and ovary. In *Neuroendocrinology of reproduction* (ed. N. T. Adler), pp. 243–78. Plenum, New York.

—— (1981b). Perinatal hormones and their role in the development of sexually dimorphic behaviour. In *Neuroendocrinology of reproduction* (ed. N. T. Adler), pp. 127–57. Plenum, New York.

Fitzsimons, J. T. (1979). *The physiology of thirst and sodium appetite*. Cambridge University Press.

Fukata, J., Nakai, Y., Endo, K., and Imura, H. (1982). Hypoglyamic induced elevation of immunoreactive β-endorphin levels in cerebrospinal fluid in the cat. *Brain Res.* **246**, 164–7.

Garris, D. R. (1979). Direct septo-hypothalamic projections in the rat. *Neurosci. Lett.* **13**, 223–37.

Gray, J. A. (1982). *The neuropsychology of anxiety*. Oxford University Press.

Grossman, S. P., Grossman, L., and Walsh, L. (1975). Functional organisation of the rat amygdala with respect to avoidance behaviour. *J. comp. Physiol. Psychol.* **88**, 829–50.

Haber, S. (1981). Social factors in evaluating the effects of biological manipulations on aggressive behaviour on non-human primates. In *Biobehavioural aspects of aggression* (ed. D. A. Hamburg and M. B. Trudeau), pp. 41–9. A. R. Liss, New York.

Hansen, S., Köhler, Ch., Goldstein, M., and Steinbusch, H. V. M. (1982). Effects of ibotenic acid-induced neuronal degeneration in the medial preoptic area and the lateral hypothalamic area on sexual behaviour in the male rat. *Brain Res.* **239**, 213–32.

——, Stanfield, E. J., and Everitt, B. J. (1981). The effects of lesions of lateral tegmental noradrenergic neurons on components of sexual behaviour and pseudo pregnancy in female rats. *Neuroscience* **6**, 1105–17.

Harlan, R. E., Gordon, J. H., and Gorski, R. A. (1979). Sexual differentiation of the brain – implications for neuroscience. In *Reviews of neuroscience* (ed. D. Schneider), Vol. 4, pp. 31–52. Raven Press, New York.

Harris, V. S. and Sachs, B. D. (1975). Copulatory behaviour in male rats following amygdaloid lesions, *Brain Res.* **86**, 514–18.

Heimer, L. and Larsson, K. (1966). Impairment of mating behaviour in male rats following lesions in the preoptic and anterior hypothalamic continuum. *Brain Res.* **3**, 248–63.

Herbert, J. (1968).Sexual preference in the rhesus monkey *Macaca mulatta* in the laboratory. *Animal Behav.* **16**, 120–8.

—— (1977). Hormones and behaviour. *Proc. R. Soc.* **B199**, 425–43.

—— (1980). Neurological concepts and methods in the study of sexual behaviour. In *Methodology in sex research* (ed. R. Green and J. Wiener), pp. 207–25. National Institute of Mental Health, Washington.

—— and Martensz, N. D. (1982). The effects of intraventricular prolactin infusions on pituitary responsiveness to thyrotropin-releasing hormone, 5-hydroxytryptophan or morphine in rhesus monkeys. *Brain Res.* **258**, 251–62.

Hofsetter, G., Gallo, R. V., and Brownfield, M. S. (1981). Presence of immunoreactive luteinizing hormone in the rat forebrain. *Neuroendocrinology* **33**, 241–5.

Kanematsu, T. and Heggelund, P. (1982). Single cell responses in cat visual cortex to visual stimulation during iontophoresis of noradrenaline. *Exp. brain Res.* **45**, 317–27.

Keverne, E. B. (1979). Sexual and aggressive behaviour in social groups of talapoin monkeys. In *Sex, hormones and behaviour* (ed. Ruth Porter and Julie Whelan), pp. 271–286. Ciba Foundation Symposium 62 (new series). Elsevier, Amsterdam.

—— and de la Riva, C. (1982). Pheromones in mice: reciprocal interaction between nose and brain. *Nature* **296**, 148–50.

Kluver, H. and Bucy, P. L. (1938). An analysis of certain effects of bilateral temporal lobotomy in the rhesus monkey with special reference to 'psychic blindness'. *J. Psychol.* **5**, 33–54.

McEwen, B. S. (1981). Cellular biochemistry of hormone action in brain and pituitary. In *Neuroendocrinology of reproduction* (ed. N. T. Adler), pp. 485–518. Plenum, New York.

McNeilly, A. S. (1980). Prolactin and the control of gonadotrophin secretion. *J. Reprod. Fertil.* **58**, 537–49.

Malsbury, C. W., Kow, L. M., and Pfaff, D. W. (1977). Effects of medial hypothalamic lesions on the lordosis response and other behaviours in female golden hamsters. *Physiol. Behav.* **19**, 223–37.

Martensz, N. D. and Herbert, J. (1982). Relationship between prolactin in the serum and cerebrospinal fluid of ovariectomised female rhesus monkeys. *Neuroscience* **7**, 2801–12.

Miczek, K. A., Brykczynski, T., and Grossman, S. P. (1974). Differential effects of lesions in the amygdala periamygdaloid cortex, and stria terminalis on aggressive behaviour in rats. *J. comp. Physiol. Psychol.* **87**, 760–71.

Mishkin, M. and Aggleton, J. (1981). Multiple functional contributions of the amygdala in the monkey. In *The amygdaloid complex* (ed. Y. Ben-Ari), pp. 409–20. Elsevier, Amsterdam.

Moyer, K. E. (1968). Kinds of aggression and their physiological basis. *Commun. Behav. Biol.* **2**, 65–87.

Neill, J. D. (1980). Neuroendocrine regulation of prolactin secretion. In *Frontiers in neuroendocrinology*, Vol. 6 (ed. L. Martini and W. F. Ganong), pp. 129–55. Raven Press, New York.

Nicholson, G., Greeley, G. H., Humm, J., Youngblood, W. W., and Kizer, J. S. (1980). Prolactin in cerebrospinal fluid: a probable site of auto regulation. *Brain Res.* **190**, 447–57.

Ottersen, O. P. (1981). The afferent connections of the amygdala of the rat as studied with retrograde transport of horseradish perioxidase. In *The amygdaloid complex* (ed. Y. Ben-Ari), pp. 91–104. Elsevier, New York.

Pedersen, C. A., Ascher, J. A., Monroe, Y. L., and Prange, A. J. (1982). Oxytocin induces maternal behaviour in virgin female rats. *Science* **216**, 648–9.

Pellegrino, L. (1968). Amgydaloid lesions and inhibition in the rat. *J. Comp. Physiol. Psychol.* **65**, 483–91.

Raisman, G. and Field, R. M. (1973). Sexual dimorphism in the neuropil of the preoptic area of the rat and its dependence on neonatal androgen. *Brain Res.* **54**, 1–12.

Robbins, T. W. and Everitt, B. J. (1982). Functional studies of the central catecholamines. *Int. Rev. Neurobiol.* **23**, 303–65.

Rolls, E. T. and Rolls, B. J. (1973). Altered food preferences after lesions in the basolateral region of the amygdala in the rat. *J. comp. Physiol. Psychol.* **83**, 248–59.

Rosenblatt, J. S., Siegel, H. I., and Mayer, A. D. (1979). Progress in the study of maternal behaviour in the rat: hormonal, non hormonal, sensory and developmental aspects. *Adv. Study Behav.* **10**, 225–311.

Rubin, B. S. and Barfield, R. J. (1980). Priming of estrus responsiveness by implants of 17-β estradiol in the ventromedial hypothalamic nucleus of female rats. *Endocrinology* **106**, 504–9.

Siegel, A. and Edinger, H. (1982). Neural control of aggression and rage behaviour. In Handbook of the hypothalamus (ed. J. Panksepp and P. J. Morgane), pp. 203–40. Dekker, New York.

Stengaard-Pedersen, K. and Larsson, L-I. (1981). Localisation and opiate receptor binding of enkephalin, CCK and ACTH and β-endorphin in the rat central nervous system. *Peptides* **2** (Suppl. 1), 3–19.

Tennent, B. J., Smith, E. R., and Dorsa, D. M. (1982). Comparison of some CNS effects of luteinizing-hormone releasing hormone and progesterone. *Horm. Behav.* **16**, 76–86.

Thorner, M. O. (1977). Hyperprolactinaemia and ovulation. In *Control of ovulation* (ed. D. B. Crighton, N. D. Haynes, G. R. Foxcroft, and S. E. Lamming), pp. 397–409. Butterworth, London.

Turner, B. H. (1973). Sensimotor syndrome produced by lesions of the amygdala and lateral hypothalamus. *J. comp. Physiol. Psychol.* **82**, 37–47.

—— (1981). The cortical sequence and terminal distribution of sensory related afferents to the amygdaloid complex of the rat and monkey. In *The amygdaloid complex* (ed. Y. Ben-Ari), pp. 51–62. Elsevier, Amsterdam.

Yodyingyuad, U., Eberhart, J. A., and Keverne, E. B. (1982). Effects of rank and novel females on behaviour and hormones in male talapoin monkeys. *Physiol. Behav.* **28**, 995–1005.

Young, W. C. (1961). The hormones and mating behaviour. In *Sex and internal secretions* (2nd edn.) (ed. W. C. Young), pp. 1173–239. Williams and Wilkins, Baltimore, Maryland.
Zuckerman, S. (1932). *The social life of monkeys and apes.* Kegan Paul, London.

4

Cortical influences on striatal function

PAUL WORMS, BERNARD SCATTON, MARIE-THÉRÈSE
WILLIGENS, ANDRÉ OBLIN, AND KENNETH G. LLOYD

INTRODUCTION

There is evidence indicating that cortical efferents to the extrapyramidal areas, possibly using glutamate as their neurotransmitter (Fonnum *et al.* 1981; McGeer *et al.* 1977), are involved in the regulation of basal ganglia function. Thus,

1. Lesions of the corticostriatal pathway enhance the behavioural effects of amphetamine (Iversen *et al.* 1971) and apomorphine (Scatton *et al.* 1982) and reduce the uptake of glutamate in the striatum (Divac *et al.* 1977; McGeer *et al.* 1977);

2. Electrical stimulation of the motor cortex stimulates the striatal release of [³H]-dopamine (DA) *in vivo* (Nieoullon *et al.* 1978);

3. *In vitro*, glutamate releases [³H]-DA and [³H]-acetylcholine from striatal slices (Giorguieff *et al.* 1977; Scatton and Lehmann 1982).

It is believed that the catalepsy resulting from dopamine (DA) receptor blockade by neuroleptics originates within the striatum. The events following DA receptor blockade, involved in the expression of this behaviour, are likely mediated by neuronal pathways utilizing – *inter alia* – acetylcholine (ACh) and γ-aminobutyric acid (GABA).

In the present study, we investigated

1. The effects of lesions of different cortical areas on neuroleptic-(and non-neuroleptic-) induced catalepsy;

2. The effects of such lesions on some of the biochemical parameters of extrapyramidal function;

3. The effects of GABAmimetic drugs on haloperidol-induced catalepsy in frontal-cortex-lesioned rats.

MATERIAL AND METHODS

Male rats of Sprague–Dawley origin (COBS strain, Charles River France: 180–200 g body weight at the time of surgery) were used for all experiments.

For surgery, rats were anaesthetized with Brietal[R] (85 mg kg⁻¹, s.c.) and were subjected to bilateral aspirations of either the frontal, parietal, or occipital cortex, or to a removal of both olfactory bulbs. So called 'sham-operated' rats were subjected to all the surgical procedures except the lesion.

Three weeks after the lesions, catalepsy induced by either haloperidol (Janssen–Lebrun Lab.), d-butaclamol HCl (Ayerst), chlorpromazine HCl (Bonapace), *cis*-flupenthixol HCl (Lundbeck Lab.), or morphine HCl (Boyer) was measured as the time (in seconds, maximum 120 s) during which the rat remained on the four-cork board (Worms and Lloyd 1979). This catalepsy was assessed every 30 min for three hours post-injection.

In some experiments the GABA mimetic drugs, progabide (Kaplan *et al.* 1980), muscimol (HCl), or aminooxyacetic acid (AOAA) were injected simultaneously with or two hours before haloperidol, in frontal-lesioned rats (three weeks post-lesion) and the catalepsy was measured as described above.

All drugs were injected intraperitoneally (i.p.) and the doses always refer to the base.

For biochemical measurements, rats were killed 10 or 21 days after the ablation of the frontal cortex. Striatum and/or substantia nigra of sham-operated and lesioned rats were dissected out in the cold. Dihydroxyphenylace-tic acid (DOPAC) and dopamine (DA) were measured according to Westerink and Korf (1977); acetylcholine (ACh) was measured according to the method of Guyenet *et al.* (1975) as modified by Scatton and Worms (1979); Choline acetyltransferase (CAT) was assayed as described by Fonnum (1975); substance P (SP) was measured by radioimmunoassay (Powell *et al.* 1973). Glutamic acid decarboxylase activity (GAD) was estimated according to the method of Lloyd *et al.* (1975). Proteins were assayed as described by Lowry *et al.* (1951).

RESULTS

Catalepsy

As shown in Table 4.1, bilateral ablations of either the frontal or the parietal cortex reduced the cataleptic state induced by haloperidol (1 mg kg^{-1}, i.p.) by 57 and 56 per cent, respectively, as compared to sham-operated rats. In contrast, bilateral ablations of the occipital cortex or of the olfactory bulbs failed to modify haloperidol-induced catalepsy (Table 4.1).

Three weeks after removal of both frontal cortices, the catalepsy induced by d-butaclamol (1 mg kg^{-1}, i.p.), chlorpromazine (12 mg kg^{-1}, i.p.) and *cis*-flupenthixol (0.8 mg kg^{-1}, i.p.) was reduced by 56, 59, and 52 per cent respec-

TABLE 4.1. *Effect of bilateral ablations of different cortical areas or olfactory bulbs on haloperidol-induced catalepsy. Data expressed as the mean ($\pm$ SEM) cumulated catalepsy time in per cent of the corresponding neuroleptic-treated, sham-operated animals*

Cortex lesion	Cumulated catalepsy time (as per cent of sham controls) after haloperidol (1 mg kg^{-1}, i.p.)			
	Frontal	Parietal	Occipital	Olfact. bulbs
Sham	100 ± 8	100 ± 13	100 ± 17	100 ± 17
Lesioned	43 ± 7†	44 ± 10*	99 ± 12	85 ± 16

*$p < 0.01$. †$p < 0.001$ vs. sham values.

TABLE 4.2. *Effect of bilateral ablation of the frontal cortex on the catalepsy induced by various neuroleptics or by morphine. Data are expressed as the mean (± SEM) catalepsy time in per cent of the corresponding neuroleptic-treated sham-operated animals*

Groups	Cumulated catalepsy time (in per cent of sham controls)			
	d-Butaclamol $(1 \text{ mg kg}^{-1}, \text{i.p.})$	Chlorpromazine $(12 \text{ mg kg}^{-1}, \text{i.p.})$	*cis*-Flupenthixol $(0.8 \text{ mg kg}^{-1}, \text{i.p.})$	Morphine $(25 \text{ mg kg}^{-1}, \text{i.p.})$
Sham	100 ± 10	100 ± 12	100 ± 9	100 ± 16
Frontal lesioned	$44 \pm 9^*$	$41 \pm 9^*$	$48 \pm 9†$	103 ± 24

$^*p < 0.01$. $†p < 0.001$ vs. sham values.

tively, whereas morphine- $(25 \text{ mg kg}^{-1}, \text{i.p.})$ induced catalepsy was not affected by this lesion (Table 4.2).

When GABAmimetics such as progabide $(200 \text{ mg kg}^{-1}, \text{i.p.})$, muscimol $(2 \text{ mg kg}^{-1}, \text{i.p.})$, or AOAA $(25 \text{ mg kg}^{-1}, \text{i.p.})$ (Fig. 4.1) were administered to rats with bilateral lesions of the frontal cortex, the catalepsy induced by haloperidol was completely re-established, being similar to that in sham-lesioned rats.

Biochemical measurement

As shown in Fig. 4.2, the striatal concentrations of DA were unchanged in frontal-lesioned rats as compared to sham-operated rats.

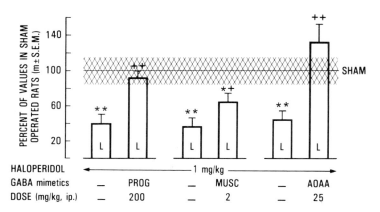

Fig. 4.1. Influence of progabide (PROG), muscimol (MUSC), and AOAA on haloperidol-induced catalepsy in frontal lesioned rats. Data are expressed as cumulated catalepsy times, in per cent (mean with SEM) of haloperidol values in sham rats (L: lesioned). $^*p < 0.05$; $^{**}p < 0.01$ vs. sham; $^+p < 0.05$; $^{++}p < 0.01$ vs. haloperidol alone in lesioned rats.

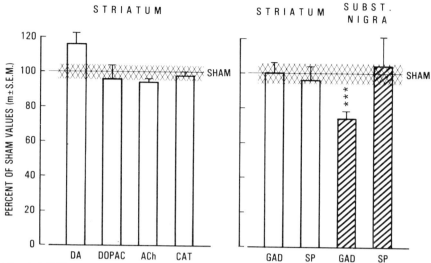

Fig. 4.2. Effect of bilateral ablation of the frontal cortex on the biochemical indices of striatal and nigral functions. Opened columns represent mean (± SEM) percentages of sham values in the striatum; hatched columns represent mean (± SEM) percentages of sham values in the substantia nigra. ***p <0.001 vs. sham values (striatum: DA = 9750 ± 350 ng/g; DOPAC = 1006 ± 47 ng/g; ACh = 38.8 ± 2.0 nmol/g; SP = 309 ± 16 ng/g; CAT = 16.1 ± 0.6 μmol/g/hour; GAD = 150 ± 6 nmol/h/mg protein. Substantia nigra: SP = 2132 ± 340 ng/g; GAD = 343 ± 14 nmol/h/mg protein). (Data from Scatton *et al.* 1982.)

Such a lesion also failed to affect CAT activity or GAD activity in the striatum, or SP levels in the substantia nigra.

However, nigral GAD activity was diminished by 23 per cent (p <0.001), in lesioned as compared to sham-operated rats (Fig. 4.2).

In the striatum, the ability of haloperidol (1 mg kg^{-1}, i.p.) to increase DOPAC levels and to decrease ACh concentrations was unaffected by bilateral lesions of the frontal cortex (Table 4.3).

TABLE 4.3. *Effect of bilateral ablation of the frontal cortex on haloperidol-induced increase in DOPAC levels and decrease in ACh content, in the striatum. Data are expressed as mean (± SEM) percentages of the control values in sham-operated rats*

Groups	Striatal DOPAC		Striatal ACh	
	Saline	Haloperidol 1 mg kg^{-1}	Saline	Haloperidol 1 mg kg^{-1}
Sham	100 ± 5	350 ± 10*	100 ± 2	52 ± 3*
Frontal lesioned	98 ± 7	336 ± 10*	93 ± 3	66 ± 3*

*p <0.01 vs. respective saline controls (DOPAC: 924 ± 46 ng g^{-1}; ACh: 41.8 ± 0.8 nmol g^{-1}).

DISCUSSION

The present study shows that a bilateral ablation of the frontal cortex markedly reduces neuroleptic-induced catalepsy in rats, three weeks post-lesion. Similarly, ablations of both parietal cortices diminish haloperidol catalepsy, whereas removal of either the occipital cortex or the olfactory bulbs does not affect this cataleptic state.

These data are consistent with those neuroanatomical findings showing that the neostriatum and the substantia nigra (SN), both areas involved in neuroleptic-induced catalepsy, receive extensive projections from the sensori-motor cortex, whereas projections from the visual cortex or from the olfactory bulbs to these areas are minimal or nonexistent (Fonnum *et al.* 1981; Grofova 1979).

The present behavioural observations are in agreement with previous data which demonstrated that lesions of the frontal cortex reduce fluphenazine catalepsy (Carter and Pycock 1980) and increase the ability of apomorphine (Scatton *et al.* 1982) and amphetamine (Iversen *et al.* 1971) to induce stereotypies in rats.

The differential effects of frontal cortex lesions on the catalepsy induced by different agents support the hypothesis that this interaction is occurring at the striatal level. Thus, the effect of neuroleptics (haloperidol, butaclamol, flupenthixol, and chlorpromazine) – which is to likely to be dependent on striatal function (Costall *et al.* 1972) – is reduced by such a lesion, whereas morphine-induced catalepsy, which does not appear to involve the striatum (Dunstan *et al.* 1981) is unaffected by ablation of the frontal cortex.

From these data, it can be hypothesized that the cortico–striatal projections (possibly glutamatergic in nature, Divac *et al.* 1977; McGeer *et al.* 1977) play an important role in the expression of catalepsy and therefore in the regulation of striatal functions. This, however, does not exclude the involvement of other subcortical areas (i.e. SN, ventromedial thalamus) which receive inputs from the frontal cortex (Fonnum *et al.* 1981; Thierry *et al.* 1982) and which appear implicated in catalepsy (Dray 1979; DiChiara *et al.* 1979).

The mechanisms whereby lesions of the sensori-motor cortex reduce neuroleptic catalepsy are as yet unclear. This behaviour appears to be related to the function of at least three neurotransmitters within the basal ganglia, DA, ACh, and GABA (Lloyd 1978; Worms and Lloyd 1980).

1. *DA*. Although glutamate (*in vitro*) and stimulation of the frontal cortex (*in vivo*) both enhance the release of newly synthesized DA in the striatum (Giorguieff *et al.* 1977; Nieoullon *et al.* 1978), it is unlikely that the frontal cortex influences catalepsy through a direct effect on nigro-striatal DAergic terminals within the striatum. Thus, neither the presynaptic indicators of striatal DA activity (levels, turn-over, synthesis) nor their modification by haloperidol (present study) or apomorphine (Scatton *et al.* 1982), are affected by these fronto-cortical lesions.

2. *ACh*. For this neurotransmitter, the situation is complex. After lesions for the frontal cortex, neither ACh levels nor the activity of CAT in the striatum, are modified as compared to sham-operated rats. Moreover, the

haloperidol-induced reduction (present study) and the apomorphine-induced increase (Scatton *et al.* 1982) in striatal ACh levels remain unchanged after this lesion. However, such a lesion was shown to reduce ACh turnover significantly within the striatum (Wood *et al.* 1979). Altogether, these findings cannot exclude an involvement of modified cholinergic activity in the cortical control of the expression of catalepsy.

3. *GABA*. In contrast, evidence indicates that GABA neurones (i.e. striato-pallidal, striato-nigral, intra-striatal) could be involved in the regulatory role of corticofugal neurones.

Thus, we report here that direct (progabide, Lloyd *et al.* 1982; muscimol, Krogsgaard-Larsen 1981) and indirect (AOAA) GABAmimetic drugs are able to restore haloperidol-induced catalepsy in rats with a bilateral lesion of the frontal cortex. Moreover, the activity of GAD is significantly reduced in the substantia nigra of lesioned rats, although this enzyme is unchanged in the striatum of the lesioned animals.

Other neurochemical findings also support this hypothesis: after ablation of the frontal cortex or lesion of the cortico–striatal pathway, the turnover of striatal GABA is decreased (Wood *et al.* 1979), the GABA content is increased in the striatum (Hassler *et al.* 1982), and [^{3}H]-GABA binding is increased in both the pallidum and the substantia nigra, pars reticulata, but unchanged in the striatum or the substantia nigra, pars compacta (Kuppersmith and Goldstein 1980). These data suggest that, after these lesions, the activity of striato-pallidal and striato-nigral GABA neurones is diminished, and these neurones could represent the target cells for the cortical projections (cf. also the neuro-anatomical findings of Somogyi *et al.* 1981).

Finally, the lack of effect of cortical lesions on nigral and striatal SP levels suggests that striato-nigral SP neurones are unlikely to play a role in the cortical influences on striatal function.

In conclusion, the present results show that, in the rat, the integrity of the sensori-motor cortex is necessary for the expression of the catelepsy induced by DA receptor blockade. The influence of the frontal cortex is likely exerted distally to striatal DAergic terminals. GABAergic neurones functioning as output pathways for the striatum may well be involved in these complex phenomena.

REFERENCES

Carter, C. J. and Pycock, C. J. (1980). Behavioural and biochemical effects of dopamine and noradrenaline depletion within the medial prefrontal cortex of the rat. *Brain Res.* **192**, 163–72.

Costall, B., Naylor, R. J., and Olley, J. E. (1972). Catalepsy and circling behaviour after intracerebral injections of neuroleptic, cholinergic and anticholinergic agents into the caudate–putamen, globus pallidus and substantia nigra of rat brain. *Neuropharmacology* **11**, 645–63.

DiChiara, G., Morelli, M., Porceddu, M. L., and Gessa, G. L. (1979). Role of thalamic γ-aminobutyrate in motor functions: catalepsy and ipsiversive turning after intrathalamic muscimol. *Neuroscience* **4**, 1453–65.

Divac, I., Fonnum, F., and Storm-Mathisen, J. (1977). High affinity uptake of gluta-mate in terminals of cortico–striatal axons. *Nature* **226**, 377–8.

Dray, A. (1979). The striatum and substantia nigra: a commentary on their relation-ship. *Neuroscience* **4**, 1407–39.

Dunstan, R., Broekkamp, C. L. E., and Lloyd, K. G. (1981). Involvement of caudate nucleus, amygdala or reticular formation in neuroleptic and narcotic catalepsy. *Pharmacol. Biochem. Behav.* **14**, 169–74.

Fonnum, F. (1975). Radiochemical assays for choline acetyltransferase and acetyl-cholinesterase. In *Research methods in neurochemistry* (ed. N. Marks and R. Rodnight), pp. 253–75. Plenum Press, New York.

——, Storm-Mathisen, J., and Divac, I. (1981). Biochemical evidence for glutamate as neurotransmitter in cortico–striatal and corticothalamic fibers in rat brain. *Neuroscience* **6**, 863–74.

Giorguieff, M. F., Kemel, M. L., and Glowinski, J. (1977). Presynaptic effect of L-glutamic acid on dopamine release in striatal slices. *Neurosci. Lett.* **6**, 77–8.

Grofova, I. (1979). Extrinsic connections of the neostriatum. In *The neostriatum* (ed. I. Divac and R. G. E. Oberg), pp. 37–51. Pergamon Press, Oxford.

Guyenet, P. G., Agid, Y., Javoy, F., Beaujouan, J. C., Rossier, J., and Glowinski, J. (1975). Effects of dopamine receptor agonists and antagonists on the activity of the neostriatal cholinergic system. *Brain Res.* **84**, 227–44.

Hassler, R., Haug, P., Nitsch, C., Kim, J. S., and Paik, K. (1982). Effect of motor and premotor cortex ablation on concentrations of amino acids, monoamines, and acetyl-choline and on the ultrastructure in rat striatum. A confirmation of glutamate as the specific corticostriatal transmitter. *J. Neurochem.* **38**, 1087–98.

Iversen, S. D., Wilkinson, S., and Simpson, B. (1971). Enhanced amphetamine responses after frontal cortex lesions in the rat. *Eur. J. Pharmacol.* **13**, 387–90.

Kaplan, J. P., Raizon, B., Desarmenien, M., Feltz, P., Worms, P., Lloyd, K. G., and Bartholini, G. (1980). New anticonvulsants: Schiff bases of GABA and GABAmide. *J. Med. Chem.* **23**, 702–4.

Krogsgaard-Larsen, P. (1981). γ-Aminobutyric acid agonists, antagonists and uptake inhibitors. Design and therapeutic aspects. *J. Med. Chem.* **24**, 1377–83.

Kuppersmith, M. J. and Goldstein, M. (1980). The effect of decortication on the basal ganglia GABA receptor. *Neurosci. Lett.* **17**, 335–7.

Lloyd, K. G. (1978). Neurotransmitter interactions related to central dopamine neurons. In *Essays in neurochemistry and neuropharmacology* (ed. M. B. H. Youdim, W. Lovenberg, D. F. Sharman, and J. R. Lagnado), pp. 131–207. Wiley, New York.

——, Arbilla, S., Beaumont, K., Briley, M., DeMontis, G., Scatton, B., Langer, S. Z., and Bartholini, G. (1982). γ-Aminobutyric acid receptor stimulation. II. Specificity of progabide (SL 76002) and SL 75102 for γ-aminobutyric acid receptor. *J. Pharma-col. exp. Ther.* **220**, 672–80.

——, Möhler, H. Heitz, P., and Bartholini, G. (1975). Distribution of glutamate decarboxylase within the substantia nigra and other brain regions from control and parkinsonian patients. *J. Neurochem.* **25**, 789–95.

Lowry, O. H., Rosebrought, N. J., Farr, A. L., and Randall, R. J. (1951). Protein measurement with the Folin phenol reagent. *J. Biol. Chem.* **193**, 265–75.

McGeer, P. L., McGeer, E. G., Scherer, U., and Singh, K. (1977). A glutamatergic cortico-striatal path? *Brain Res.* **128**, 369–73.

Nieoullon, A., Cheramy, A., and Glowinski, J. (1978). Release of dopamine evoked by electrical stimulation of the motor and visual areas of the cerebral cortex in both caudate nuclei and substantia nigra in the rat. *Brain Res.* **145**, 69–84.

Powell, D., Leeman, S., Tregear, G. W., Niall, H. D., and Potts. J. T. (1973). Radio-immunoassay for substance P. *Nature* **241**, 252–4.

Scatton, B. and Lehmann, J. (1982). N-methyl-D-aspartate-type receptors mediate striatal ^{3}H-acetylcholine release evoked by excitatory amino-acids. *Nature* **297**, 422–4.

—— and Worms, P. (1979). Tolerance to increase in striatal acetylcholine concentrations after repeated administration of apomorphine dipivaloyl ester. *J. Pharm. Pharmacol.* **31**, 861–3.

——, ——, Lloyd, K. G., and Bartholini, G. (1982). Cortical modulation of striatal function. *Brain Res.* **232**, 331–43.

Somogyi, P., Bolam, J. P., and Smith, A. D. (1981). Monosynaptic cortical input and local axon collaterals of identified striato–nigral neurons. *J. comp. Neurol.* **195**, 567–84.

Thierry, A. M., Chevalier, G., Ferron, A., and Glowinski, J. (1982). Electrophysiological identification of some efferent pathways of the rat medial prefrontal cortex. *Neurosci. Lett.* (Suppl. 10), S478–S479.

Westerink, B. H. C. and Korf, J. (1977). Rapid concurrent automated fluorimetric assay of noradrenaline, dopamine, 3,4-dihydroxyphenylacetic acid, homovanillic acid and 3-methoxytyramine in milligram amounts of nervous tissue after isolation on Sephadex G-10. *J. Neurochem.* **29**, 697–706.

Wood, P. L., Moroni, F., Cheney, D. L., and Costa, E. (1979). Cortical lesions modulate turnover rates of acetylcholine and gamma-aminobutyric acid. *Neurosci. Lett.* **12**, 349–54.

Worms, P. and Lloyd, K. G. (1979). Predictability and specificity of behavioral screening tests for neuroleptics. *Pharmacol. Ther.* **5**, 445–50.

—— and —— (1980). Biphasic effects of direct, but not indirect, GABAmimetics and antagonists on haloperidol-induced catalepsy. *Naunyn-Schmiedeberg's Arch. Pharmacol.* **311**, 179–84.

5

Amygdaloid-kindled seizures in the rat: pharmacological studies

J. M. STUTZMANN, C. GARRET, J. C. BLANCHARD, AND
L. JULOU

INTRODUCTION

In 1966, Herberg and Watkins reported that after induction of convulsions in the rat by a first electrical hypothalamic stimulation, the convulsive threshold, after an initial rise for at least two hours, clearly decreased for a longer period. One year later, Goddard (1967) showed, in the same species, that a short-duration and low-intensity subcortical or cortical electrical stimulation was able to induce modifications of the electroencephalogram (EEG) characterized by an afterdischarge which consisted of spikes and multiple spikes and waves associated with a few behavioural signs.

Moreover, taking into account Herberg and Watkins's observation of a delayed convulsive threshold decrease following a first electrical stimulation, Goddard studied the effect of a daily stimulation in different subcortical areas and observed that under such conditions the EEG afterdischarge, primarily localized in the stimulated structure, progressively increased both in amplitude and duration and spread to all the cerebral structures. In 1969, Goddard *et al.* described more precisely the methodology to be used and called this phenomenon *kindling*, confirming, as stated in Goddard (1967), that this phenomenon is more easily triggered from the limbic system and particularly from the amygdaloid complex. Thereafter, kindling studies were extended to different species and different structures. Many animal species, including reptiles and amphibians can be kindled (Morell and Tsuru 1976; Rial and Gonzalez 1976); however the rat, the rabbit, the cat, and the monkey are the species most often used (Racine 1978).

It was also found that kindling could be elicited from various cerebral structures: the globus pallidus, the pyriform cortex, the olfactive cortex, the frontal cortex, the entorhinal cortex, the olfactive bulb, the septum, the preoptic area, the striatum, and the hippocampus (Goddard *et al.* 1969; Wake and Wada 1975; Corcoran *et al.* 1976; Racine 1978). But some structures cannot be kindled and, in particular, in the rat, the occipital cortex is refractory to kindling.

In this chapter, we shall recall the main characteristics of the amygdaloid kindling in the rat. We shall then emphasize the interest of this phenomenon as a model for the study of the anticonvulsant properties of drugs. Our commentary will be illustrated by data from the literature and by personal results obtained during the course of the pharmacological study of zopiclone (ZC), a hypnotic agent, and suriclone (SC), a minor tranquillizer, both belonging to the original family of cyclopyrrolones.

MAIN PARAMETERS OF AMYGDALOID KINDLING IN THE RAT

The first stimulation (2 seconds duration; biphasic impulsions of 0.1 ms, 60 Hz, 50 to 1000 μamp) of the amygdala induces, together with minor behavioural signs, a localized and short afterdischarge (Fig. 5.1). Thereafter, with the repetition of this daily stimulation, this afterdischarge increases in amplitude and in duration and spreads, first, to the more directly connected structures (contralateral amygdala, hippocampus, pyriform cortex) and progressively to all of the brain (Le Gal La Salle 1980).

We must mention that, between stimulation periods, isolated spontaneous spikes frequently occur during and after the development of the kindling effect. These spikes appear first in the stimulated amygdala and reach the contralateral amygdala followed by the hippocampus and the cortex (Wada *et al.* 1974; Tanaka *et al.* 1976, Fitz and McNamara 1979).

At the same time as the EEG modifications, behavioural symptoms appear and progress resulting finally in a generalized convulsive seizure. According to Racine (1972), five successive behavioural stages can be distinguished (Fig. 5.1)

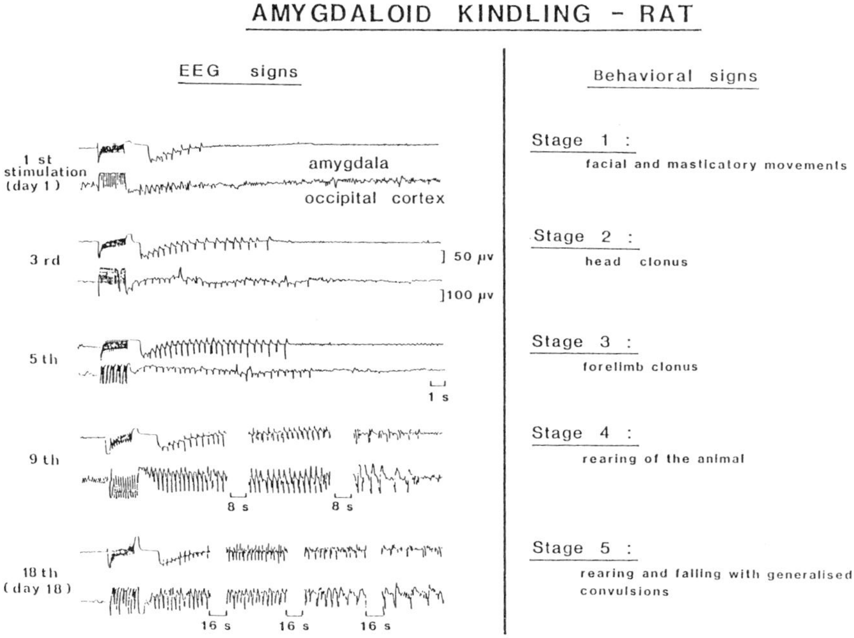

Fig. 5.1. Progression of the EEG and behavioural signs during the installation of the kindling phenomenon.

Stage 1. Facial and masticatory movements;
Stage 2. Head clonus;
Stage 3. Forelimb clonus;
Stage 4. Rearing of the rat;
Stage 5. Rearing and falling with generalized convulsions.

It is also noteworthy that, once an animal has been kindled to the point of generalized convulsions (stage 5), the epileptic seizure can be triggered electrically, even after several weeks without any stimulation (Goddard *et al.* 1969; Wada *et al.* 1974). As we shall elaborate later on, this characteristic appears to be of interest for pharmacological studies. It must be mentioned that when the animal has been stimulated several times (at least 100 stimulations), spontaneous seizures may occur on very rare occasions (Wada *et al.* 1974; Pinel and Rovner 1978).

Finally, we have to underline that it has been clearly shown that kindled epileptic seizures do not result from nervous-tissue damage or glial-tissue damage (Goddard and Douglas 1976; Brotchi *et al.* 1978). Thus, kindling differs completely from other epileptic models such as the penicillin or cobalt-powder focus or the kainic-acid model.

THE USE OF AMYGDALOID KINDLING AS A MODEL FOR THE STUDY OF ANTICONVULSANT PROPERTIES OF DRUGS

The effects of various pharmacological classes of drugs have been reviewed by different authors (Babington and Wedeking 1973; Bowyer and Winters 1981; Albertson *et al.* 1980; Chapter 6, this volume). Anti-epileptic agents and compounds which possess, among other properties, an anticonvulsant effect and can eventually be used as anti-epileptics, represent one type of drug which has been specially studied in amygdaloid kindling (Wada *et al.* 1976; Ashton and Wauquier 1979; Albertson *et al.* 1981*a*).

We shall illustrate the protective effect of anticonvulsant drugs on amygdaloid-kindled seizures (AKS) by presenting our own results concerning two well-known families of drugs used as anticonvulsants: barbiturates and benzodiazepines. In addition, we shall present the results that we have obtained with two cyclopyrrolone derivatives, zopiclone (ZC) and suriclone (SC) (Fig. 5.2). Finally the activity of the benzodiazepine antagonist Ro 15-1788 against the previous drugs will be considered.

Concerning classical anti-epileptic drugs, several studies have shown that valproic acid, carbamazepine, and ethosuximide are effective against AKS (Albertson *et al.* 1980, 1981*b*; Leviel and Naquet 1977). For phenytoin, opposite results have been found; Racine *et al.* (1975) observed that this drug facilitated the development and duration of AKS whilst Albertson *et al.* (1980) found that it was able to protect against seizures.

In our own studies we have evaluated the effect of barbiturates, benzodiazepines, and cyclopyrrolones on the AKS by measuring at different times (30 min to 24 h) the mean EEG seizure duration after oral drug administration of the compounds.

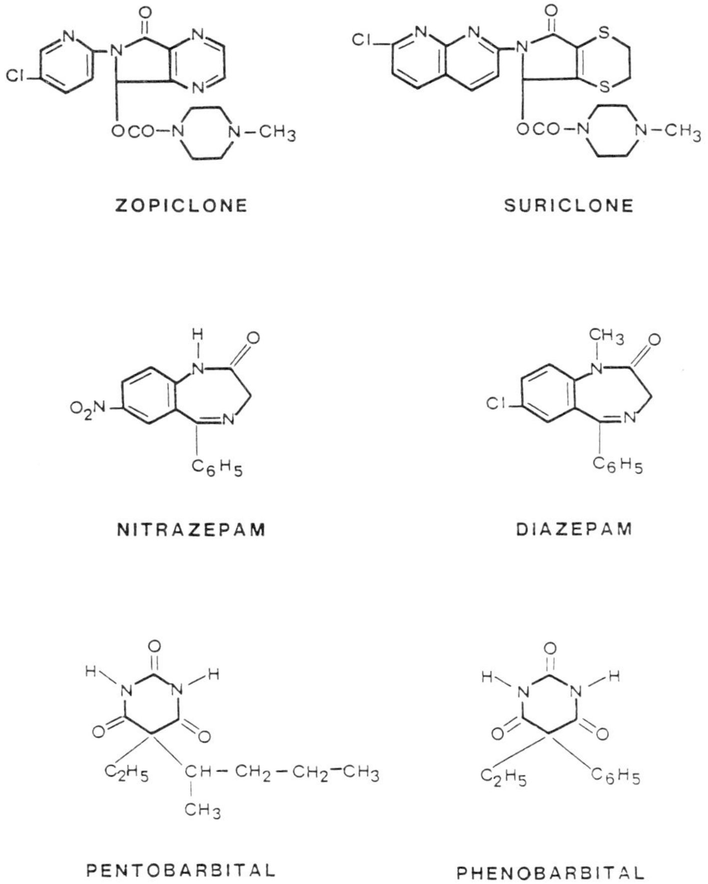

ZOPICLONE

SURICLONE

NITRAZEPAM

DIAZEPAM

PENTOBARBITAL

PHENOBARBITAL

Fig. 5.2. Chemical structures of the different cyclopyrrolones, benzodiazepines, and barbiturates.

Barbiturates

The two barbiturates, pentobarbital (PB) and phenobarbital (PhB), that we have studied do not possess at 2.5 mg kg^{-1} p.o. any effect against AKS duration (Figs. 5.3 and 5.4). PB, at 20 mg kg^{-1} p.o., is shown to have a marked activity (-68 per cent) at 30 min after administration only (Fig. 5.5). PhB, at the same dose, possesses the same protective effect but a longer duration of activity (up to 4 h 30 min) than that of PB (Fig. 5.6). With the two barbiturates a significant decrease in behavioural manifestations can be seen only from the dose of 20 mg kg^{-1} p.o. (in the two cases, two of the five tested rats do not exceed stage 2).

These effects of barbiturates have already been described by many other authors, using different routes of administration (Babington and Wedeking 1973; Ashton and Wauquier 1979; Albertson *et al.* 1980).

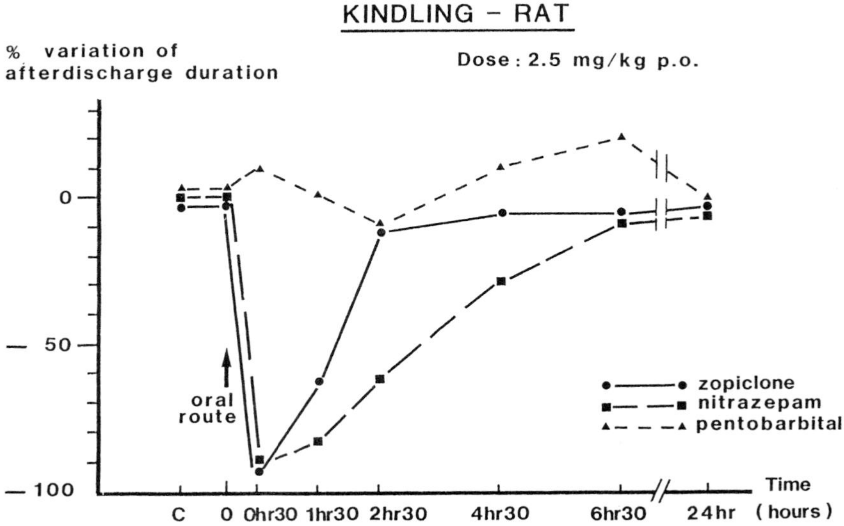

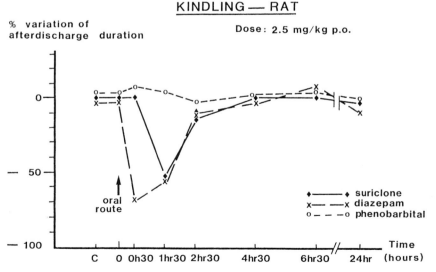

Fig. 5.3. Modification of afterdischarge duration measures on four to six rats per dose. C: Control period = 48 h, 24 h, and 1 h 30 min before drug administration. Treatment period = 0 h 30 min, 1 h 30 min, 2 h 30 min, 4 h 30 min, 6 h 30 min, and 24 h after drug administration.

Fig. 5.4. Modification of afterdischarge duration measured on four to six rats per dose. C: Control period = 48 h, 24 h, and 1 h 30 min before drug administration. Treatment period = 0 h 30 min, 1 h 30 min, 2 h 30 min, 4 h 30 min, 6 h 30 min, and 24 h after drug administration.

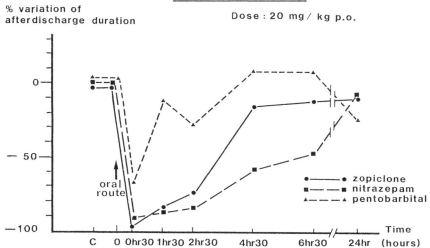

Fig. 5.5. Modification of afterdischarge duration measured on four to six rats per dose. C: Control period = 48 h, 24 h, and 1 h 30 min before drug administration. Treatment period = 0 h 30 min, 1 h 30 min, 2 h 30 min, 4 h 30 min, 6 h 30 min, and 24 h after drug administration.

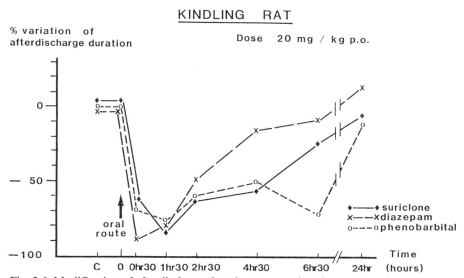

Fig. 5.6. Modification of afterdischarge duration measured on four to six rats per dose. C: Control period = 48 h, 24 h, and 1 h 30 min before drug administration. Treatment period = 0 h 30 min, 1 h 30 min, 2 h 30 min, 4 h 30 min, 6 h 30 min, and 24 h after drug administration.

Benzodiazepines

Nitrazepam (NZ), at 2.5 mg kg^{-1} p.o., is shown to have a significant and maximal effect against AKS (-92 per cent) and behavioural manifestations from 30 min to 2 h 30 min after the drug administration. Six and a half hours later, the seizure duration is close to that of the control value (Fig. 5.3). At the highest dose of 20 mg kg^{-1} p.o., this same marked effect of NZ is of at least 6 h 30 min duration (Fig. 5.5).

Diazepam (DZ), at the dose of 2.5 mg kg^{-1} p.o., presents a clear protective activity against AKS at 30 min and 1 h 30 min (respectively, -70 and 57 per cent of protection). The control seizure duration recovers 2 h 30 min after the drug administration (Fig. 5.4). At the highest dose of 20 mg kg^{-1} p.o., the drug effect is confirmed against AKS, but the duration of activity is much greater, up to 4 h 30 min (Fig. 5.6).

We have studied other benzodiazepines such as clonazepam (CNZ), chlordiazepoxide (CDZ), bromazepam (BMZ), and clorazepate dipotassium (CLZ). All of them exert a clear and significant protective activity against AKS and their ED$_{50}$ values will be presented thereafter together with those of the other compounds.

Several authors have found similar results when studying benzodiazepines such as DZ, CDZ, CNZ, and oxazepam in this model. DZ seems to be the most investigated drug and it has been shown that it could also act on the development of the kindling seizure (Wise and Chinerman 1974; Racine *et al.* 1975; Le Gal La Salle, personal communication). Furthermore, according to Racine *et al.* (1973), DZ would be more potent in blocking AKS than in blocking cortical frontal kindled seizures.

Cyclopyrrolones

ZC and SC, exert, like benzodiazepines, a strong activity against convulsions induced by various agents such as electroshock, pentylenetetrazole, bicuculline, strychnine, isoniazid, 3-mercaptopropionic acid, and allylglycine (Bardone *et al.* 1978; Julou *et al.* 1982; Garret *et al.* 1982). Moreover ZC and SC possess a high affinity for the so-called BZD receptors (Blanchard *et al.* 1979; Blanchard and Julou 1982). In man, ZC proved to be very interesting as a hypnotic agent, being as active as NZ (Duriez *et al.* 1979), and SC proved very interesting as a new anxiolytic agent (Basset *et al.* submitted for publication).

At the doses of 2.5 mg kg^{-1} p.o. and 20 mg kg^{-1} p.o., ZC induces a significant and maximal reduction in the AKS duration at 30 min (respectively, -93 and -98 per cent) which disappears 2 h 30 min after the drug administration at the lowest dose and 4 h 30 min after the highest dose (Figs. 5.3 and 5.5). At the same time, the behavioural manifestations are reduced or abolished.

This protective effect of ZC against AKS is also found with SC. At 2.5 mg kg^{-1} p.o. the duration of activity of SC is about 1 h 30 min and, at 20 mg kg^{-1} p.o., about 6 h 30 min (Figs. 5.4 and 5.6).

Comparison of the potency of the different barbiturates, benzodiazepines, and cyclopyrrolones

The precise comparison of the potency of the different drugs studied against AKS necessitates the determination of $ED_{50}s$. In Table 5.1, we present the $ED_{50}s$ after 30 min and 1 h 30 min of ZC and SC compared with those of the six different benzodiazepines and the two barbiturates that we have examined.

All the compounds are more active or as active after 30 min than after 1 h 30 min, with the exception of SC which is three times more active after 1 h 30 min.

TABLE 5.1. *Antikindling effect in rat of cyclopyrrolones, benzodiazepines, and barbiturates*

	ED_{50} mg kg^{-1} p.o.*	
Compound	30 min after administration	1 h 30 min after administration
Zopiclone	0.6	2.2
Suriclone	6.6	2.2
Clonazepam	0.3	0.75
Nitrazepam	0.4	1.2
Diazepam	1.3	1.9
Chlordiazepoxide	1.95	2.5
Bromazepam	2.2	2.8
Clorazepate	5.8	13.0
Pentobarbital	9.5	40.0
Phenobarbital	16.0	17.0

*The ED_{50} which corresponds to the dose which reduces by 50 per cent the mean afterdischarge duration is graphically determined with at least four doses.

At the time of maximal effect, the ED_{50} of ZC is 0.6 mg kg^{-1} p.o., a value close to that of NZ and CNZ (0.4 and 0.3 mg kg^{-1} p.o., respectively); the ED_{50} of SC is 2.2 mg kg^{-1} p.o., a value very similar to that of CDZ or BMZ (1.95 and 2.2 mg kg^{-1} p.o., respectively), slightly superior to that of DZ (1.3 mg kg^{-1} p.o.), but clearly inferior to that of CLZ (5.8 mg kg^{-1} p.o.).

With regard to the $ED_{50}s$, the duration of activity of ZC appears to be shorter than that of SC.

Barbiturates, PB and PhB, possess $ED_{50}s$ clearly greater than those of the cyclopyrrolones or benzodiazepines.

Study of the antagonist effect of Ro 15-1788 on the protection of barbiturates, benzodiazepines, and cyclopyrrolones

The recent discovery that an imidazobenzodiazepine could antagonize the sedative, anticonvulsant, and anticonflict effects of benzodiazepines (Bonetti *et al.* 1981; Hunkeler *et al.* 1981; Darrach *et al.* 1981; Lloyd *et al.* 1981) led us to study this benzodiazepine antagonist against the protective effects of barbiturates, benzodiazepines, and cyclopyrrolones in the kindling phenomenon.

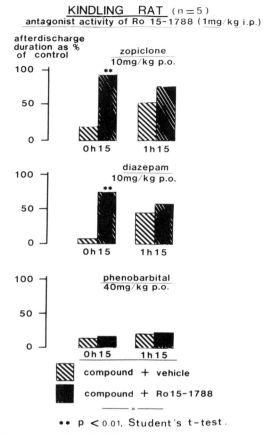

Fig. 5.7. Antagonist activity of Ro 15-1788. Ro 15-1788 is administered 1 h 15 min after drug treatment and kindling test is induced 15 min and 1 h 15 min after Ro 15-1788 administration. $(\times - \times) = p \leqslant 0.01$ (Student's t-test).

As shown in Fig. 5.7, Ro 15-1788 (1 mg kg^{-1} i.p.) antagonizes the anti-kindling effect of ZC (10 mg kg^{-1} p.o.) and DZ (10 mg kg^{-1} p.o.). The protective effect against AKS of ZC and DZ (-83 and -94 per cent, respectively when the drugs are administered alone) is reduced to -6 and -26 per cent respectively, when Ro 15-1788 is given 1 h 15 after ZC and DZ and 15 min before the test. The antagonist activity of Ro 15-1788 against ZC and DZ is less marked when the test is performed 1 h 15 after Ro 15-1788 administration, which indicates that Ro 15-1788 activity probably lasts a short time ($\leqslant 2$ h).

It must be underlined that, on the contrary, Ro 15-1788 is not able to antagonize PhB protection.

CONCLUSION

By using a standardized technique for amygdaloid kindling which permits the determination of the ED$_{50}$ of compounds, we have been able to quantify the activity of the two cyclopyrrolone derivatives (zopiclone, a hypnotic agent, and suriclone, an anxiolytic agent) by comparison with various benzodiazepines and two barbiturates.

This study enables us to underline the interest of the kindling model to pharmacology.

1. Once kindling has been achieved, this phenomenon is reproducible in the same animal and from one animal to another, which can reduce the number of animals required to study a test substance. This represents a marked advantage over the experimental lesioned epilepsies where the extent of the lesion cannot always be controlled exactly.

2. The protective activity of a compound can be evaluated both by EEG and behavioural signs.

3. Above all, kindling is a model which does not only allow the determination of the activity of the compound on the duration of the generalized convulsive seizure, as in this present study, but also an evaluation of the activity of the compound on the installation of kindling upon repeated treatment.

4. Finally, it should be noted that kindling can be obtained from various brain structures; in particular, certain authors suggest that amygdaloid kindling represents the only experimental model for temporal-lobe epilepsy in man (Sato *et al.* 1977; Adamec *et al.* 1981). Hence, using the same compound, one can evaluate activity with regard to different types of kindling. Thus, at the same dose level, DZ apparently provides more effective protection against kindled seizures of amygdaloid origin than against those of frontal cortical origin (Racine *et al.* 1973).

The research presented here has also enabled us to confirm the similarity of the pharmacological activity of the 2 cyclopyrrolones by comparison with benzodiazepines, an activity which is clearly different from that of the barbiturates.

With respect to AKS in the rat, orally administered ZC and SC exert a marked protective effect against EEG and behavioural epileptic manifestations which is comparable to that of the benzodiazepines. This effect generally appears at dose levels of the order of 0.3 to 2 mg kg^{-1} p.o.; the activity of ZC is similar to that of NZ and CNZ, whereas the activity of SC is closer to that of DZ, CDZ, and BMZ. The results obtained with benzodiazepines are in agreement with those which have already been cited in various papers (Babington and Wedeking 1973; Albertson *et al.* 1980). It is interesting to note that the duration of action of ZC is shorter than that of SC.

The activity of the barbiturates appears at much higher dose levels (9.5 and 16 mg kg^{-1} p.o.); this finding is also in conformity with those of other researchers (Babington and Wedeking 1973; Albertson *et al.* 1980).

The use of Ro 15-1788, which is able to displace the benzodiazepines from their brain receptor sites and which possesses an antagonistic effect with regard to the sedative, anticonvulsant, and anticonflict properties of benzodiazepines (Hunkeler *et al.* 1981), has enabled us to clearly differentiate between the cyclopyrrolones and barbiturates. Ro 15-1788 is capable of inhibiting the protective activity of the cyclopyrrolones against AKS as for the benzodiazepines, but this is not the case with the barbiturates. The results which we have obtained are in agreement with those recently published by Albertson *et al.* (1982) regarding the antagonistic effect of Ro 15-1788 against diazepam in this amygdaloid-kindling model in the rat.

Our results suggest that the protective effect of the benzodiazepines and cyclopyrrolones with respect to amygdaloid kindling probably involves the recognition of common receptor sites and that, on the contrary, the mechanism of action of barbiturates is different.

ACKNOWLEDGEMENTS

We are grateful to Mr M. Cochon, and Mr M. Roux, and Mrs J. Lafforgue, and Mrs M. Lucas for their technical assistance.

We very much thank Mr Long for the translation of this chapter.

REFERENCES

Adamec, R. E., Stark-Adamec, C., Perrin, R., and Livingston, K. E. (1981). What is the relevance of kindling for human temporal lobe epilepsy? In *Kindling* (ed. J. A. Wada), Vol. 2, pp. 303–14. Raven Press, New York.

Albertson, T. E., Bowyer, J. F., and Paule, M. G. (1982). Modification of the anti-convulsant efficacy of diazepam by Ro 15-1788, in the kindled amygdaloid seizure model. *Life Sci.* **31**, 1597–601.

——, Peterson, S. L., and Stark, L. G. (1980). Anticonvulsant drugs and their antago-nism on kindled amygdaloid seizures in rats. *Neuropharmacology* **19**, 643–52.

——, ——, and —— (1981*a*). The anticonvulsant effects of diazepam and pheno-barbital in prekindled and kindled seizures in rats. *Neuropharmacology* **20**, 597–603.

——, ——, ——, and Baselt, R. C. (1981*b*). Valproic acid serum levels and protection against kindled amygdaloid seizures in the rat. *Neuropharmacology* **20**, 95–7.

Ashton, D. and Wauquier, A. (1979). Behavioral analysis of the effects of 15 anti-convulsants in the amygaloid kindled rat. *Psychopharmacology* **65**, 7–13.

Babington, R. and Wedeking, P. (1973). The pharmacology of seizures induced by sensitization with low intensity brain stimulation. *Pharmacol. Biochem. Behav.* **1**, 461–7.

Bardone, M. C., Ducrot, R., Garret, C., and Julou, L. (1978). Benzodiazepine-like effects of R.P. 27 267 a dihydro-6,7 5H-pyrrolo [3,4-b]pyrazine derivative. Abstracts of the 7th International Congress of Pharmacology, Paris, p. 743. Pergamon, Oxford.

Basset, P., Durand, G., Forette, B., Jean-Louis, P., Lecourt-Bakouche, M. C., Lormeau, G., Pilate, G., and Rives, H. Etude en double insu contre placebo d'un nouvel anxiolytique: la suriclone. *Therapie (Paris)*. (In press.)

Blanchard, J. C., Boireau, A., Garret, C., and Julou, L. (1979). In vitro and in vivo inhibition by zopiclone of benzodiazepine binding to rodent brain receptors. *Life Sci.* **24**, 2417–20.

—— and Julou, L. (1982). Specific rat brain binding of suriclone, a cyclopyrrolone derivative with anxiolytic properties, *Abstracts of the 13th C.I.N.P. (Collegium Internationale Neuro-Psychopharmacologium)*, Jerusalem, p. 55.

Bonetti, E. P., Schaffner, R., Cumin, R., and Pieri, L. (1981). A selective benzodiaze-pine antagonist: Ro 15-1788. *Experientia* **37** (6), 666.

Bowyer, J. F. and Winters, W. D. (1981). The effects of various anesthetics on amygda-loid kindled seizures. *Neuropharmacology* **20**, 199–209.

Brotchi, J., Tanaka, T., and Leviel, V. (1978). Lack of activated astrocytes in the kindling phenomenon. *Exp. Neurol.* **58**, 119–25.

Corcoran, M. E., Urstad, H., Mac Caughran, J. A., Jr., and Wada, J. A. (1976). Frontal lobe and kindling in the rat. In *Kindling* (ed. J. A. Wada), pp. 215–28. Raven Press, New York.

Darragh, A., Lambe, R., Brick, I., and Downie, W. (1981). Reversal of benzodiazepine-induced sedation by intravenous Ro 15-1788. *Lancet* **8254**, 1042.

Duriez, R., Barthelemy, C., Rives, H., Courkaret, J., and Gregoire, J. (1979). Traitement des troubles du sommeil par la zopiclone. Essais cliniques en double insu contre placebo. *Therapie* **34**, 317–25.

Fitz, J. G. and McNamara, J. O. (1979). Spontaneous interictal spiking in the awake kindled rat. *Electroenceph. clin. Neurophysiol.* **47**, 592–6.

Garret, C., Bardone, M. C., Blanchard, J. C., Stutzmann, J. M., and Julou, L. (1982). Pharmacological studies of suriclone (R. P. 31,264), a cyclopyrrolone derivative with anxiolytic properties. *Abstracts of the 13th C.I.N.P. Congress,* Jerusalem, p. 244.

Goddard, G. V. (1967). Development of epileptic seizures through brain stimulation at low intensity. *Nature* **214**, 1020–1.

—— and Douglas, R. M. (1976). Does the engram of kindling model the engram of normal long term memory? In *Kindling* (ed. J. A. Wada&, pp. 1–180. Raven Press, New york.

——, Mc Intyre, D. C., and Leech, C. K. (1969). A permanent change in brain function resulting from daily electrical stimulation. *Neurology* **25**, 295–330.

Herberg, L. J. and Watkins, P. J. (1966). Epileptiform seizures induced by hypothalamic stimulation in the rat: resistance to fits following fits. *Nature* **209**, 515.

Hunkeler, W., Möhler, H., Pieri, L., Polc, P., Bonetti, E. P., Cumin, R., Schaffner, R., and Haefely, W. (1981). Selective antagonists of benzodiazepines. *Nature* **290**, 514–16.

Julou, L., Bardone, M. C., Blanchard, J. C., Garret, C., and Stutzmann, J. M. (1982). Pharmacological studies on zopiclone. *Abstracts of 13th C.I.N.P. Congress,* Jerusalem, p. 838.

Le Gal La Salle, G. (1980). Le complexe amygdalien: approches neurochimiques et fonctionnelle chez le rat. Thèse Doctorat ès Sciences, Paris.

Leviel, V. and Naquet, R. (1977). A study of the action of valproic acid on the kindling effect. *Epilepsia* **18**, 229–34.

Lloyd, K. G., Bovier, P., Broekkamp, C. L., and Worms, P. (1981). Reversal of the antiaversive and anticonvulsant actions of diazepam, but not of progabide, by a selective antagonist of benzodiazepine receptors. *Eur. J. Pharmacol.* **75**, 77–8.

Morrell, F. and Tsuru, N. (1976). Kindling in the frog: development of spontaneous epileptiform activity. *Electroenceph. clin. Neurophysiol.* **40**, 1–11.

Pinel, J. P. and Rovner, L. I. (1978). Experimental epileptogenesis: kindling induced epilepsy in rats. *Exp. Neurol.* **58**, 190–202.

Racine, R. (1972). Modification of seizure activity by electrical stimulation. II-Motor seizure. *Electroenceph. clin. Neurophysiol.* **32**, 281–94.

—— (1978). Kindling: the first decade. *Neurosurgery* **3**, 234–52.

——, Burnham, W., Gartner, J., and Levitan, D. (1973). Rates of motor seizure development in rats subjected to electrical brain stimulation: strain and interstimulation interval effects. *Electroenceph. clin. Neurophysiol.* **35**, 553–6.

——, Livingston, K., and Joaquin, A. (1975). Effects of procaine hydrochloride, diazepam and phenylhydantoin on seizure development in cortical and subcortical structures in rats. *Electroenceph. clin. Neurophysiol.* **4**, 355–65.

Rial, R. V. and Gonzales, J. (1976). The effect of diaphenylhydantoin in the prevention of the threshold descent in the electroshock on the reptilian telencephalon. Abstracts of First European Regional Conference on Epilepsy, Warsaw.

Sato, M., Wake, A., Nakashima, T., and Otsuki, S. (1977). A study of temporal lobe seizures with kindled cat preparations. *Electroenceph. clin. Neurophysiol.* **43** (No. E 009).

Tanaka, T., Lange, H., and Naquet, R. (1976). Sleep, subcortical stimulation and kindling in the cat. In *Kindling* (ed. J. A. Wada), pp. 117–33. Raven Press, New York.

Wada, J. A., Osawa, T., Sato, M., Wake, A., Green, J. R., and Troupin, A. S. (1976). Acute anticonvulsant effects of diphenylhydantoin, phenobarbital and carbamazepine. A combined electroclinical and serum level study in amydaloid kindled cats and baboons. *Epilepsia* **17**, 77–88.

——, Sato, M., and Corcoran, M. E. (1974). Persistent seizure susceptibility and recurrent spontaneous seizures in kindled cats. *Epilepsia* **15**, 465–78.

Wake, A. and Wada, J. A. (1975). Frontal cortical kindling in cats. *Can. J. neurol. Sci.* **2**, 493–9.

Wise, R. A. and Chinerman, J. (1974). Effects of diazepam and phenobarbital on electrically-induced amygdaloid seizures and seizure development. *Exp. Neurol.* **45**, 355–63.

6

Anticonvulsants and the limbic system

BRIAN MELDRUM

INTRODUCTION

Discussion of the action of anticonvulsant drugs on the limbic system suggests that there are pharmacological features of limbic seizures not shown by seizures involving the whole cerebral hemispheres. This view is supported by the widespread belief that some anticonvulsants (e.g. diphenylhydantoin and carbamazepine) are preferentially effective in seizures originating in the temporal lobes. However it may be argued that the preferential anticonvulsant action of these compounds is against partial seizures of all kinds, irrespective of whether their focal origin is in neocortex or palleocortex. There is also some evidence that other classes of anticonvulsant drug are as effective as diphenylhydantoin and carbamazepine in complex partial seizures. Thus it is necessary to look rather closely at the concept of limbic seizures and at the evidence that such seizures are pharmacologically different from seizures which primarily involve the neocortex.

LIMBIC SEIZURES

The special clinical features of seizures involving the limbic system have been recognized for a little over 100 years (see Daly 1982). Preferred designations for the syndrome have sometimes stressed the altered mental state ('dreamy states' or psychomotor epilepsy) and sometimes the anatomical substrate ('temporal lobe seizures'). The poorly descriptive term, 'complex partial seizures', is now the recommended nomenclature for focal seizures with disturbance of consciousness. The use of stereotactic EEG recordings sometimes allows the definition of the site of origin of limbic seizures, which is commonly in the hippocampus, the amygdala, or temporal cortex. However, it is never possible using electrophysiological techniques in man to define precisely the maximal extent of involvement of the limbic system in seizure discharge. However, improvements in imaging techniques (such as positron emission tomography and nuclear magnetic resonance) may permit an exact neuroanatomical delineation of the structures involved in a 'limbic seizure'. The use of autoradiography of microtome sections of the whole brain (with isotopically labelled deoxyglucose or other tracer) permits the exact delineation of the structures involved in limbic seizures in experimental animals.

This chapter considers evidence for involvement of particular neurotransmitters in limbic seizures and critically evaluates the assumption that some anticonvulsant drugs are selectively active against complex partial seizures.

Spread of experimentally induced limbic seizures

Focal electrical stimulation of the amygdala or hippocampus can trigger local afterdischarges or sustained seizure activity that spreads to other limbic structures or becomes generalized. Recurrent focal stimulation of the amygdala can 'kindle' seizure activity so that focal seizures are induced by a stimulus initially subconvulsant (Goddard *et al.* 1969). These models have been studied in the rat using [^{14}C]-deoxyglucose and autoradiography (Sokoloff 1981). The enhanced glucose utilization associated with seizure activity (Borgström *et al.* 1976; Evans and Meldrum 1984) allows the identification of structures secondarily involved in seizure activity. Afterdischarges initiated electrically in the hippocampus activate the lateral septum (Kliot and Poletti 1979). Afterdischarges in the amygdala during kindling are associated with enhanced glucose utilization in the ipsilateral hippocampus, entorhinal and pyriform cortex, the medial septal area and the 'frontal limbic field (sulcal area)' (Engel *et al.* 1978*a*).

Limbic seizures can also be induced by drugs injected focally into the hippocampus or amygdala. A vast range of compounds is potentially suitable for this but most attention has been devoted to experiments with kainic acid. As seizure activity is prolonged with this excitotoxic agent, it is possible to study either glucose utilization or secondary pathology arising as a consequence of seizure activity. Both techniques identify the same structures (Ben Ari *et al.* 1980; Lothman and Collins 1981).

In the rat, and in primates, seizures induced by the focal injection of kainic acid into one amygdala rapidly spread to the ipsilateral hippocampus, and subsequently to the contralateral hippocampus and amygdala (Ben Ari *et al.* 1980; Menini *et al.* 1980). In marmosets and baboons, seizure activity can continue for many hours in these structures with little further spread. However in rats it readily spreads to involve a variety of other structures including some thalamic nuclei (mediodorsal), entorhinal cortex, claustrum, putamen, and substantia nigra.

In man, sustained seizure activity confined to the limbic system has been recognized as a clinical syndrome for 25 years (Gastaut *et al.* 1956). The leading clinical feature of complex partial status epilepticus is a prolonged period of confusion or altered consciousness, which may be followed by permanent defects including memory impairment (Engel *et al.* 1978*b*; Markland *et al.* 1978; Wieser 1980; Treiman and Delgado-Escueta 1983).

Systemic administration of certain drugs induces seizures that either originate in or predominantly involve the limbic system as demonstrated either by electrophysiological monitoring, or [^{14}C]-deoxyglucose utilization, or by the later pathological effects. Some of these regionally selective convulsants are listed in Table 6.1. The most substantial evidence concerns kainic acid. Interestingly, systemic administration induces limbic seizures of similar extent to those which occur after focal injections into the amygdala or hippocampus, both in rats and primates. Thus seizures in the marmoset can last several hours yet remain largely confined to the hippocampus (see Fig. 6.1), whereas in the rat seizures involve the limbic core (hippocampus, amygdala, septum, and

TABLE 6.1 *Systemic drug treatments initiating limbic seizures*

	Receptors	Reference
Kainic acid	Dicarboxylic amino acid	Collins *et al.* (1980)
Dipiperidinoethane		Olney *et al.* (1980)
Lidocaine		Campos and Cavalheiro (1980)
		Post (1981)
Cholinomimetics	Cholinergic (muscarinic)	Turski *et al.* (1983 *a,b,*)

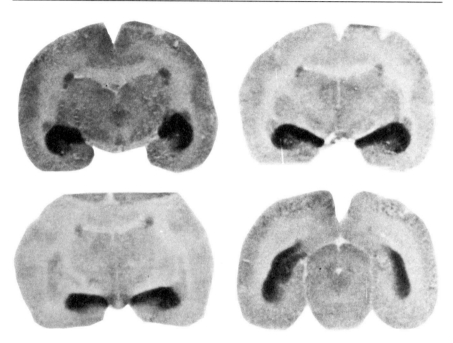

Fig. 6.1. Autoradiographs of coronal sections of a marmoset brain showing a limbic seizure induced by the intravenous injection of kainic acid, 10 mg kg^{-1}. Facial signs of limbic seizure activity were evident after 26 min. Eleven min later ^{14}C-deoxyglucose was injected i.v., and the animal sacrificed after a further 40 min. The extent of seizure activity, indicated by the enhanced glucose utilization associated with seizure activity, is seen in the amygdala and hippocampus. (M. C. Evans and B. S. Meldrum).

subiculum) but also many associated structures including entorhinal cortex. thalamus, nucleus accumbens, and ventral putamen (see illustrations in Collins *et al.* 1980).

As emphasized by Iversen (Chapter 1, this volume), anatomical methods identify structures that provide afferents to core limbic structures (hippocampus and amygdala) or receive efferents from them. There are no criteria for deciding whether these structures should be regarded as coming within the limbic system. Study of limbic seizures offers perhaps the only operational criterion – i.e. those structures that are normally invaded by seizure activity that initially involves hippocampus and amygdala.

NEUROTRANSMITTERS AND LIMBIC SEIZURES

Experiments with focal or systemic injections of convulsant agents have provided important clues as to the involvement of particular neurotransmitters in limbic seizures. Most important are the excitatory and inhibitory amino acids and drugs modifying their synaptic activity. Acetylcholine is probably more important for seizure threshold than the catecholamines or indoleamines. Excitatory amino acids (Fig. 6.2) are of the greatest significance. Glutamate and aspartate are probably the neurotransmitters at several different synaptic sites within the hippocampus (Table 6.2). They are released by electrical or high-potassium stimulation of hippocampal slices. The principal excitatory input to the hippocampus (the perforant path, from entorhinal cortex to stratum moleculare of dentate gyrus) is probably glutamatergic. The neurotransmitter of the mossy-fibre system (from dentate granule cells to pyramidal neurones in CA_{3-4}) remains uncertain, but is probably an excitatory amino acid.

The output transmitter from CA_3 pyramidal neurones to ipsilateral CA_1 pyramidal neurones (Schaffer collaterals) and to contralateral hippocampus (commissural fibres) is also probably glutamate, aspartate, or a related amino acid.

Local application of dicarboxylic amino acids can induce burst discharges, particularly in the CA_3 region. Kainic acid given intracerebroventricularly or systemically induces seizure activity most readily in the CA_3 region of the hippocampus. Although this is consistent with observations that these neurones are highly susceptible to burst firing (Traub 1982), it also implies that they are well endowed with 'kainic acid receptors'. The endogenous agent that acts on kainic acid receptors has yet to be identified. However focal application of excitatory amino acids acting on any of the presently recognized selective receptor sites (i.e. preferring quisqualic acid, or N-methyl-D-aspartate, or kainic acid) in hippocampal slices will induce burst firing in CA_1 or CA_3 neurones. Thus among possible aetiologies for seizures arising focally within the hippocampus we must consider overactivity at synapses utilizing excitatory amino acids. This could take three forms – either a release of an absolute or relative excess of the neurotransmitter, or an increase in the sensitivity or effectiveness of the postsynaptic receptors, or a failure of the uptake system that inactivates the transmitter. Mechanisms that could yield these effects include local neuronal degeneration, followed either by postsynaptic receptor supersensitivity or the sprouting of the remaining terminals to occupy vacated areas on the dendrites.

The phenomenon of long-term potentiation involves a sustained increase in postsynaptic responses in hippocampal neurones (granule or pyramidal) following particular patterns of electrical stimulation of afferent pathways (Bliss and Lømo 1973). Mechanisms leading to an increase in the sensitivity of postsynaptic glutamate receptors have been described (Baudry and Lynch 1980).

In terms of *anticonvulsant* mechanisms we can look for either diminished

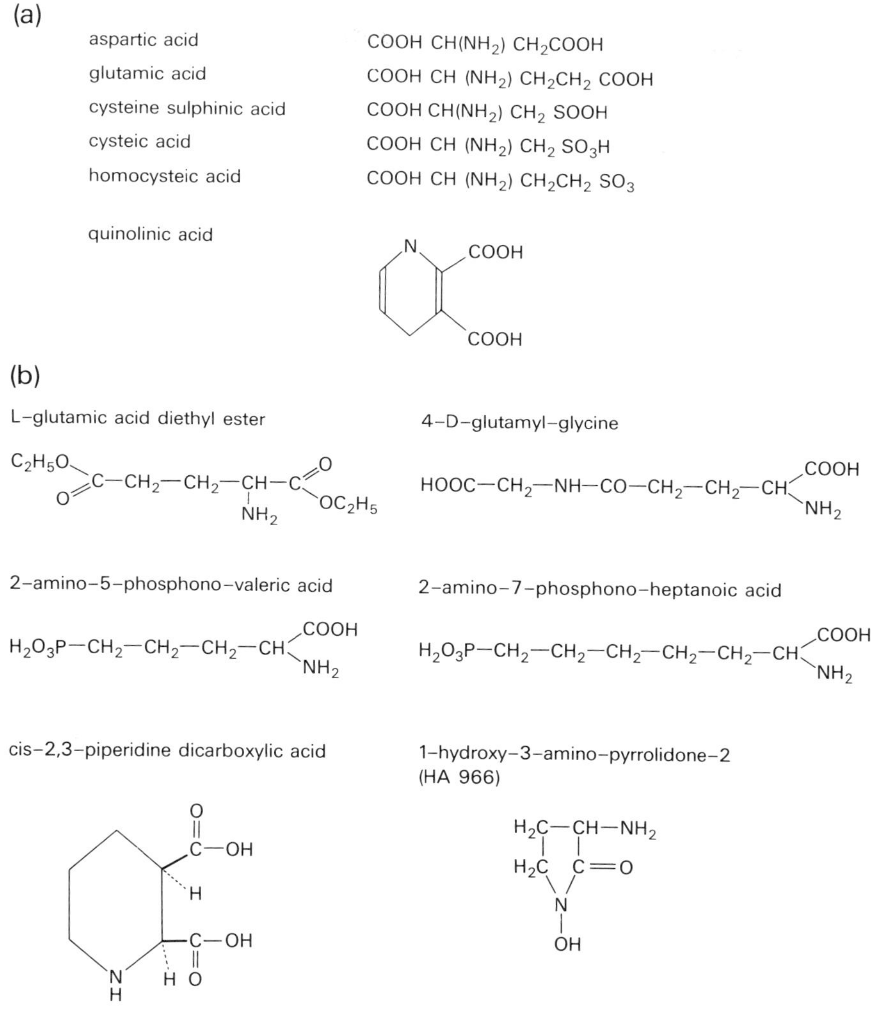

(a)

aspartic acid	COOH CH(NH$_2$) CH$_2$COOH
glutamic acid	COOH CH (NH$_2$) CH$_2$CH$_2$ COOH
cysteine sulphinic acid	COOH CH(NH$_2$) CH$_2$ SOOH
cysteic acid	COOH CH (NH$_2$) CH$_2$ SO$_3$H
homocysteic acid	COOH CH (NH$_2$) CH$_2$CH$_2$ SO$_3$

quinolinic acid

(b)

L-glutamic acid diethyl ester

4-D-glutamyl-glycine

$$HOOC-CH_2-NH-CO-CH_2-CH_2-CH \begin{smallmatrix} COOH \\ NH_2 \end{smallmatrix}$$

2-amino-5-phosphono-valeric acid

$$H_2O_3P-CH_2-CH_2-CH_2-CH \begin{smallmatrix} COOH \\ NH_2 \end{smallmatrix}$$

2-amino-7-phosphono-heptanoic acid

$$H_2O_3P-CH_2-CH_2-CH_2-CH_2-CH_2-CH \begin{smallmatrix} COOH \\ NH_2 \end{smallmatrix}$$

cis-2,3-piperidine dicarboxylic acid

1-hydroxy-3-amino-pyrrolidone-2
(HA 966)

Fig. 6.2. (a) Molecular formulae of amino acids with an excitatory action that occur in the hippocampus. (b) Molecular formulae of some compounds that act as antagonists of the excitatory amino acids in (a).

synaptic release of excitatory amino acids or diminished postsynaptic action. Several anticonvulsants (e.g. barbiturates, benzodiazepines) act on presynaptic membranes to diminish neurotransmitter release (see discussion below). Sodium valproate decreases hippocampal aspartate content (see below).

Some amino-acid analogues that act postsynaptically to antagonize excitation due to glutamate or aspartate are anticonvulsant in animal models of reflex epilepsy (Croucher *et al.* 1982). The most potent such agent, 2-amino-7-

TABLE 6.2. *Excitatory synapses of the hippocampus*

Synapse	Synaptic blockade	Receptor subtype	Transmitter	Reference
Perforant path (lat. entorhinal cortex → ap. dend. granule cells)	L-2-APB	(*Not* NMDA)	glutamate	Ganong and Cottman (1982) Nadler and Smith (1981)
Mossy fibre (granule cells → ap. dend. CA_3 pyramid. neu.)			glutamate	Crawford and Conner (1973)
Schaffer collaterals (CA_3 pyramid. neur. → ap. dend. pyramid. neur.)	D-α-AS DGG (not APV, GDEE, DAA)	(Not NMDA)	glutamate	Koerner and Cottman (1982) Collingridge *et al.* (1983)
Commissural fibres (CA_3, CA_4 pyramid. neu. → bas. + ap. dend. CA_1 & CA_3			Aspartate glutamate	Nitsch and Okada (1979)
Fornix–septum (CA_1 + CA_3 pyramid. neur. → lat. septal neurons)			glutamate	Walaas and Fonnum (1980)
Subiculum (CA_1 pyramid. neurons → pyramid. neur. subiculum)			glutamate	
Septo–hippocampal (Medial septum → bas., dend. CA_2 + CA_3 hilar polymorph cells + nicotine ap. dend. granule cells)	Atropine	Muscarinic	Acetylcholine	Fibiger (1982) Ropert and Krnjević (1982)

NMDA = N-methyl-D-aspartic acid
DAA   = D-α-amino adipate
L-2 APB = L-2-amino-4-phosphonobutyric acid
D-α AS = D-α-amino suberate

DGG = γ-D-glutamyl glycine
APV = 2-amino-5-phosphonovaleric acid
GDEE = Glutamic acid diethyl ester

phosphonoheptanoic acid, blocks seizures induced by N-methyl-D-aspartate, but not seizures induced by kainic acid (Czuczwar and Meldrum 1982). Its effect on electrically-induced limbic seizures is not yet known.

Inhibitory amino acids

Glycine, γ-aminobutyric acid (GABA), and taurine are all found in the hippocampus. GABA is synthesized in intrinsic inhibitory interneurones, that can be identified immunocytochemically by the use of antibodies to the enzyme synthesizing GABA from glutamic acid (glutamic acid decarboxylase) (Ribak *et al.* 1978). These include pyramidal-basket or short-axon cells occurring in the innermost aspect of the granule-cell layer, interspersed in the pyramidal-cell layer, and lying free in the stratum oriens of stratum radiatum. These supply GABAergic (type II, symmetrical) synapses to the cell bodies of granule cells and pyramidal cells.

There is evidence for two kinds of GABA receptors on hippocampal pyramidal neurones (Alger and Nicoll 1982). The GABA receptors on the soma membrane produce hyperpolarizing responses and benzodiazepines selectively enhance this response to GABA. Receptors on the dendrites produce depolarizing responses to GABA, and pentobarbitone preferentially enhances these responses.

Local paroxysmal activity can be induced in the pyramidal-cell layer (CA$_1$ or CA$_3$) by any compounds that block GABAergic inhibition. Thus a failure or insufficiency of the intrinsic GABAergic system could be an aetiological factor in complex partial seizures. Selective damage to the polymorphic neurones on the innermost aspect of the dentate granule-cell layer has been described as a consequence of sustained limbic seizures induced by kainic acid (Sloviter and Damiano 1981). A similar pattern of damage can be induced by sustained electrical stimulation of the perforant path in the rat (Sloviter 1983). In these two situations electrophysiological evidence has been provided to show that pathology developing in these inhibitory neurones correlates with failure of the recurrent inhibitory input to dentate granule cells. A similar selective neuronal pathology is apparent after sustained seizures induced by bicuculline or allyglycine (Atillo *et al.* 1983; Griffiths *et al.* 1983). A selective loss of intrinsic GABAergic inhibitory neurones as a result of prolonged seizure activity leading to a lowered hippocampal seizure threshold could explain the clinical observations that prolonged febrile convulsions in infancy sometimes precede the development of temporal lobe epilepsy in adolescence (Falconer 1974) and that generalized seizures in childhood and adolescence may be followed by complex partial seizures in adult life.

Numerous pharmacological procedures for enhancing GABAergic inhibition are available (Meldrum 1982). Several of them can suppress paroxysmal discharges in the hippocampus. The probable importance of enhanced GABAergic inhibition in the anticonvulsant action of benzodiazepines and other drugs is discussed below.

Acetylcholine

The cholinergic septo–hippocampal pathway was first identified histochemically on the basis of staining for acetylcholinesterase (Shute and Lewis 1967). Its properties have subsequently been investigated by a variety of anatomical and physiological techniques (Fibiger 1982: Ropert and Krnjević 1982). Terminals containing acetylcholinesterase are found predominantly on basal and apical dendrites of CA_3 and CA_2 pyramidal neurones, around hilar polymorph cells, and on apical dendrites of dentate granule cells. Acetylcholine or cholinergic agents applied in the vicinity of CA_1 or CA_3 pyramidal neurones enhance firing induced by synaptic activation by acting presynaptically to decrease GABAergic inhibition and by acting postsynaptically to decrease membrane K^+ conductance (Ben Ari *et al.* 1981; Ropert and Krnjević 1982). Hippocampal pyramidal cells in tissue culture show paroxysmal depolarization shifts in response to acetylcholine (Gähwiler and Dreifuss 1982).

Transplants of rat hippocampus into the anterior chamber of the eye show seizure-like activity in response to superfusion of cholinomimetics or endogenously released acetylcholine (Freedman *et al.* 1979). These discharges can be suppressed by atropine or by barbiturates.

In the whole animal, the systemic or intrahippocampal administration of cholinomimetic agents can produce limbic seizures that are sufficiently sustained to induce hippocampal damage similar to that seen after kainic acid (Turski *et al.* 1983*a*, *b*). However, there is at present no evidence for excessive cholinergic activity as a cause of complex partial seizures in man, nor is there any evidence that atropine or other muscarinic antagonists have a therapeutic action. It remains possible that some subgroup of patients might respond to such therapy.

Noradrenaline

The noradrenergic input to the hippocampus from cell bodies in the brainstem (locus ceruleus) was first identified by histofluorescence but has subsequently been studied by other methods including selective lesions and dopamine-β-hydroxylase immunocytochemistry (Swanson and Hartman 1975: Loy *et al.* 1980). The most dense terminations are in the hilar region of the dentate gyrus in the rat. A similar distribution or noradrenergic agonist-binding sites is found in the human hippocampus (Biegon *et al.* 1982).

The major effect of locally applied noradrenaline or of locus ceruleus stimulation on unit activity or population responses is inhibition (Segal and Bloom 1976; Mueller *et al.* 1981). This is a modulatory function, depending on enhancement of other inhibitory action. The proportion of noradrenergic terminals in the hippocampus is low (perhaps 0.3 per cent, Storm-Mathisen 1977), and receptor-binding studies indicate that the total number of noradrenergic receptor sites in the rat hippocampus is about one-tenth the number of sites revealed by ligands for cholinergic receptors or for benzodiazepine-binding sites (Dooley and Bittiger 1982).

Epileptiform activity induced by penicillin in intraocular hippocampal transplants is reduced by locally applied noradrenaline or by activation of adrenergic inputs (Freedman *et al.* 1979). Systemic administration of clonidine or other α_2 noradrenergic agonists prevents reflex epilepsy in rodents (Horton *et al.* 1980). Whether α_2 noradrenergic agonists have a therapeutic action in temporal-lobe epilepsy remains to be established.

Serotonin

The serotoninergic innervation of the hippocampus derives from cell bodies in the brainstem raphe nuclei (Azmitia and Segal 1978; Moore and Halaris 1975). As with the noradrenergic input, terminals are densest in the hilar region of the dentate gyrus.

The physiological action of serotonin on hippocampal pyramidal neurones is inhibitory, acting directly to increase membrane potassium conductance and to hyperpolarize dendrites and soma (Segal 1980; Segal and Gutnick 1980).

Both anticonvulsant and proconvulsant effects of serotonin agonists or precursors have been described. In photosensitive epilepsy in baboons, *Papio papio*, myoclonic responses are suppressed by L-5-hydroxytryptophan or various agents acting at 5HT receptor sites (Wada *et al.* 1972; Meldrum and Naquet 1971). In rodents a syndrome characterized by tremors, head jerks, and sometimes forelimb myoclonus is induced by 5-hydroxytryptophan and a range of 5HT agonists, acting at $5HT_2$ receptors (Corne *et al.* 1963; Peroutka *et al.* 1981). This syndrome in rodents appears to involve the limbic serotoninergic system, and close interactions with noradrenergic and cholinergic systems can be demonstrated (Handley and Brown 1982).

The relevance of these observations to limbic seizures in man remains to be explored. However autoradiographs of human hippocampus using tritiated serotonin show a moderate density of 5HT-binding sites in the dentate gyrus and a high density in the subiculum (Biegon *et al.* 1982).

Peptides

Immunocytochemical methods have recently permitted the detailed mapping of neurones in the amygdala and hippocampus which contain neuropeptides (Roberts *et al.* 1982, 1983: See also Chapter 15) (methionine-enkephalin, somatostatin, cholecystokinin, substance P, neurotensin and vasoactive polypeptide). The functional significance of these and other neuropeptides remains to be established (Snyder 1980). Cholecystokinin is a potent excitant agent and could be significant in epileptogenesis. Cell bodies containing cholecystokin are found in the amygdala and in the various regions of the hippocampal formation (pyramidal cell layer, dentate gyrus, and subiculum). An apparent anticonvulsant action of cholecystokinin against pentylenetetrazol and penicillin in mice has been described (Zetler 1980) but experiments to evaluate any involvement of cholecystokinin in limbic seizures have yet to be performed.

The only peptidergic systems to have undergone pharmacological investigation in relation to epilepsy are the enkephalin and β-endorphin systems. Intracerebroventricular injection of low doses of β-endorphin in rats induces seizure activity in the hippocampus and amygdala and clinical signs associated with limbic seizures (Henricksen *et al.* 1978). This appears to be due to a 'disinhibitory' action of β-endorphin, decreasing GABAergic inhibition within the hippocampus. The μ-receptor antagonist, naloxone, prevents the epileptogenic effect of β-endorphin.

Rather high doses of methionine or leucine–enkephalin given intracerebroventricularly produce transient electrocortical signs of seizure activity, that have been compared to those seen in petit mal (Frenk *et al.* 1978; Snead and Bearden 1980). Rather high doses of naloxone suppress these electrical seizures.

In experimental primates intracerebroventricular injections of enkephalins or β-endorphin do not induce seizure activity (Meldrum *et al.* 1979). Similarly focal injections of metenkephalin or a potent synthetic enkephalin analogue, FK 33,824, into the amygdala or hippocampus are not epileptogenic (Meldrum *et al.* 1981). In contrast FK 33,824, which acts as an agonist at μ-receptors, is anticonvulsant both in baboons with photosensitive epilepsy (Meldrum *et al.* 1979) and in rats exposed to the convulsant, flurothyl (Cowan *et al.* 1981).

In patients with temporal-lobe epileptic foci and implanted depth electrodes, low doses of naloxone (0.8–2.4 mg, i.v.) are without effect on focal limbic discharges (Montplaisir *et al.* 1981). However further experiments are required to evaluate the possible involvement of enkephalins and endorphin in limbic seizures.

Purinergic transmission

Adenosine and adenosine triphosphate are involved in neurotransmission in the hippocampus, possibly as neuromodulators rather than as specific neurotransmitters.

Adenosine or its derivatives (prelabelled *in vivo*) is released following specific electrical stimulation of the perforant path in hippocampal slices (Lee *et al.* 1982). Endogenously released adenosine depresses synaptic transmission in the hippocampus (Dunwiddie and Hoffer 1980; Schubert and Mitzdorf 1979). In the peripheral autonomic system adenosine has been shown to act presynaptically to reduce the release of acetylcholine or noradrenaline (Vizi 1979; Burnstock and Brown 1981; Stone 1981) and similar central actions probably involve these and other neurotransmitters, including excitatory amino acids.

There is a high-affinity system for uptake of adenosine into neurones and synaptosomes (Bender *et al.* 1980), which is potently inhibited by benzodiazepines (Phillis *et al.* 1981).

Interictal spiking induced by penicillin in *in vitro* hippocampal slices can be diminished by exogenously applied adenosine (1–10 μM) (Dunwiddie 1980).

ANTICONVULSANT DRUG ACTION

The anticonvulsants whose mechanism of action will be discussed here are diphenylhydantoin, carbamazepine, phenobarbitone, sodium valproate, and the 1,4-benzodiazepines, diazepam and clonazepam. In the principal limbic-seizure models in experimental animals, i.e. focal or systemic kainic-acid- and amygdala-kindled seizures, benzodiazepines and sodium valproate are as effective as diphenylhydantoin and carbamazepine (Ashton and Wauquier 1979). In man it is generally considered that the primary drugs for treating complex partial seizures are carbamazepine and diphenylhydantoin. However the other drugs are undoubtedly effective in a proportion of cases of complex partial seizures, and definitive comparative studies of the available drugs, given as monotherapy are not available. Studying previously untreated adults with partial seizures, Turnbull *et al.* (1982) found that sodium valproate, 600–1000 mg daily in 20 patients gave results that were not significantly different from those observed in 17 patients receiving phenytoin 300 mg daily.

Enhancement of GABAergic transmission

Potent convulsant effects of GABA antagonists, and anticonvulsant effects of GABA agonists are seen in hippocampal slices studied *in vitro*. Of all the anticonvulsant drugs the most detailed evidence for an action dependent on enhancement of GABAergic transmission concerns the benzodiazepines. There is slightly weaker evidence concerning barbiturates and equivocal evidence concerning sodium valproate and hydantoin.

Benzodiazepines

Studies of the binding to brain membrane preparations of radioactively-labelled ligands have established reciprocal interactions between GABA (or GABA agonists) and benzodiazepines, i.e. the binding of benzodiazepines is enhanced in the presence of GABA or GABA agonists (Braestrup *et al.* 1979; Karobath *et al.* 1979) and the (low-affinity) binding of GABA is enhanced by benzodiazepines (Skerritt *et al.* 1982*a*). It is also clear that both types of compound bind to the same receptor complex (Chang and Barnard 1982; Olsen 1982).

Studies of cultured spinal neurones show that the inhibitory action of iontophoretically-applied GABA is enhanced when benzodiazepines are present in concentrations comparable to those obtained *in vivo* during anticonvulsant therapy (MacDonald and Barker 1979). Noise analysis of membrane potential shows that benzodiazepines increase the number of chloride channels opened by a given dose of GABA (Study and Barker 1981).

Benzodiazepines given systemically to rats enhance recurrent inhibition (due to GABAergic interneurones) in the hippocampus (Wolf and Haas 1977).

Benzodiazepines also enhance presynaptic inhibition on primary afferent pathways, as assessed by spinal reflexes or dorsal root potentials (Haefely *et al.* 1981).

Of all the effects of anticonvulsant drugs on synaptic transmission so far described, enhancement by benzodiazepines of postsynaptic inhibition mediated by GABA is the most fully documented. It shows a good correlation with anticonvulsant action when different benzodiazepines are compared, and occupation of a relatively low proportion of benzodiazepines sites produces the full anticonvulsant effect (Duka *et al.* 1979). This does not exclude the possibility that other actions of benzodiazepines may also be relevant to anti-epileptic action.

Barbiturates

A similar range of data indicates that barbiturates enhance GABAergic transmission. Thus in studies of binding to brain-membrane sites, an interaction of barbiturates with GABA-binding is seen that involves a site closely related to but distinct from the GABA recognition site and the benzodiazepine-binding site (Olsen 1982). Using cultured spinal neurones, enhancement of the effect of exogenous GABA is observed, and can be shown by noise analysis to be due to prolongation of the open time of chloride channels (Study and Barker 1981). Recurrent inhibition in the hippocampus is also enhanced by systemically administered barbiturates (Wolf and Haas 1977). However, comparing a variety of barbiturates, there may be a better correlation between the sedative and anaesthetic actions of barbiturates and the action on GABA-mediated inhibition than there is with anticonvulsant action. Thus other effects of barbiturates are likely to be of major importance in the anticonvulsant action of phenobarbitone.

Sodium valproate

A recent comprehensive review of the mechanisms of action of sodium (Na) valproate concludes that the view that it acts by inhibiting the further metabolism of GABA is not substantiated (Chapman *et al.* 1982*b*). In binding studies Na valproate shows no affinity for GABA-receptor sites or for the benzodiazepine-binding sites, but at high concentrations, it inhibits the binding of dihydropicrotoxin (Ticku and Davis 1981), which implies action at the site which is sensitive to barbiturate action on GABA receptors (Olsen 1982). Enhancement of the inhibitory action of GABA is seen in the presence of high concentrations of NA valproate (3–10 mM) *in vitro* (MacDonald and Bergey 1979; Harrison and Simmonds 1982). A postsynaptic action of Na valproate to enhance the effect of GABA could, via a negative feedback mechanism, explain the increase in the concentration of GABA seen in nerve terminals after valproate (Gale and Iadarola 1980). However the effects of Na valproate on brain GABA concentration are most evident in the neocortex and substantia nigra, and are absent in the hippocampus (see Fig. 6.3).

Hydantoin

Diphenylhydantoin does not modify GABA binding in brain membrane preparations either directly or indirectly (although interactions with benzodiazepine-binding sites have been described) (Gallager and Mallorga 1980; Shah *et al.* 1981). Effects of diphenylhydantoin on GABA-mediated inhibition vary according to the preparation studied; in cultured mouse spinal neurones, concentrations of 5–10 μg ml^{-1} are required to enhance inhibitory responses to GABA (MacDonald 1983).

Carbamazepine

There is no evidence that carbamazepine significantly effects binding of ligands to the GABA-receptor complex, or that it enhances GABA-mediated inhibition.

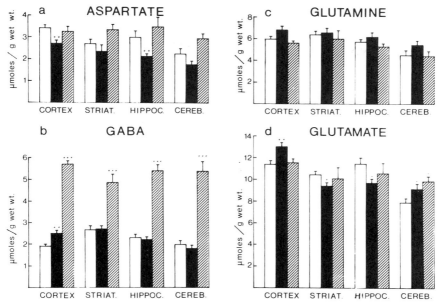

Fig. 6.3. Effects of sodium valproate or of γ-vinyl-GABA neurotransmitter on amino acids in different brain regions. The histograms show the concentration of aspartate, GABA, glutamine, and glutamate in the cortex, striatum, hippocampus, and cerebellum in control rats (open bars), and in rats given Na valproate, 400 mg kg^{-1}, i.p. (solid bars) or in rats given γ-vinyl-GABA 1 g kg^{-1} (hatched bars). The values are expressed as mean ± SEM, and statistical significance between an experimental group and the corresponding control group is indicated as *$p < 0.05$; **$p < 0.01$; ***$p < 0.001$. Note that after NA valproate, GABA is significantly increased only in the cortex, whereas the decrease in aspartate concentration in the hippocampus is highly significant. (From Chapman *et al.* (1982).)

However carbamazepine may exert an inhibitory influence on synaptic activity in the hippocampus through an action on adenosine receptors. Carbamazepine appears to be unique among anticonvulsant drugs in potently inhibiting the high-affinity binding of phenylisopropyl-adenosine (a stable ligand for purinergic receptors) to brain-membrane preparations (Skerritt *et al.* 1982*b*).

Excitatory amino-acid transmitters

The extremely important role played by dicarboxylic amino acids, and related compounds in the hippocampus makes it likely that effects on these systems are critical for anticonvulsant drug action. However these effects have been much less exhaustively studied than those on inhibitory transmission. Indeed the possibility of preventing seizures by specifically blocking excitatory amino-acid transmission has only recently been established (Croucher *et al.* 1982; Czuczwar and Meldrum 1982).

Benzodiazepines

There is limited evidence for two actions of benzodiazepines one presynaptic, the other postsynaptic.

In rat hippocampal slices preloaded with radiolabelled glutamate or cysteine sulphinic acid, release of the amino acid (induced by high concentration of potassium) is decreased by benzodiazepines but not by barbiturates or hydantoin (Baba *et al.* 1983). This effect of benzodiazepines can be reproduced by GABA and is blocked by bicuculline, so it appears to be due to action at a GABA/benzodiazepine receptor complex, possibly presynaptic on the excitatory terminals.

Synaptic release of aspartate from rat olfactory cortex slices is depressed by chlordiazepoxide (Collins 1981).

The convulsant β-carboline, DMCM, which acts via the high-affinity benzodiazepine-binding site, enhances the release of preloaded aspartate in rat forebrain slices (Kerwin and Meldrum 1983). This provides further evidence for a control of excitatory amino-acid release by the GABA/benzodiazepine receptor complex.

Postsynaptically the excitatory action of glutamate is diminished in spinal cord and cortex by local application of benzodiazepines (Evans *et al.* 1977; Assumpcao *et al.* 1979). However, the excitatory action of glutamate on cultured spinal neurones is not modified in the presence of concentrations of benzodiazepines that potently enhance GABAergic inhibition (MacDonald and Barker 1979).

Barbiturates

Studies employing many different preparations (including sympathetic ganglia and spinal cord) have shown that barbiturates impair excitatory transmission. Barbiturates reduce calcium entry into synaptic terminals (Heyer and MacDonald 1982) and decrease stimulated release of aspartate from brain slices (Willow *et al.* 1980; Waller and Richter 1980).

Postsynaptically, barbiturates decrease the excitatory action of glutamate or homocysteate (Lambert and Flatman 1981; Heyer and MacDonald 1982).

Actions of barbiturates on excitatory transmission are important for their sedative and anaesthetic action and probably contribute to some anticonvulsant effects.

Sodium valproate

In rodents Na valproate induces a fall in brain-aspartate concentration, the time course of which correlates with the anticonvulsant effect (Schechter *et al.* 1978). In the hippocampus the change in aspartate is more prominent than any effect on GABA concentration (Fig. 6.3). Comparison of various branched chain fatty acids, that are analogues of valproic acid, shows that the anticonvulsant effect correlates better with actions on aspartate than on GABA (Chapman *et al.* 1983). The changes in aspartate concentration probably arise from a direct action of Na valproate on intermediary metabolism. Electrophysiological studies showing changes in excitatory transmission in the hippocampus after Na valproate are not yet available.

Hydantoin

Phenytoin at relatively high concentrations (2 μg ml^{-1}) acts presynaptically to reduce the release of many different neurotransmitters (including acetylcholine, noradrenaline, GABA, and glutamate). This effect appears to depend on reduced calcium entry (MacDonald 1983). Inhibition of a calcium–calmodulin-dependent protein-kinase in nerve terminal membranes by hydantoin appears to be a closely related phenomenon (DeLorenzo 1980).

Hydantoin applied iontophoretically in the hippocampus antagonizes the postsynaptic excitatory action of glutamate and of acetylcholine (Matthews and Connor 1977), but it has not been shown that pharmacologically relevant concentrations of hydantoin also do this.

SUMMARY

Many drugs acting on inhibitory or excitatory transmission induce seizures that begin in the hippocampus and may remain confined to the limbic system, even when they continue for several hours. Sustained limbic seizures may be followed by pathology similar to that found in patients with complex partial seizures.

Anticonvulsant drugs modify synaptic transmission in the hippocampus in many different ways. The effects which appear to be most relevant to anticonvulsant action are

1. A postsynaptic action to enhance the inhibitory action of GABA – shown most potently by benzodiazepines and barbiturates;

2. A presynaptic action to diminish the release of excitatory amino-acid neurotransmitters (shown by hydantoin, benzodiazepines, and barbiturates);

3. A post synaptic action to decrease the excitatory effect of amino-acid neurotransmission (shown by barbiturates and benzodiazepines).

Pharmacological understanding of the interactions of drugs and neurotransmitters in the limbic system is at a primitive stage, and further study is required of interactions with amino acids, monoamines, peptides, and purines.

ACKNOWLEDGEMENTS

The author thanks the Wellcome Trust, the Medical Research Council, and the National Fund for Research into Crippling Diseases for financial support. Thanks are also due to Drs G. Anlezark, Chapman, and M. C. Evans and Mr M. Croucher for their collaboration.

REFERENCES

Alger, B. E. and Nicoll, R. A. (1982). Pharmacological evidence for two kinds of GABA receptor on rat hippocampal pyramidal cells studied *in vitro*. *J. Physiol.* **328**, 125–41.

Ashton, D. and Wauquier, A. (1979). Behavioral analysis of the effects of 15 anticonvulsants in the amygdaloid kindled rat. *Psychopharmacology* **65**, 7–13.

Assumpcao, J. A., Bernardi, N., Brown, J., and Stone, T. W. (1979). Selective antagonism by benzodiazepines of neuronal responses to excitatory amino acids in the cerebral cortex. *Br. J. Pharmacol.* **67**, 563–8.

Atillo, A., Söderfeldt, B., Kalimo, H., Olsson, Y., and Siesjö, B. K. (1983). Pathogenesis of brain lesions caused by experimental epilepsy. Light- and electron-microscopic changes in the rat hippocampus following bicuculline-induced status epilepticus. *Acta Neuropathol., Berl.* **59**, 11–24.

Azmitia, E. C. and Segal, M. (1978). An autoradiographic analysis of the differential ascending projections of the dorsal and median raphe nuclei in the rat. *J. comp. Neurol.* **179**, 641–68.

Baba, A., Okumura, S., Mizuo, H., and Iwata, H. (1983). Inhibition by diazepam and γ-aminobutyric acid of depolarization-induced release of ^{14}C-cysteine sulfinate and ^{3}H-glutamate in rat hippocampal slices. *J. Neurochem.* **40**, 280–4.

Baudry, M. and Lynch, G. (1980). Regulation of hippocampal glutamate receptors: evidence for the involvement of a calcium-activated protease. *Proc. Nat. Acad. Sci.* **77**, 2298–302.

Ben-Ari, Y., Krnjević, K., Reinhardt, W., and Ropert, N. (1981). Intracellular observations on the disinhibitory action of acetylcholine in the hippocampus. *Neuroscience* **12**, 2475–84.

——, Tremblay, E., Ottersen, O. P., and Meldrum, B. S. (1980). The role of epileptic activity in hippocampal and 'remote' cerebral lesions induced by kainic acid. *Brain Res.* **191**, 79–97.

Bender, A. S., Wu, P. H., and Phillis, J. W. (1980). The characterization of (3H) adenosine uptake into rat cerebral cortical synaptosomes. *J. Neurochem.* **35**, 629–40.

Biegon, A., Rainbow, T. C., Mann, J. J., and McEwen, B. S. (1982). Neurotransmitter receptor sites in human hippocampus: a quantitative autoradiographic study. *Brain Res.* **247**, 379–82.

Bliss, T. V. P. and Lømo, T. (1973). Long-lasting potentiation of synaptic transmission in the dentate area of the anaesthetised rabbit following stimulation of the perforant path. *J. Physiol.* **232**, 331–56.

Borgström, L., Chapman, A. G., and Siesjö, B. K. (1976). Glucose consumption in the cerebral cortex of rat during bicuculline-induced status epilepticus. *J. Neurochem.* **27**, 971–3.

Braestrup, C., Nielsen, M., Krogsgaard-Larsen, P., and Falch, E. (1979). Partial agonists for brain GABA/benzodiazepine receptor complex. *Nature* **280**, 331–3.

Burnstock, G. and Brown, C. M. (1981). An introduction to purinergic receptors. In *Purinergic receptors* (ed. G. Burnstock), pp. 1–45. Chapman & Hall, London.

Campos, C. J. R. and Cavalheiro, E. A. (1980). The paradoxical effect of lidocaine on an experimental model of epilepsy. *Arch. int. Pharmacodyn. Ther.* **243**, 66–73.

Chang, L-R. and Barnard, E. A. (1982). The benzodiazepine/GABA receptor complex: Molecular size in brain synaptic membranes and in solution. *J. Neurochem.* **39**, 1507–18.

Chapman, A. G., Riley, K., Evans, M. C., and Meldrum, B. S. (1982*a*). Acute effects of sodium valproate and γ-vinyl.GABA on regional amino acid metabolism in the rat brain. *Neurochem. Res.* **7**, 1089–105.

——, Keane, P. E., Meldrum, B. S., Simiand, J. and Vernieres, J. C. (1982*b*). Mechanism of anticonvulsant action of valproate. *Prog. Neurobiol.* **19**, 315–59.

——, Meldrum, B. S., and Mendes, E. (1983). Acute anticonvulsant activity of structural analogues of valproic acid and changes in brain GABA and aspartate content. *Life Sci.* **32**, 2023–31.

Collingridge, G. L., Kehl, S. J., and McLennan, H. (1983). Excitatory amino acids in synaptic transmission in the Schaffer collateral–commissural pathway of the rat hippocampus. *J. Physiol.* **334**, 33–46.

Collins, G. G. S. (1981). The effects of chlordiazepoxide on synaptic transmission and amino acid neurotransmitter release in slices of rat olfactory cortex. *Brain Res.* **224**, 389–404.

Collins, R. C., McLean, M., and Olney, J. (1980). Cerebral metabolic responses to systemic kainic acid: ^{14}C-deoxyglucose studies. *Life Sci.* **27**, 855–62.

Corne, S. J., Pickering, R. W., and Warner, B. T. (1963). A method for assessing the effects of drugs on the central actions of 5-hydroxytryptamine. *Br. J. Pharmacol.* **20**, 106–20.

Cowan, A., Tortella, F. C., and Adler, M. W. (1981). A comparison of the anticonvulsant effects of two systematically active enkephalin analogues in rats. *Eur. J. Pharmacol.* **71**, 117–21.

Crawford, I. L. and Connor, J. D. (1973). Localization and release of glutamic acid in relation to the hippocampal mossy fibre pathway. *Nature* **244**, 442–3.

Croucher, M. J., Collins, J. F., and Meldrum, B. S. (1982). Anticonvulsant actions of excitatory amino acid antagonists. *Science* **216**, 899–901.

Czuczwar, S. J. and Meldrum, B. (1982). Protection against chemically-induced seizures by 2-amino-7-phosphonoheptanoic acid. *Eur. J. Pharmacol.* **83**, 335–8.

Daly, D. D. (1982). Complex partial seizures. In *A textbook of epilepsy* (2nd edn.) (ed. J. Laidlaw and A. Richens), pp. 131–46. Churchill Livingstone, Edinburgh.

DeLorenzo, R. J. (1980). Phenytoin: calcium- and calmodulin-dependent protein phosphorylation and neurotransmitter release. In *Antiepileptic drugs: mechanisms of action* (ed. G. H. Glaser, J. K. Penty, and D. M. Woodbury), pp. 399–414. Raven Press, New York.

Dooley, D. J. and Bittiger, H. (1982). Characterization of neurotransmitter receptors in the rat hippocampal formation. *J. Neurochem.* **38**, 1621–6.

Duka, T., Hoellt, V., and Herz, A. (1979). In vivo receptor occupation by benzodiazepines and correlations with the pharmacological effect. *Brain Res.* **179**, 147–56.

Dunwiddie, T. V. (1980). Endogenously released adenosine regulates excitability in the *in vitro* hippocampus. *Epilepsia* **21**, 541–8.

—— and Hoffer, B. J. (1980). Adenine nucleotides and synaptic transmission in the *in vitro* rat hippocampus. *Br. J. Pharmacol.* **69**, 59–68.

Engel, J., Wolfson, L., and Brown, L. (1978*a*). Anatomical correlates of electrical and behavioral events related to kindling. *Ann. Neurol.* **3**, 538–44.

——, Ludwig, B. I., and Fetell, M. (1978*b*). Prolonged partial complex status epilepticus: EEG and behavioural observations. *Neurology, Minneapolis* **28**, 863–9.

Evans, R. H., Francis, A. A., and Watkins, J. C. (1977). Differential antagonism by chlorpromazine and diazepam of frog mononeurone depolarization induced by glutamate-related amino acids. *Eur. J. Pharmacol.* **44**, 325–30.

Evans, M. C. and Meldrum, B. S. (1984). Regional brain glucose metabolism in chemically-induced seizures in the rat. *Brain Res.* (In press.)

Falconer, M. A. (1974). Mesial temporal (Ammon's horn) sclerosis as a common cause of epilepsy. Aetiology, treatment and prevention. *Lancet* **ii**, 767–70.

Fibiger, H. C. (1982). The organization and some projections of cholinergic neurons of the mammalian forebrain. *Brain Res. Rev.* **4**, 327–88.

Freedman, R., Taylor, D. A., Seiger, A., Olson, L., and Hoffer, B. J. (1979). Seizures and related epileptiform activity in hippocampus transplanted to the anterior chamber of the eye: modulation by cholinergic and adrenergic input. *Ann. Neurol.* **6**, 291–95.

Frank, H., Urca, G., and Liebeskind, J. C. (1978). Epileptic properties of leucine and methionine–enkephalin: comparison with morphine and reversibility by naloxone. *Brain Res. Arch.* **147**, 327–37.

Gähwiler, B. H. and Dreifuss, J. J. (1982). Multiple actions of acetylcholine on hippocampal pyramidal cells in organotypic explant cultures. *Neuroscience* **7**, 1243–56.

Gale, K. and Iadarola, M. J. (1980). Seizure protection and increased nerve terminal GABA: delayed effects of GABA-transaminase inhibition. *Science* **208**, 288–91.

Gallager, D. W. and Mallorga, P. (1980). Diphenylhydantoin: pre- and post-natal administration alters diazepam binding in developing rat cerebral cortex. *Science* **208**, 64–6.

Ganong, A: H. and Cotman, C. W. (1982). Acidic amino acid antagonists of lateral perforant path synaptic transmission: agonist–antagonist interactions in the dentate gyrus. *Neurosci. Lett.* **34**, 195–200.

Gastaut, H., Roger, J., and Roger, A. (1956). Sur la signification de certaines fugues épileptiques: état de mal temporal. *Rev. Neurol., Paris* **94**, 298–301.

Goddard, G. V., McIntyre, D. C., and Leech, C. K. (1969). A permanent change in brain function resulting from daily electrical stimulation. *Exp. Neurol.* **25**, 296–330.

Griffiths, T., Evans, M. C., and Meldrum, B. S. (1983). Intracellular calcium accumulation in rat hippocampus during seizures induced by bicuculline or L-allylglycine. *Neuroscience* **10**, 385–95.

Haefely, W., Pieri, L., Polc, P., and Schaffner, R. (1981). General pharmacology and neuropharmacology of benzodiazepine derivatives. *Handbook exp. Pharmacol.* **55**, 8–262.

Handley, S. L. and Brown, J. (1982). Effects on the 5-hydroxytryptamine-induced head-twitch of drugs with selective actions on alpha$_1$ and alpha$_2$-adrenoceptors. *Neuropharmacology* **21**, 507–10.

Harrison, N. L. and Simmonds, M. A. (1982). Sodium valproate enhances responses to GABA receptor activation only at high concentrations. *Brain Res.* **250**, 201–4.

Henriksen, S. J., Bloom, F. E., McCoy, F., Ling, N., and Guillemin, R. (1978). β-endorphin induces non-convulsive limbic seizures. *Proc. Nat. Acad. Sci., Wash.* **75**, 5221–5.

Heyer, E. J. and Macdonald, R. L. (1982). Barbiturate reduction of calcium-dependent action potentials: correlation with anaesthetic action. *Brain Res.* **236**, 157–71.

Horton, R., Anlezark, G., and Meldrum, B. (1980). Noradrenergic influences on sound-induced seizures. *J. Pharmacol. exp. Ther.* **214**, 437–42.

Karobath, M., Placheta, P., Lippitsch, M., and Krogagaard-Larsen, P. (1979). Is stimulation of benzodiazepine receptor binding mediated by a novel GABA receptor? *Nature* **278**, 748–9.

Kerwin, R. W. and Meldrum, B. S. (1983). Effect on cerebral ^{3}H-D-aspartate release of 3-mercaptopropionic acid and methyl 6,7-dimethoxy-4-ethyl-β-carboline-3-carboxylate. *Eur. J. Pharmacol.* **89**, 265–9.

Kliot, M. and Poletti, C. E. (1979). Hippocampal after-discharges: differential spread of activity shown by the ^{14}-C-deoxyglucose technique. *Science* **204**, 641–3.

Koerner, J. F. and Cotman, C. W. (1982). Response of Schaffer collateral-CA$_1$ pyramical cell synapses of the hippocampus to analogues of acidic amino acids. *Brain Res.* **251**, 105–15.

Lambert, J. D. C. and Flatman, J. A. (1981). The interaction between barbiturate anaesthetics and excitatory amino acid responses on cat spinal neurones. *Neuropharmacology* **20**, 227–40.

Lee, K., Schubert, P., Brigkoff, V., Sherman, B., and Lynch, G. (1982). A combined *in vivo/in vitro* study of the presynaptic release of adenosine derivatives in the hippocampus. *J. Neurochem.* **38**, 80–3.

Lothman, E. W. and Collins, R. C. (1981). Kainic acid induced limbic seizures: metabolic, behavioral, electroencephalographic and neuropathological correlates. *Brain Res.* **218**, 299–318.

Loy, R., Koziell, D. A., Lindsey, J. D., and Moore, R. Y. (1980). Noradrenergic innervation of the adult rat hippocampal formation. *J. comp. Neurol.* **189**, 699–710.

Macdonald, R. L. (1983). Mechanisms of anticonvulsant drug action. In *Recent advances in epilepsy* (ed. T. Pedley and B. S. Meldrum), Vol. 1. Churchill-Livingstone, Edinburgh.

—— and Barker, J. L. (1979). Enhancement of GABA-mediated post-synaptic inhibition in cultured mammalian spinal cord neurons: a common mode of anticonvulsant action. *Brain Res.* **167**, 323–36.

—— and Bergey, G. K. (1979). Valproic acid augments GABA-mediated postsynaptic inhibition in cultured mammalian neurons. *Brain Res.* **170**, 558–62.

Markland, O. N., Wheeler, G. L., and Pollack, L. P. (1978). Complex partial status epilepticus (psychomotor status). *Neurology, Minneapolis* **28**, 189–96.

Matthews, W. D. and Connor, J. D. (1977). Actions of iontophoretic phenytoin and medazepam on hippocampus neurons. *J. Pharmacol. exp. Ther.* **201**, 613–21.

Meldrum, B. (1982). Pharmacology of GABA. *Clin. Neuropharmacol.* **5**, 293–316.

——, Menini, Ch., Stutzmann, J. M., and Naquet, R. (1979). Effects of opiate-like peptides, morphine and naloxone in the photosensitive baboon, *Papio papio. Brain Res.* **170**, 333–48.

——, ——, Naquet, R., Riche, D., and Silva-Comte, C. (1981). Absence of seizure activity following focal cerebral injection of enkephalin in a primate. *Regulatory Peptides* **2**, 383–90.

—— and Naquet, R. (1971). Effects of psilocybin, dimethyltryptamine, mescaline and various lysergic derivatives on the EEG and on photically induced epilepsy in the baboon (*Papio papio*). *Electroenceph. clin. Neurophysiol.* **31**, 563–72.

Menini, C., Meldrum, B. S., Riche, D., Silva-Comte, C., and Stutzmann, J. M. (1980). Sustained limbic seizures induced by intra-amygdaloid kainic acid in the baboon: symptomatology and neuropathological consequences. *Ann. Neurol.* **8**, 501–9.

Montplaisir, J., Saint-Hilaire, J. M., Walsh, J. T., Lavardiere, M., and Bouvier, G. (1981). Naloxone and focal epilepsy: a study with depth electrodes. *Neurology* **31**, 350–2.

Moore, R. Y. and Halaris, A. E. (1975). Hippocampal innervation by serotonin neurons of the midbrain raphe in the rat. *J. comp. Neurol.* **164**, 171–84.

Mueller, A. L., Hoffer, B. J., and Dunwiddie, T. V. (1981). Noradrenergic responses in rat hippocampus: evidence for mediation by γ and β receptors in the in vitro slice. *Brain Res.* **214**, 113–26.

Nadler, J. V. and Smith, E. M. (1981). Perforant path lesion depletes glutamate content of fascia dentata synaptosomes. *Neurosci. Lett.* **25**, 275–80.

Nitsch, C. and Okada, Y. (1979). Distribution of glutamate in layers of the rabbit hippocampal fields CA1, CA3, and the dentate area. *J. Neurosci. Res.* **4**, 161–7.

Olney, J. W., Fuller, T. A., Collins, R. C., and de Gubareff, T. (1980). Systemic dipiperidinoethane mimics the convulsant and neurotoxic actions of kainic acid. *Brain Res.* **200**, 231–5.

Olsen, R. W. (1982). Drug interactions at the GABA receptor–ionophore complex. *Ann. Rev. Pharmacol. Toxicol.* **22**, 245–77.

Peroutka, S. J., Lebovitz, R. M., and Snyder, S. H. (1981). Two distinct central serotonin receptors with different physiological functions. *Science* **212**, 827–30.

Phillis, J. W., Wu, P. H., and Bender, A. S. (1981). Inhibition of adenosine uptake into rat brain synaptosomes by the benzodiazepines. *Gen. Pharmacol.* **12**, 67–70.

Post, R. M. (1981). Lidocaine-kindled limbic seizures: behavioural implications. In *Kindling* (ed. J. A. Wada), Vol. 2, pp. 149–60. Raven Press, New York.

Ribak, C. E., Vaughn, J. E., and Saito, K. (1978). Immunocytochemical localization of glutamic acid decarboxylase in neuronal somata following colchicine inhibition of axonal transport. *Brain Res.* **140**, 315–32.

Roberts, G. W., Woodhams, P. L., Polak, J. M., and Crow, T. J. (1982). Distribution of neuropeptides in the limbic system of the rat: the amygdaloid complex. *Neuroscience* **7**, 99–131.

——, ——, ——, and —— (1984). Distribution of neuropeptides in the limbic system of the rat: the hippocampus. *Neuroscience* **11**, 35–77.

Ropert, N. and Krnjević, K. (1982). Pharmacological characteristics of facilitation of hippocampal population spikes by cholinomimetics. *Neuroscience* **7**, 1863–77.

Schechter, P. J., Tranier, Y., and Grove, J. (1978). Effect of dipropylacetate on amino acid concentrations in mouse brain: correlation with anticonvulsant activity. *J. Neurochem.* **31**, 1325–7.

Schubert, P. and Mitzdorf, U. (1979). Analysis and quantitative evaluation of the depressive effect of adenosine on evoked potentials in hippocampal slices. *Brain Res.* **172**, 186–90.

Segal, M. (1980). The action of serotonin in the rat hippocampal slice preparation. *J. Physiol.* **303**, 423–39.

—— and Bloom, F. E. (1976). The action of norepinephrine in the rat hippocampus IV:

The effects of locus coeruleus stimulation on evoked hippocampal unit activity. *Brain Res.* **107**, 513–25.

—— and Gutnick, M. J. (1980). Effects of serotonin on extracellular potassium concentration in the rat hippocampal slice. *Brain Res.* **195**, 389–401.

Shah, D. S., Chambon, P., and Guidotti, A.(1981). Binding of ³H-5,5-diphenylhydantoin to rat brain membranes. *Neurophrmacology* **20**, 1115–9.

Shute, C. C. D. and Lewis, P. R. (1967). The ascending cholinergic reticular system: neocortical, olfactory and subcortical projections. *Brain* **90**, 497–522.

Skerritt, J. H., Willow, M., and Johnston, G. A. R. (1982*a*). Diazepam enhancement of lower affinity GABA binding to rat brain membranes. *Neurosci. Lett.* **29**, 663–6.

——, Davies, L. P., and Johnston, G. A. R. (1982*b*). A purinergic component in the anticonvulsant action of carbamazepine. *Eur. J. Pharmacol.* **82**, 195–7.

Sloviter, R. S. (1983). "Epileptic" brain damage in rats induced by sustained electrical stimulation of the perforant path. I. Acute electrophysiological and light microscopic studies. *Brain Res. Bull.* **10**, 675–97.

—— and Damiano, B. P. (1981). On the relationship between kainic acid-induced epileptiform activity and hippocampal neuronal damage. *Neuropharmacology* **20**, 1003–11.

Snead, O. C. and Bearden, L. J. (1980). Anticonvulsants specific for petit mal antagonize epileptogenic effect of leucine Enkephalin. *Science* **210**, 1031–3.

Snyder, S. H. (1980). Brain peptides as neurotransmitters. *Science* **209**, 976–83.

Sokoloff, L. (1981). Localization of functional activity in the central nervous system by measurement of glucose utilization with radioactive deoxyglucose. *J. Cerebr. Blood Flow Metabol.* **1**, 7–36.

Stone, T. W. (1981). Physiological roles for adenosine and adenosine 5'-triphosphate in the nervous system. *Neuroscience* **6**, 523–55.

Storm-Mathisen, J. (1977). Localization of transmitter candidates in the brain: the hippocampal formation as a model. *Prog. Neurobiol.* **8**, 119–81.

Study, R. E. and Barker, J. L. (1981). Diazepam and (–) pentobarbital: fluctuation analysis reveals different mechanisms for potentiation of γ-aminobutyric acid responses in cultured neurons. *Proc. Nat. Acad. Sci.* **78**, 7180–4.

Swanson, L. W. and Hartman, B. K. (1975). The central adrenergic system. An immunofluorescence study of the location of cell bodies and their efferent connections in the rat utilizing dopamine-β-hydroxylase as a marker. *J. comp. Neurol.* **163**, 467–505.

Ticku, M. K. and Davis, W. C. (1981). Effect of valproic acid on (³H) diazepam and (³H) dihydropicrotoxinin binding sites at the benzodiazepine–GABA receptor-ionophore complex. *Brain Res.* **223**, 218–22.

Traub, R. D. (1982). Stimulation of intrinsic bursting in CA₃ hippocampal neurons. *Neuroscience* **7**, 1233–42.

Treiman, D. M. and Delgado-Escueta, A. V. (1983). Complex partial status epilepticus. In *Advances in neurology,* Vol. 34: Status epilepticus (ed. A. V. Delgado-Escueta, C. G. Wasterlain, D. M. Treiman, and R. J. Porter), pp. 69–81. Raven Press, New York.

Turnbull, D. M., Rawlins, M. D., Weightman, D., and Chadwick, D. W. (1982). A comparison of phenytoin and valproate in previously untreated adult epileptic patients. *J. Neurol. Neurosurg. Psychiat.* **45**, 55–9.

Turski, W. A., Cavalheiro, E. A., Schwarz, M., Czuczwar, S. J., and Turski, L. (1983*a*). Limbic seizures produced by pilocarpine in rats: behavioural, electroencephalographic and neuropathological aspects. *Behav. Brain Res.* **9**, 315–35.

——, ——, Turski, L., and Kleinrok, Z. (1983*b*). Intrahippocampal bethaechol in rats: behavioural, electroencephalographic and neuropathological correlates. *Behav. Brain Res.* **7**, 361–70.

Vizi, E. S. (1979). Presynaptic modulation of neurochemical transmission. *Prog. Neurobiol.* **12**, 181–290.

Wada, J. A., Balzamo, E., Meldrum, B. S., and Naquet, R. (1972). Behavioural and electrographic effects of L-5-hydroxytryptophan and D,L-parachlorophenylalanine on epileptic Senegalese baboon (*Papio papio*). *Electroenceph. clin. Neurophysiol.* **33**, 520–6.

Walaas, I. and Fonnum, F. (1980). Biochemical evidence for glutamate as a transmitter in hippocampal efferents to the basal forebrain and hypothalamus in the rat brain. *Neuroscience* **5**, 1691–9.

Waller, M. B. and Richter, J. A. (1980). Effects of pentobarbital and Ca^{++} on the resting and K^+-stimulated release of several endogenous neurotransmitters from rat midbrain slices. *Biochem. Pharmacol.* **29**, 2189–98.

Wieser, H. G. (1980). Temporal lobe or psychomotor status epilepticus: a case report. *Electroenceph. clin. Neurophysiol.* **48**, 558–72.

Willow, M., Bornstein, J. C., and Johnston, G. A. R. (1980). The effects of anaesthetic and convulsant barbiturates on the efflux of (^{3}H) D-aspartate from brain mini-slices. *Neurosci. Lett.* **18**, 185–90.

Wolf, P. and Haas, H. L. (1977). Effects of diazepines and barbiturates on hippocampal recurrent inhibition. *Naunyn-Schmiedeberg's Arch. Pharmacol.* **299**, 211–18.

Zetler, G. (1980). Anticonvulsant effects of caerulein and cholecystokinin octapeptide, compared with those of diazepam. *Eur. J. Pharmacol.* **65**, 297–300.

7

Limbic system disorders in man

MICHAEL R. TRIMBLE

INTRODUCTION

The naming of the limibic lobe is traditionally attributed to Broca, who, in defining the comparative anatomy of an area of cortex which he referred to as 'le grand lobe limbique', included in its description some parts of the hippo-campus, and the subcallosal and the cingulate gyrae (Broca 1878). The close connections of these brain areas to the olfactory apparatus of the brain was noted, and this led to the adoption of the term 'rhinencephalon' which assumed that such parts of the neuroanatomy were intimately related with the reception of smell. A number of contemporary anatomists, including Ferrier and Hiss objected to this latter name, but it remained in popular use until the middle of this century, when the concept of the limbic system became more fully elabor-ated, initially by Papez (1937) and then by Maclean (1970). The former drew attention to the distinction between activities of the medial cortex, with the hippocampus and cingulate cortex participating in hypothalamic activity, and the lateral cortex which mediated general sensory circuits. The so-called 'Papez' circuit consisted of the hippocampal formation, the mammillary bodies, the anterior thalamic nuclei, and the cortex of the gyrus cinguli, which was thought to be central to affective experience forming, in Papez's own words, 'a harmonious mechanism which may elaborate the functions of central emotion, as well as participate in emotional expression'.

These ideas were given much impetus by the discovery of changes in affect and behaviour in animals who were subject to lesions and stimulation of these nuclei and their tracts (see Herbert, Chapter 3, this volume) and also by the growth of knowledge in biological psychiatry which, in particular, concentrated on disturbances of the limbic system as a possible explanation for psychiatric illness.

Critics of the theories of limbic system function, such as Brodal (1969), note that brain research increasingly is leading towards more holistic ideas with regards to brain function, making it difficult to separate one functionally dif-ferent region of the brain from another. Although Brodal recommended that the term 'limbic system' should be abandoned, Maclean (1970) noted the intimate links between the phylogenetic development of the brain and the corresponding behaviour of the species. He pointed out that what is distinctive to the mammalian brain is the marked expansion and differentiation of prim-itive cortex into limbic structures, which have similar features in all mammals. In marked contrast to the neo-cortex, this cortex has strong connections with the hypothalamus, and directly therefore influences autonomic and hormonal activity.

Accepting the reservations of Brodal (1969), in clinical practice, it is not uncommon for lesions and diseases that affect certain parts of the central nervous system to present clinically in their own characteristic ways. In clinical neurology the essence of the clinical examination, since it was introduced in its modern form at the turn of this century, has been to delineate clinical signs and symptoms which will reflect disturbances in one or another part of the central nervous system.

Symmetrical inequality, whether of reflexes, tone, or sensation, is given prime place, since in clinical practice, prior to the development of CT scanning at least, one of its most important aspects was the identification of structural lesions which would lead one side of the central nervous system to perform inappropriately. Indeed the history of neurology over the last two centuries may be seen as the continued identifaction of the structural bases for such clinical signs and symptoms; what may have markedly retarded progress in psychiatry has been the lack of the equivalent findings *in vivo* or at post-mortem, in patients with psychiatric disability.

Two highly significant events in this century have led to a restructuring of our ideas and to the clinical acceptance of the limbic system, not as an isolated unity, but as one aspect of the brain, which, when dysfunctioning, is likely to lead to clinical signs and symptoms that can be detected and from which limbic system disturbances may be inferred. The first of these was the occurrence of acute influenza epidemics in the second decade of this century which led to the condition of encephalitis lethargica, most elegantly described by Von Economo (1931). The sequelae of this to be discussed further below, included a wide range of psychiatric disturbances. So striking was this disorder and its complications that Von Economo wrote at the end of his monograph

this short review may have demonstrated that encephalitis lethargica has not only set much of our neurological knowledge on a partly new base, but has also fundamentally influenced our appreciation of normal and pathological psychological manifestations. The dialectic combinations and psychological constructions of many ideologists will collapse like houses of cards, if they do not in future take into account these new basic facts. Every psychiatrist who wishes to probe into the phenomena of disturbed motility and changes of character, the psychological mechanism of mental inaccessibility of the neuroses, etc., must be thoroughly acquainted with the experiences gathered from encephalitis lethargica . . . encephalitis lethargica can scarcely again be forgotten.

The second significant event was the discovery, particularly following the use of the electroencephalogram in the 1930s and 1940s, that some areas of the limbic system, such as parts of the temporal lobe, not only were subject to their own pathologies, but also that stimulations and lesions of these areas in patients led to disturbance of behaviour and affects analogous to those described in animals. In particular the discovery by Gibbs (1951) that the psychiatric complications of epilepsy were more prevalent in those with temporal-lobe disturbances led to a continuing search for the clinical correlations of limbic system disease in man.

It is here proposed to briefly review some of these, with full acknowledgement of the following two facts. First, a clear distinction needs to be drawn between diseases of function and diseases of structure. Although it has been argued elsewhere (Trimble 1982) that the term 'functional' has little value in the sense it is often used in (for example, as a polite eponym for psychiatric illness), if used correctly, in the historical sense of meaning disturbed function within the nervous system, then it can be seen that functional disturbance can arise in one of two ways. Either damage to a structure may lead to alteration of function or functional disturbance, in the sense of increased or decreased activity of various pathways in a system, may occur *de nouveau*, or secondary to the administration of, for example, drugs which alter such activity. While this statement seems obvious, it is worthwhile remembering that it was only in the 1950s that neuro-transmitters were finally identified in the central nervous system, and only in the last two decades that psychiatric disturbances and neurotransmitter abnormalities have been linked in an aetiological fashion. As Dr Iversen (Chapter 1) has pointed out, the concepts of the anatomy of the limbic system have radically altered in recent years, and the links between it and neuronal groupings such as the basal ganglia provide us with the neurochemical and neuroanatomical basis for understanding some of the clinical phenomena we see. It is conceivable that many will turn out to be purely functional, in the sense of being the result of specific neurochemical systems undergoing alteration of activity leading to the clinical phenomena. On the contrary, many structural lesions have little respect for neurochemical boundaries, and therefore the clinical patterns that follow structural lesions not only will differ qualitatively from those seen following functional changes, but may also differ in their evolution and their prognosis.

The second fact to consider is that attempts to localize psychiatric illness to areas within the central nervous system have failed in the past, and are likely to continue to do so. The on going debate between the localizers, who feel that various parts of the brain are responsible for specific functions, and the holists, who subscribe more to equipotentiality of nervous structures, finds an acceptable outlet in an alternative concept. This suggests that, while the brain acts holistically in the mediation of behaviour, destruction or overactivity of one part of it will lead, as pointed out many years ago by Hughlings Jackson (see Taylor 1931), to the clinical consequences of change of activity of that particular part of the brain acting in combination with the now altered activity of the rest of the central nervous system.

LIMBIC-SYSTEM DISORDERS

Temporal-lobe disorders

Encephalitis

The disease once referred to as 'acute necrotizing encephalitis', in which destruction of the temporal lobe occurred in particular with an emphasis on the limbic part of the hemispheres (see Fig. 7.1), has been shown by virological studies to be due to a herpes simplex virus. The illness often starts with a non-specific prodromal stage, which is followed by a variety of neurological signs

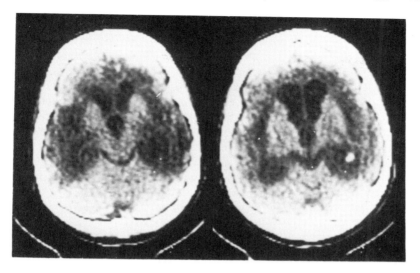

Fig. 7.1. CT scan of a patient who had herpes encephalitis showing gross bilateral destruction of fronto–temporal limbic structures. The patient had a dense amnesia, visual agnosia, and outbursts of irritability and aggression.

and symptoms. The mortality is high (around 70 per cent) and, in survivors, disturbances of behaviour are common. The most illuminating account of these is given by Hierons *et al.* (1978). They had clinico-pathological data on 10 patients who survived from 3 to 39 years. The most affected areas of the brain were the anterior part of the temporal lobe, the uncus and the amygdaloid nucleus, the hippocampus and dentate fascia, the insula and the parahippo-campal, posterior orbital, and cingulate gyri. The salient clinical features were an amnesia, sometimes going back for many years, with an inability to lay down new memory traces, aphasic disturbances, and a wide variety of behaviour patterns which were 'a little reminiscent of one or other feature of the Kluver–Bucy Syndrome'. Aggressive outbursts, periods of apathy and depression, and episodes of restlessness and overactivity were seen in nearly all patients, although no constant pattern of disturbances of affect occurred.

In addition to the limbic encephalitis of herpes, a similar clinical and path-ological picture has been observed in a number of other cases. Corsellis *et al.* (1968) noted its presentation with carcinoma. The presenting clinical features were convulsive episodes followed by a permanent amnesic disorder. In addition to this, the patients also had disturbances of affect – usually anxiety and depression – and sometimes hallucinations occurred. Again at post-mortem, the amygdaloid nucleus, the hippocampus, the fornix, and the mammiliary bodies were the areas of the brain most affected, and the pathological picture was very similar to that noted with encephalitis. In some of these cases, complete Kluver–Bucy syndromes have been reported (Marlow *et al.* 1975).

Other viruses known to affect limbic structures include those responsible for rabies, and subacute sclerosing pan encephalitis (SSPE). In the former, periods of irrational and often overactive behaviour occur, and, in particular, hydrophobia is said to be a distinguishing feature of the clinical state. SSPE is due to chronic infection of the central nervous system with a measles virus, and is associated with progressive intellectual deterioration over a short course of time in addition to which behaviour disturbances, mainly affective in nature, are seen. During the course of this condition, florid psychiatric presentations resembling schizophrenia-like conditions have also been recorded (Koehler and Jakumeit 1976).

Epilepsy

The association between epilepsy and psychopathology has been discussed at length elsewhere (Trimble 1981; Reynolds, Chapter 8, this volume). Although subject to much controversy, the central issues have been whether patients with certain forms of epilepsy for example, temporal-lobe epilepsy, are more prone to the development of psychopathology and whether certain psychiatric disturbances, such as specific personality types or the schizophreniform psychoses, are related to the disturbance caused by epilepsy itself, its treatment, to social deprivation, or to the stigmatization that patients with epilepsy suffer from.

With regards to the first issue, there is a growing body of evidence that the psychopathological profile seen in those with temporal-lobe epilepsy does differ from those with generalized epilepsies. One of the early reports was that of Nuffield (1961) who found that children with temporal-lobe epilepsy had more aggressive behaviour disturbances than those who had three-per-second spike and wave abnormalities who tended to display neurotic behaviour patterns. In adults, Gibbs (1951) suggested that the presence of long-standing abnormalities in the temporal lobe structures led to an increased prevalence of personality disorders and psychiatric syndromes in epilepsy, a hypothesis which has recently been subject to much investigation. A summary of some of the work in this field is given elsewhere (Trimble 1983). In studies that have attempted to measure psychopathology using standardized and validated rating scales (Trimble and Perez 1980), results have included a large number of negative findings. However, more recent data provides some positive indicators of profile differences between temporal-lobe and generalized epileptic patients, further clarified when specific rating scales, designed specially to assess potential interictal behaviour changes that are thought to be related to temporal-lobe epilepsy, are used. Paranoia, in particular is one trait observed more frequently in several studies of temporal-lobe epileptic populations.

Clinically, patients with temporal-lobe epilepsy do not all display psychopathology however, and some efforts have been made towards identifying which groups of patients are most likely to be affected. With regards to the development of a schizophreniform psychosis, risk factors are becoming identified. These include (Stevens 1982) multiple spikes or sphenoidal spikes recorded on the electroencephalogram, automatisms or visceral aurae, an abnormal neurological examination, and an age of onset in late childhood. In addition, Ounsted and Lindsay (1982) suggest that psychopathology in temporal-lobe epilepsy is commoner with left-sided lesions, which led them to comment 'viewed developmentally, limbic seizures and their associated lesions might be seen as contributing to an increasing noisiness within those territories where perversion of function leads to psychotic behaviour'.

Further work has suggested that the presentation of psychosis in epilepsy is intimately linked with the type of epilepsy, and the site of the epileptic discharge. The early speculations of Flor-Henry (1969) that left-sided lesions predominately were associated with a schizophrenic psychosis have received support in a prospective study in which epileptic patients with chronic psychosis, occurring in a setting of clear consciousness, were examined using the Wing Present State Examination (PSE) (Trimble and Perez 1982). In 24 patients, a PSE category of nuclear schizophrenia, which essentially is based on the identification of Schneiderian first rank symptoms, was found in 11. All of these had an EEG diagnosis of temporal-lobe epilepsy, in contrast to other forms of psychosis which were seen in both generalized and temporal-lobe groupings. Further examination of the profiles of the psychotic temporal-lobe epilepsy patients indicated that left temporal lesions were significantly linked to the presentation of nuclear schizophrenia. Thus it appears that a psychosis, which phenomenologically resembles schizophrenia, occurs in some patients with temporal-lobe epilepsy, and that first rank symptoms are particularly common with left-sided lesions. In contrast to process schizophrenia, however, the personalities of these patients are often well preserved, and they maintain their ability to adequately form stable relationships with their families and peers. As a result many do not end up in psychiatric hospitals and are maintained well in the community.

In view of the risk factors outlined above, it is possible that it is patients with chronic 'limbic epilepsy', that is those with allocortical as opposed to neocortical lesions, that are most prone to these presentations. Further indication of this comes from recent examination of the anterior pituitary hormone output following epileptic seizures in a variety of patient groups (Dana Hari and Trimble 1984). Patients with partial seizures originating in the temporal lobes were examined for alteration of prolactin, LH, and FSH levels 20 minutes following a seizure. Patients were independently evaluated as either having psychopathology associated with their epilepsy or not, and the neurohormonal changes in the two groups noted. Those with psychopathology had greater increases above baseline values postictally of their LH and prolactin levels, than those without psychopathology. In addition, baseline FSH values were lower in those with psychopathology. These data suggest that patients prone to the development of psychopathology in epilepsy may show greater facility of spread of the epileptic seizure through limbic-system structures, thus leading to alteration of hormone release.

Whatever the mechanism of the development of psychiatric syndromes, in particular the psychoses in epilepsy, explanations have to take into account recent work using positron-emission tomography in epilepsy. Trimble and colleagues (Bernardi *et al.* 1982), using radioactively labelled oxygen-15 and computed axial tomography, have examined a group of patients with epileptic psychosis and compared cerebral blood flow and metabolism in various areas

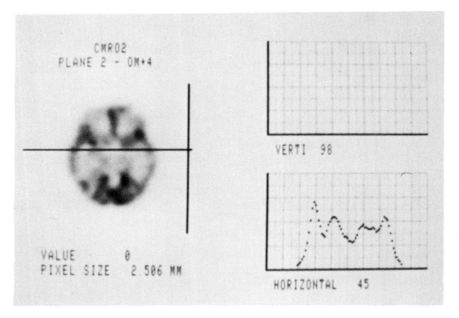

Fig. 7.2. A scan of the regional cerebral metabolic rate for oxygen utilization, taken using positron-labelled oxygen in a patient with epilepsy. Slice level is OM + 4. The scan shows an area of reduced metabolism in the right temporal cortex, with similarly reduced values in the basal ganglia region. This patient had mesial temporal sclerosis at operation.

of the limbic system to that of an age-matched control group. These data indicate that, interictally in these patients while they are psychotic, hypometabolism and low blood flow are seen in temporal-lobe structures bilaterally. In addition, similar down regulation of activity occurs in the region of the head of the caudate nucleus (see Fig. 7.2). As similar low levels of activity in basal ganglia structures have been reported in schizophrenic patients using flurodeoxyglucose (Buchsbaum *et al.* 1982), the changes noted in the epileptic patients may not purely be a consequence of, for example, the receiving of anticonvulsant medication, and may have direct relevance for our understanding of the development of these clinical syndromes.

Other temporal-lobe pathologies

Of importance for further understanding of the relationship between limbic-system lesions and behaviour is the clinical information on disturbances which follow destruction of the temporal lobes, by either tumours, trauma, or iatrogenically following surgical lesions. Davison and Bagley (1969), in their extensive review of psychotic states associated with organic disorders, indicate a significantly high proportion of temporal-lobe, hypophyseal and suprasellar, and supratentorial tumours in patients who present with psychosis. No specific features of the psychiatric presentations are commented on, and remission of the psychosis following surgical removal of a tumour occurred only in some cases. Malamud (1975) in a large survey of 245 tumour cases noted that 3 per cent had mental symptoms and 11 per cent received a clinical diagnosis of 'purely functional disorder'. All of these cases involved one or other part of the limbic system, and of eight with medial temporal tumours, four were diagnosed as schizophrenia, and three as affective disorder. This series comprised gliomas, craniopharyngiomas, colloid cysts, and haemangiomas of varying degrees of malignancy.

Psychiatric disturbances following cerebral trauma are relatively common. Often this takes the form of a neurosis with accompanying depression, although more severe psychiatric symptoms including psychoses have been reported. It is important to note the experimental literature, which suggests that, following a blow to the head, the relative movement of the brain in relationship to the skull leads to stresses and strains at specific areas within the central nervous system. In particular, the frontal and temporal lobes are prone to stress due to their being relatively restricted in the skull by the rigid anterior and temporal fossae (Pudenz and Sheldon 1946).

Hillbom (1960), examining 415 cases of head injury, noted a higher than expected occurrence of schizophrenia-like states in these patients, and commented that psychiatric disturbances generally were commoner in patients with left-sided lesions. Temporal-lobe trauma was linked with psychosis. Lishman (1968) studied the psychiatric morbidity of 670 patients with pene-trating brain injury, and again noted an increased association of psychiatric disability with left-hemisphere, particularly temporal, lesions. Davison and Bagley (1969) documented data on 40 cases of psychosis from the literature on head injury and noted features likely to be associated with its development. The strongest statistical associations were with the occurrence of epilepsy and a lesion in the temporal lobes.

Amnesic syndromes

One of the temporal-lobe syndromes that has been well defined is that of amnesia. Since the description of Korsakoff's psychosis in the last century, amnesic syndromes associated with alcoholic encephalopathies have been well described. This memory disorder comprises an inability to recall information acquired before the illness (retrograde amnesia) and an inability to acquire new information, referred to as anterograde amnesia. Clinically these patients are able to repeat digits normally, thus having a short-term memory which is intact, and in pure cases the amnesic disturbances occur in the absence of impairment of cognitive abilities in other fields. Other symptoms seen in Korsakoff's psychosis include mood disturbances, especially apathy and an air of detach-ment, lack of initiative and insight, and lack of spontaneity. Confabulation may occur, but not in all cases.

The anatomical basis for the Korsakoff's state has yet to be clearly defined, but limbic system involvement is suggested from post-mortem examinations which indicate that the mammillary bodies are involved in particular but, in addition, similar clinical impairments have occurred following destruction of hippocampal–limbic system circuits. These include bilateral temporal lobectomy, carried out in the early days of surgery for temporal-lobe epilepsy, occlusion of the posterior cerebral arteries with subsequent hippocampal infarction, anoxic damage, subarachnoid haemorrhage, head trauma, and meningitis. Although there is some argument as to whether damage to the medial temporal structures by such diseases leads to a consistent pattern of amnesia and also as to which elements of the temporal lobe need to be maximally involved for the amnesia to occur, it does seem clear that integrity of limbic system links between medial temporal and forebrain structures are essential for the adequate laying-down of memory traces, and that interference with the system from a variety of causes may lead to an amnesic state.

Frontal-lobe syndromes (see Fig. 7.3)

Since the early well known case of Phineas Gage, the American who, in 1878 had his frontal lobes shattered by a metal bar, which in its wake, led to marked personality changes such that he was said to be 'no longer Gage', the frontal-lobe syndromes have been well defined. As noted by Dr Iversen (Chapter 1), the frontal lobes, or at least part of them, form intimate links with the limbic system, with dopaminergic pathways existing which connect frontal cortex to areas of the mesolimbic system (Berger *et al.* 1976). The observations of Jacobsen (1935), followed by the careful testing of patients who received head injuries during the Second World War or received prefrontal lobotomy for the treatment of psychiatric illness, have led to the acknowledgement of specific neuropsychological deficits associated with frontal-lobe damage. One of the most pertinent is in the field of attention (Fuster 1980) with patients showing distractability and poor concentration. Patients present with poor memory, and it can be seen that execution of sustained behavioural acts is interfered with. The memory deficit is sometimes referred to as 'forgetting to remember'.

The thinking of patients with frontal-lobe lesions appears concrete, and in

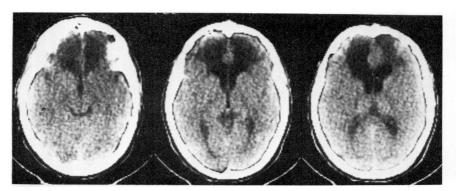

Fig. 7.3. CT scan of a patient following a closed head injury with bilateral frontal-lobe damage. The patient presented with episodes of unacceptable social behaviour (stealing his wife's jewellery and selling it) and on examination showed neuropsychological impairments typical of frontal-lobe damage.

their behaviour they may show perseveration and stereotypy of their responses. The perseveration, with inability to switch from one line of thought to another, leads to difficulty in performing arithmetical calculations; serial sevens, for example, get performed incorrectly, as are carry-over subtractions.

An aphasic disorder is seen following frontal-lobe lesions which differs from the classical Brocas's aphasia. Luria (1968) referred to it as 'dynamic aphasia'. The patients have well preserved motor speech and no anomia; repetition is intact, but they are unable to 'propositionize', and active speech is severely disturbed. He suggested this was due to a disturbance of the predictive function of speech, that which takes part in formulating the structure or scheme of a sentence. This syndrome is very similar to transcortical motor aphasia, which may be produced by a wide variety of pathologies including tumours, intra-cerebral haematomas, or occlusion of the anterior cerebral artery.

Other features of the prefrontal syndrome include reduced activity, in partic-ular with regard to spontaneous movement, sometimes associated with bouts of restless rather aimless unco-ordinated behaviour, and disturbance of affect. Most characteristic is apathy, with blunting of the emotions, the patient showing indifference to the world about him. This can resemble the picture of a retarded depression. In contrast, euphoria and disinhibition is also described. The euphoria is not that of a manic illness however, having an empty quality to it, and the disinhibition can lead to markedly abnormal behaviour which may be associated with outbursts of irritability and aggression. '*Witzelsucht*' has also been described, in which some patients show inappropriate facetiousness with a tendency to pun.

Some authors have distinguished between lesions of the lateral frontal cortex, most closely linked with the motor structures of the brain, which lead to disturbances of movement and action with perseveration and inertia, and lesions of the orbital–medial areas, those most intimately linked with the reticular formation, which lead to the disinhibition and changes of affective life. The terms pseudodepressed and pseudopsychopathic have been used to describe these two types respectively (Blumer and Benson 1975). Since a similar clinical picture to the pseudodepressed picture is seen in some patients with subcortical dementia, it has been suggested that the clinical pattern results from destructive pathology of the prefrontal cortex and its connection to subcortical structures, in particular the basal ganglia and thalamus. In contrast the pseudo-psychopathic picture results from injury to the orbital frontal lobe or pathways transversing it.

In some patients paroxysmal behaviour disorders are recorded, which are short-lived and include episodes of confusion, sometimes associated with hallucinations. Again these may reflect transient disturbances of fronto–limbic connections. Finally, following massive frontal-lobe lesions, the so-called apathico–akinetico–abulic syndrome may occur, in which the patient lies passively unaroused, unable to complete tasks or listen to commands.

Various mechanisms have been suggested for these changes in behaviour and cognition that follow frontal lobe lesions. Luria emphasized the connections between the prefrontal zones and the brainstem, thalamic structures, and other cortical zones, implying that the frontal lobes play a vital part in organizing and programming intellectual acts and checking performance (Luria 1973). Teuber (1964) suggested that the frontal lobes anticipate 'sensory stimuli' which result from behaviour, thus preparing the brain for events that are about to occur. The expected results of an action are compared to the actual experience, and regulation of activity occurs.

Other limbic-system lesions

A variety of subcortical structures are involved in disease processes that are linked with the limbic system. In particular the basal ganglia structures, the thalamus, and areas of the limbic midbrain, such as the ventral tegmental area (VTA), should be briefly considered.

It is now accepted that an area of the corpus striatum, namely the ventral striatum, has intimate links with limbic structures and may be referred to as part of the limbic forebrain (see Iversen, Chapter 1, this volume). The importance of these areas of the brain in animal models of behaviour has become clarified, and it is not surprising that, in man, abnormalities in such regions of the brain have been defined in psychiatric illness, and disturbances of behaviour have been noted following tumours and lesions. For example, tumours of the septal region have been reported to produce rage attacks and increased irritability (Valenstein and Heilman 1979) and stimulation of similar areas leads to pleasant, sometimes sexually arousing sensations (Heath 1964). Disorders of the basal ganglia such as Parkinson's disease and Huntington's chorea are commonly associated with psychiatric disability. Thus, in Parkinson's disease depression is common, often occurring in patients before the onset of the motor disability. Cognitive deficits are seen, and in one study a 40 per cent incidence of dementia was found which was related to the severity and duration of the illness (Celesia and Wanamaker 1972). In Huntington's chorea, in which destruction of corpus striatum occurs, early symptoms include irritability, emotional lability, and increased excitability. Relatives complain of personality changes which include irresponsible and sometimes promiscuous behaviour, but also include loss of interest in the home, work, or family life. The depression which characterizes Huntington's chorea is psychotic, particularly with persecutory delusions and a high frequency of suicide or attempted suicide is reported.

In contrast to Parkinson's disease, a schizophreniform psychosis occurs with markedly increased frequency in Huntigton's chorea (see Trimble 1981) and the dementia is almost universal. The latter has some specific characteristics resembling so-called 'subcortical' dementia.

Other basal-ganglia disorders where psychopathology is commonly encountered include Wilson's disease, in which abnormal accumulations of copper are seen in the putamen, caudate nucleus, and globus pallidus; Sydenham's chorea, and the Gilles de la Tourette Syndrome (Trimble 1981).

Thalamic lesions have been reported as leading to aphasia (Benson 1979), dementia (Smythe and Stern 1938), and to play a role in the amnesic syndrome especially that secondary to alcoholism. The acknowledged prominent subcortical afferents which come to the prefrontal cortex from the mediodorsal nucleus of the thalamus, the latter which receives input from the mesencephalic reticuar formation, from the amygdala, the prepiriform cortex, and the inferior temporal cortex (Fuster 1980), suggest that the thalamic–frontal connection may be of principal importance in regulating behaviour.

Lesions of the VTA and interrelated limbic nuclei may also be responsible for behaviour disturbances in man. Earlier it was emphasized that encephalitis lethargica was one of the turning points in the history of neuropsychiatry. The pathology in that condition was of microscopic foci of inflammation which mainly attacked the grey matter of the midrain and basal ganglia. Von Economo (1931) in his descriptions noted that it was the region of the aqueduct, the tegmental region, and the posterior wall, floor, and the basal parts of the lateral walls of the third ventricle which were predominantly affected. The next most important sites were the hypothalamic region, the thalamus, the lenticular nuclei, and the substantia nigra. Such disorders are not confined to the history books however, and similar cases are still occasionally described (Hunter and Jones 1966). In addition the suggestion that encephalitis may be responsible for some cases of schizophrenia has arisen from several sources. Tyrrell *et al.* (1979) reported that the cerebrospinal fluid from some patients with acute psychosis induced cytopathic effects in human fibroblast cultures, which might have suggested the presence of a virus. Stevens (1982) has described periventricular fibrillary gliosis in schizophrenic brains, in particular in the septum, hypothalamus, the medial thalamus, and the midbrain tegmentum.

The descriptions of behaviour in animals following lesions of the mesencephalic tegmentum leads to the question as to whether there is a similar syndrome in man. Case reports of patients with upper brainstem lesions and subsequent psychiatric disturbances are to be found in the literature (Trimble and Cummings 1981) and the close anatomical association between the VTA and nuclei which regulate eye movement, suggests that a variety of lesions affecting this part of the brain may lead to clinical pictures where there is a combination of eye movement abnormalities and psychopathology. Representative pathological conditions include haemorrhage, particularly in the mesencephalic branch of the basilar artery, and brainstem tumours. Disturbances of eye movements and blinking in schizophrenic patients suggest another link between the limbic system and psychopathology worthy of further study (Stevens 1978).

CONCLUSIONS

This chapter has mainly concerned itself with the clinical aspects of limbic lesions in man. Predominant has been the discussion of temporal-lobe syndromes, frontal-lobe disorders, and brainstem and thalamic disturbances, all of which lead to disturbed behaviour. The recognition that certain parts of the brain have intimate connections with holistic, as opposed to isolated, acts of behaviour has been a major step forward in the development of ideas in behavioural neurology. In particular, the occurrence of personality changes and schizophreniform disorders in diseases of the temporal lobe, the occurrence of apathetic akinetic conditions sometimes associated with disinhibited and fatuous behaviour interlinking with frontal-lobe changes, and the production of a variety of behaviour syndromes, including schizophreniform pictures, with midbrain and tegmental lesions suggest that, even within the limbic system, different patterns of behaviour may be consequent on which particular part of the axis is involved.

In addition to the siting of the pathology, the type of the pathology and the age at which the individual is affected must also play an important role in the consequent behavioural disorder. The essential feature of limbic system disturbances seems to be that they do not lead to traditional neurological localizing signs, and, with some exceptions, for example, the eye-movement abnormalities discussed in relation to VTA lesions, the disturbances of behaviour are the predominant aspects of the clinical presentation. Why many physicians should find the concept of the limbic system and the consequences of its aberrant functioning so difficult to accept is not entirely clear. However, as our data continues to grow, the borders between neurology and psychiatry begin to shrink, and the possibility that limbic system dysfunction will be identified in a wide variety of psychiatric disorders is now a real possibility. Already subtle neuro-endocrine disturbances have been noted in patients with depression (Carrol 1982) emphasizing that, as techniques for investigation of the central nervous system become more refined, the involvement of those areas of the brain hitherto difficult to explore because of their medial situation, being far removed from the surgeon's scalpel or the investigator's stimulating probe, will become clear.

REFERENCES

Benson, D. F. (1979). *Aphasia, alexia and agraphia*. Churchill Livingstone, Edinburgh.
Berger, B., Thierry, A., Tassin, J. P., and Moyne, M. A. (1976). Dopaminergic innervation of the rat prefrontal cortex: A fluorescence histochemical study. *Brain* **106**, 133–45.
Bernadi, S., Trimble, M. R., Lammertsma, A. A., Frackowiak, R. S. J., Roberts, J. K. A., and Jones, T. (1982). Cerebral blood flow and oxygen consumption in epilepsy investigated with positron emission tomography. *Br. J. clin. Pract., Suppl.* **18**, 21–2.
Blumer, D., and Benson, D. F. (1975). Personality changes with frontal and temporal lobe lesions. In *Psychiatric aspects of neurological diseases* (ed. D. F. Benson and D. Blumer), pp. 151–69. Grune and Stratton, New York.
Broca, P. (1878). Anatomie comparee des circonvolutions cerebrals: la grand lobe limbique et la scizzure limbique dans la serie des mammiferes. *Rev. Anthorpol.* **Ser. 2**, 385–498.
Brodal, A. (1969). *Neurological anatomy in relation to clinical medicine*. Oxford University Press, London.
Buchsbaum, M. S., Ingvar, D. H., Kessler, R., Waters, R. N., Cappelletti, J., van Kammen, D. P., King, A. C., Johnson, J. L., Manning, R. G., Flynn, R. W., Mann, L. S., Bunney, W. E., and Sokoloff, L. (1982). Cerebral glucography with positron tomography. *Arch. gen. Psychiat.* **39**, 251–60.
Carroll, B. J. (1982). The dexamethasone suppression test for melancholia. *Br. Psychiat.* **140**, 292–304.
Celesia, C. G., and Wanamaker, W. M. (1972). Psychiatric disturbances in Parkinson's disease. *Dis. nerv. Syst.* **33**, 577–83.

Corsellis, J. A. N., Goldberg, G. J., and Norton, A. R. (1968). "Limbic encephalitis" and its association with carcinoma. *Brain* **91**, 481–96.

Dana Hari, J. and Trimble, M. R. (1984). Prolactin and gonadotrophin changes following partial seizures in epileptic patients with and without psychopathology. *Biol. Psych.* [In press].

Davison, K., and Bagley, C. R. (1969). Schizophrenia-like psychosis associated with organic disorders of the central nervous system: a review of the literature. In *Current problems in neuropsychiatry* (ed. R. N. Herrington), pp. 113–84. Headley Brothers, Kent.

Flor-Henry, P. (1969). Psychosis and temporal lobe epilepsy. *Epilepsia* **10**, 363–95.

Fuster, J. M. (1980). *The prefrontal cortex.* Raven Press, New York.

Gibbs, F. A. (1951). Ictal and non-ictal psychiatric disorders in temporal lobe epilepsy. *J. nerv. ment. Dis.* **113**, 522–8.

Heath, R. G. (1964). Pleasure response of human subjects to direct stimulation of the brain: physiologic and psychodynamic considerations. In *The role of pleasure in behaviour* (ed. R. G. Heath). Harper & Row, New York.

Hierons, R., Janota, I., and Corsellis, J. A. N. (1978). The late effects of necrotising encephalitis of the temporal lobes and limbic areas: a clinico-pathological study of 10 cases. *Psychol. Med.* **8**, 21–42.

Hillbom, E. (1960). After effects of brain injuries. *Acta psychiat. Neurol. Scand., Suppl.* **35**, 558–9.

Koehler, K. and Jakumeit, N. (1976). Subacute sclerosing panencephalitis presenting as Leonhard's Speech Prompt Catatonia. *Br. J. Psychiat.* **129**, 29–31.

Lishman, W. A. (1968). Brain damage in relation to psychiatric disability after head injury. *Br. J. Psychiat.* **114**, 373–410.

Luria, A. R. (1968). The mechanism of dynamic aphasia. *Foundations Language* **4**, 296–307.

—— (1973). *The working brain.* (Translated by D. Haigh.) Basic Books, New York.

Maclean, P. D. (1970). The triune brain, emotion and scientific bias. In *The neurosciences, second study programme* (ed. F. O. Schmitt), pp. 336–49. Rockefeller University Press, New York.

Malamud, N. (1975). Organic brain disease mistaken for psychiatric disorder: A clinicopathologic study. In *Psychiatric aspects of neurological disease* (ed. D. F. Benson and D. Blumer), pp. 287–307. Grune & Stratton, New York.

Marlow, W. B., Mancall, E. L., and Thomas, J. J. (1975). Complete Kluver–Bucy syndrome in man. *Cortex* **11**, 53–9.

Nuffield, E. (1961). Neurophysiology and behaviour disorders in epileptic children. *J. ment. Sci.* **107**, 348–58.

Ounsted, C. and Lindsay, J. (1982). The long-term outcome of temporal lobe epilepsy in childhood. In *Epilepsy and psychiatry* (ed. E. H. Reynolds and M. R. Trimble), pp. 185–215. Churchill Livingstone, Edinburgh.

Papez, J. W. (1937). A proposed mechanism of emotion. *Arch. Neurol. Psychiat.* **38**, 725–743.

Pudenz, R. H., and Sheldon, C. A. (1946). The lucite calvarium – a method for the direct observation of the brain. *Neurosurgery* **3**, 487–505.

Smythe, G. E. and Stern, K. (1938). Tumours of the thalamus – a clinico-pathological study. *Brain* **61**, 339–74.

Stevens, J. R. (1978). Disturbances of ocular movements and blinking in schizophrenia. *Neurol. Neurosurg. Psychiat.* **40**, 1024–30.

—— (1982). Risk factors for psychopathology in individuals with epilepsy. In *Temporal lobe epilepsy, mania, schizophrenia and the limbic system.* (ed. W. P. Koella and M. R. Trimble), pp. 56–80. Karger, Basle.

Taylor, J. (1931). *Selected writings of John Hughlings Jackson,* Vol. 11. Staples Press, London.

Teuber, H. L. (1964). The riddle of frontal lobe functions in man. In *The frontal granular cortex and behaviour* (ed. J. M. Warren and K. Akert), pp. 410–44. McGraw Hill, New York.

Trimble, M. R. (1981). *Neuropsychiatry.* J. Wiley & Sons, Chichester.

—— (1982). Functional diseases. *Br. med. J.* **285,** 1768–70.

—— (1983). Interictal behaviour and temporal lobe epilepsy. In *Recent advances in epilepsy,* Vol. 1 (ed. B. S. Meldrum and T. Pedley). Churchill Livingstone, Edinburgh.

—— and Cummings, J. L. (1981). Neuropsychiatric disturbances following brain stem lesions. *Br. J. Psychiat.* **138,** 56–9.

—— and Perez, M. M. (1980). Quantification of psychopathology in adult patients with epilepsy. In *Epilepsy and behaviour 1979* (ed. B. M. Kulig, A. Meinardi, and G. Stores), pp. 118–26. Swets & Zeitlinger Lisse.

—— and —— (1982). The phenomenology of the chronic psychoses of epilepsy. In *Temporal lobe epilepsy, mania, schizophrenia and the limbic system* (ed. W. P. Koella and M. R. Trimble), pp. 98–105. Karger, Basle.

Tyrrell, D. A. J., Perry, R. P., Crow, T. J., Johnstone, E. C., and Ferrier, I. N. (1979). Possible virus in schizophrenia and some neurological disorders. *Lancet* **i,** 839–41.

Valenstein, E. and Heilman, K. M. (1979). Emotional disorders resulting from lesions of the central nervous system. In *Clinical neurophysiology* (ed. K. M. Heilman and E. Valenstein), pp. 413–38. Oxford University Press, New York.

Von Economo, C. (1931). *Encephalitis lethargica. Its sequelae and treatment.* [Translated by K. O. Newman.] Oxford University Press, London.

8

Temporal-lobe epilepsy, behaviour, and anticonvulsant drugs

E. H. REYNOLDS

INTERICTAL PSYCHOLOGICAL DISORDERS IN TEMPORAL-LOBE EPILEPSY

In 1948 Gibbs *et al.* first reported that patients with psychomotor epilepsy had a high incidence of severe personality disorders and suggested that it was 'reasonable to attribute much of what has been called the epileptic personality to disturbances in the anterior temporal area'. With this publication we entered what Guerrant *et al.* (1962) called the 'era of psychomotor peculiarity' in which the concept gradually evolved that patients with psychomotor seizures or temporal-lobe epilepsy (TLE) are particularly susceptible to personality or psychiatric disorders of various types. The concept has been fuelled by a wealth of experimental evidence showing the importance of the limbic system and temporal lobes in the emotional and behavioural life of animals and man, and by clinical interest in some of the overlapping psychopathological manifestations of what are now called 'complex partial seizures' and certain psychiatric disorders such as schizophrenia, in which the limbic system has also been implicated (Koella and Trimble 1982).

In the last 34 years an enormous largely conflicting literature has accumulated on this subject (Reynolds and Trimble 1981; Koella and Trimble 1982). Stevens (1982) has recently summarized nine controlled studies including three of her own, which have failed to show any increase in psychopathology in patients with psychomotor, temporal-lobe (limbic) epilepsy compared with matched generalized or focal non-limbic epileptic or control populations; these studies used objective or projective psychological tests or blind psychiatric interviews.

The latest entrants to the debate were Bear and Fedio (1977) who concluded that everyone was using the wrong tools for the task, and devised their own behavioural-personality rating scale (BFS), based on review of the literature and clinical observation. They then showed that 27 out-patients with TLE differed significantly in their personality profile from neurological and normal controls (as well as demonstrating significant differences between right and left TLE). Bear *et al.* (1982) have now reported that 10 hospitalized temporal-lobe epileptics differed significantly from 10 matched generalized or non-temporal focal epileptics, with the former showing in particular circumstantiality, religious and philosophic interests, and deepened affects. This study was undertaken in a psychiatric hospital and surprisingly the authors do not tell us the psychiatric diagnosis in their epileptic groups, even though they imply that the reasons for admission in two such groups are usually different. Mungas (1982) has recently reported that a large percentage of variance in the traits on the Bear and Fedio rating scale (BFS) can be accounted for by the presence or absence of psychiatric illness. It is clearly desirable to undertake such a study in a neurological setting and this has now been done by Hermann and Riel (1981) who compared 14 matched temporal and generalized epileptic patients. The former showed significant differences only in four of the 18 traits on the BFS, i.e. increased sense of personal destiny, dependence, paranoia, and philosophical interest. None of the other traits approached significance and some of the trends were in the opposite direction to that predicted. These rather inconclusive or conflicting results with respect to the BFS are illustrative of much of the long-running literature on the whole subject of TLE and psychopathology.

It is rather surprising to a neurologist that striking differences in personality should be expected to be demonstrated with such small groups of patients. After all TLE is very common. In the admittedly retrospective study of 666 cases of TLE recruited from all the departments of a district general hospital, 57 per cent were mentally normal and only 6 per cent were thought to have severe psychological disorders (Currie *et al.* 1971).

It is worth considering possible reasons for the continuing controversy in this difficult area in order to see a way forward to the resolution of such a prolonged debate. To begin with there has been a rather loose, or even absent, definition of terms. Psychomotor seizures have imperceptibly evolved into TLE, then complex partial seizures, and now limbic epilepsy. These terms are not synonymous but have been used interchangeably or with varying diagnostic criteria. One might also suspect that there are limbic discharges in generalized seizures. More precise definition and quantification of psychopathology and a clear distinction between personality and other psychiatric disorders is also required, as Trimble and Perez (1982) have stressed. The second problem is sampling bias. The frequency of TLE in adult hospital clinics is insufficiently appreciated. Three major studies of the application of the ILAE seizure classification to adult neurological clinics in different countries suggest that 60–75 per cent have partial seizures (Alving 1978). The majority of the latter are temporal-lobe seizures, and possibly 50–70 per cent of patients at such a clinic will have TLE (Stevens 1982). If psychiatrists find a high incidence of TLE among their epileptic psychiatric populations, this may not be of momentous significance but may simply reflect the incidence of this type of epilepsy in adult clinics in general. A third problem is that the EEG has again tended to confuse as much as to illuminate the subject. Whereas many authors have included temporal spiking in their diagnostic criteria, some have included other (e.g. slow-wave) temporal phenomena, and yet others have avoided EEG criteria. Further it should not be forgotten that 'all that spikes is not fits' (Stevens 1977) and that, in some psychiatric disorders unassociated with epilepsy, temporal spikes or slow activity have been noted (Hill 1952). The latter observation however has also reinforced the search for the temporal-lobe basis of the psychopathology associated with TLE. Finally, although it is widely recognized that many other factors contribute to psychological disorders in epilepsy (Pond 1981; Reynolds 1981) including brain damage, chronic drug therapy, and powerful psychosocial influences, such factors are frequently not controlled for studies in TLE and psychopathology. This is especially true of anticonvulsant therapy. Partial seizures are more difficult to control than generalized seizures (Rodin 1968; Shorvon and Reynolds 1982), which is one reason why patients with TLE accumulate in adult clinics, where they have been shown by Rodin *et al.* (1976) to end up on more drugs than patients with other seizure types. There is increasing evidence of the adverse effects of such polytherapy on mental function.

ANTICONVULSANTS AND MENTAL SYMPTOMS

Although epileptic patients traditionally have been treated with multiple drugs over many years, there has been a surprising failure to study the effect of such treatment on mental function until the last few years (Trimble and Reynolds 1976). With the advent of blood-level monitoring of the drugs in the 1960s it soon became apparent that the older concept of barbiturate or hydantoin toxicity – exemplified by nystagmus, ataxia, and later confusion – should be considerably broadened to include other psychological states, such as depression, behaviour disorders, or intellectual deterioration (Reynolds 1970). Not infrequently, such states were associated with high drug concentrations in the absence of more conventional signs of toxicity, such as nystagmus, and had thus led to misinterpretation of the true aetiology until the drug concentration had been measured. Particular attention was focused on phenytoin, which was reported by several authors to produce a syndrome of reversible subacute or chronic encephalopathy, resulting in impairment of intellectual function and/or memory, with or without nystagmus and ataxia, sometimes associated with other neurological signs such as dyskinesias or even a rise in cerebrospinal fluid (CSF) protein. Thus, this syndrome was especially likely to be overlooked in brain-damaged or mentally retarded children in whom the classical signs of toxicity were more often absent and in whom the diagnosis was sometimes mistakenly interpreted as an unknown progressive degenerative disease. It was also apparent that the correlation between blood concentrations and these toxic psychological disturbances was not necessarily a close one, for many other patients have apparently toxic levels of phenytoin (or phenobarbitone) without any obvious immediate ill effects. As with all metabolic insults to the nervous system, there are many predisposing factors which determine whether neuro-psychiatric symptoms develop and, if so, the nature of this reaction.

Mental impairment with nontoxic levels of anticonvulsant drugs

Although impairment of mental function with toxic levels of phenytoin or other drugs is perhaps understandable, slightly more disconcerting is the possibility that subtle impairment of cognitive function and behaviour, leading eventually to mental illness, may occur with more modest or 'therapeutic' levels of some anticonvulsants. In a study of 57 epileptic out-patients, Reynolds and Travers (1974) found that, after exclusion of overt drug toxicity, gross cerebral lesions, or mental illness preceding the onset of epilepsy, patients with intellectual deterioration, psychiatric illness, personality change, or psychomotor slowing had significantly higher levels of phenytoin or phenobarbitone than those without such changes. Furthermore, the mean blood levels of the two drugs in the patients with mental changes were within the optimum or 'therapeutic' range. These observations were not simply the reflection of higher anticonvulsant prescribing for more severe epilepsy, as similar differences were noted in those patients with infrequent seizures. Similar findings were reported by Trimble and Corbett (1980) in a study of 312 epileptic children in a residential hospital school. Those children with a fall in IQ of between 10 and 40 points in at least one year had significantly higher levels of phenytoin and primidone, with a similar trend for phenobarbitone. Again the mean blood drug values were within the optimum range. Carbamazepine was also included in this study and no differences in blood level were found between mentally impaired and non-impaired groups.

The suspicion that there may be a causal relationship between relatively higher drug levels and psychological impairment is strengthened by the reported change in mental state that can occur following reduction or cessation of hydantoin or barbiturate drugs (Shorvon and Reynolds 1979; Fishbacher 1982). Particular improvement has been noted in alertness, concentration, drive, mood, and sociability. Further, many patients may be unaware of the adverse effects of the drugs until they are withdrawn, particularly if they have been on the drugs from early life and thus have no concept of how they would feel without drug therapy. Further support for the view that some drugs subtly affect mental function is now being derived from psychometric investigations. Studies in normal volunteers have shown that phenobarbitone may impair sustained attention and various measures of perceptual-motor performance with low or 'therapeutic' blood levels (Hutt *et al.* 1968) and that low levels of phenytoin can impair some aspects of psychomotor performance (Ideström *et al.* 1972). In a double-blind crossover study with placebo, Trimble *et al.* (1980) showed that a two-week administration of phenytoin or carbamazepine could impair some measures of immediate or delayed visual or verbal memory or decision-making. In general, the differences from placebo were significant for phenytoin but not for carbamazepine. There were also significant correlations between impaired performance and phenytoin blood levels.

Prospective psychometric studies in chronic epileptic patients have shown greater impairment of function associated with higher than with lower drug levels and a significant improvement in some measures of performance on reducing polytherapy (Thompson 1981). Changes were observed in mental and motor speed and in measures of attention, concentration, and memory. Interestingly, this improvement sometimes takes three to six months to become fully apparent.

Although most of the evidence has so far incriminated barbiturate and hydantoin drugs it is also true that these are the most widely used anticonvulsants. Carbamazepine does seem to be relatively free of the adverse effects discussed above, and it may be particularly useful in the treatment of epilepsy associated with psychiatric disorders (Reynolds 1982). Comparative psychometric studies of phenytoin and carbamazepine in epileptic patients (Dodrill and Troupin 1977) or normal volunteers (Trimble *et al.* 1980) certainly suggest that the latter drug has less adverse effects. However the suggestion that carbamazepine has positive psychotropic properties, especially in view of its structural relationship to tricyclic antidepressants, remains an interesting but as yet unproven possibility (Parnas *et al.* 1979; Post 1982). The newest drug, sodium valproate, has been less studied. Claims that it, too, is free from harmful psychological effects have been premature. It is certainly capable occasionally of causing an encephalopathy and is frequently responsible for drowsiness, especially in combination with other drugs (Thompson and Trimble 1981). It should also be stressed that the adverse psychological complications of anticonvulsant drugs may often stem more from the traditional, but largely unnecessary, use of polytherapy, than from the effects of any individual drug (Reynolds and Shorvon 1981). Nevertheless, there is a clear need to define more carefully and on a long-term basis the relative impact of the various major anticonvulsants, utilized on their own with blood-level monitoring, in both new and chronic patients with epilepsy (Tomlinson *et al.* 1982).

Mechanisms

If, as the above evidence suggests, anticonvulsant drugs can impair mental performance, ultimately contributing along with many other factors, to cognitive, psychiatric, or personality change, how may these effects be mediated? Chronic adverse effect of the drugs on many tissues and metabolic systems have been recognized in the last 15 years or more (Reynolds 1975), and some of these may well be relevant to mental function. Table 8.1 lists those which have been or could be implicated and there may well be others we have not yet recognized. Each of these is a complex subject in its own right, which I have discussed elsewhere (Reynolds 1975, 1981). I will refer here only to one very recent observation which could have some relationship to the so-called 'psychotropic effect' of carbamazepine or at least to the differences in the adverse mental effects of this compared to other anticonvulsant drugs.

In animal models, several anticonvulsants influence the level and turnover of serontoin (5HT) and there has been interest in the relationship of these findings to the therapeutic and toxic effects of the drugs. As plasma tryptophan levels are reported to influence the turnover of 5HT, my colleagues and I (Pratt *et al.*

TABLE 8.1. *Mechanisms of anticonvulsant effects on mental function*

Neuropathological damage
Folate deficiency
Monoamine metabolism
Hormone metabolism

TABLE 8.2. *Plasma tryptophan concentrations (95 per cent confidence limits) in epileptic patients subdivided by drug therapy, untreated epileptics, and normal volunteers. (From Pratt et al. 1984)*

Group	Clinic visits (Patients)	Whole-plasma tryptophan (µg ml)	Bound-plasma tryptophan (µg ml)	Free-plasma tryptophan (µg ml)	Bound/free tryptophan ratio
Epileptics					
Carbamazepine	28(20)	12.65(11.83,13.52)	9.86(9.20,10.57)	2.70(2.38,3.05)	3.65(3.22,4.13)
Diphenylhydantoin	53(38)	8.60†(8.20,9.03)	7.41‡(7.05,7.79)	1.15†(1.05,1.26)	6.44(5.89,7.05)
Phenobarbitone	8(7)	8.86(7.82,10.04)	7.82(6.87,8.90)	1.01†(0.80,1.27)	7.76(6.15,9.79)
Untreated	16(15)	9.17(8.40,10.01)	7.67*(7.00,8.41)	1.32(1.12,1.55)	5.83(4.95,6.88)
Normal volunteers	34(33)	10.12(9.52,10.75)	8.76(8.23,9.33)	1.33(1.19,1.49)	6.59(5.89,7.38)

All comparisons with carbamazepine group are significant ($p < 0.001$) except with bound-plasma tryptophan for phenobarbitone group ($p < 0.005$) and normal volunteers ($p < 0.025$); other comparisons with normal volunteers $*p < 0.025$; †$p < 0.05$; ‡$p < 0.001$.

1983) have examined the influence of anticonvulsants, in particular phenytoin and carbamazepine, on whole, free, and bound plasma tryptophan concentrations in chronic epileptic patients and also in newly diagnosed epileptic patients before and after the introduction of drug therapy.

As illustrated in Table 8.2, we have found that carbamazepine elevates whole and free tryptophan levels whereas phenytoin (and phenobarbitone) has the opposite effect. The rise in plasma-tryptophan concentration is related to the blood levels of carbamazepine and there is a similar trend relating the fall in plasma tryptophan to phenytoin levels. The mechanisms of these opposing effects of the two drugs on plasma tryptophan are unknown. It is interesting however that carbamazepine is related to tricyclic antidepressant drugs which have recently been shown to inhibit liver tryptophan pyrrolase (Badawy and Evans 1981). The possibility that carbamazepine has a similar effect requires investigation.

Clinical studies suggest that there is little difference in the anti-epileptic efficacy of phenytoin and carbamazepine. It is therefore unlikely that the opposite effects of these two drugs on plasma tryptophan could be related to the anticonvulsant action of either drug. However, as discussed there are clear differences in the effects of the two drugs on mental function; in addition, there are claims that carbamazepine has an independent psychotropic effect. It is possible that the divergent effects we have observed on tryptophan metabolism are related to the different influences of the two drugs on mental function.

REFERENCES

Alving, J. (1978). Classification of the epilepsies. An investigation of 1508 consecutive adult patients. *Acta neurol. scand.* **58**, 205–12.

Badaway, A.A.-B. and Evans, M. (1981). Inhibition of rat liver tryptophan pyrrolase activity and elevation of brain tryptophan concentration by administration of antidepressants. *Biochem. Pharmacol.* **30** (11), 1211–16.

Bear, D. M. and Fedio, E. (1977). Quantitative analysis of interictal behaviour in temporal lobe epilepsy. *Arch. Neurol.* **34**, 454–67.

——, Levin, K., Blumer, D., Chetham, D., and Ryder, J. (1982). Interictal behaviour in hospitalised temporal lobe epileptics: relationship to idiopathic psychiatric syndromes. *J. Neurol. Neurosurg. Psychiat.* **45**, 481–8.

Currie, S., Heathfield, W. G., Henson, R. A., and Scott, D. F. (1971). Clinical course and prognosis of temporal lobe epilepsy. A survey of 666 patients. *Brain* **94**, 173–90.

Dodrill, C. B. and Troupin, A. S. (1977). Psychotropic effects of carbamazepine in epilepsy: a double blind comparison with phenytoin. *Neurology* **27**, 1023–8.

Fischbacher, E. (1982). Effect of reduction of anticonvulsants on well being. *Br. med. J.* **285**, 423–4.

Gibbs, E. L., Gibbs, F. A., and Fuster, B. (1948). Psychomotor epilepsy. *Arch. Neurol. Psychiat.* **60**, 331–9.

Guerrant, J., Anderson, W. W., Fischer, A., Weinstein, M. R., Jaros, R. M., and Deskins, A. (1962). *Personality in epilepsy.* Thomas, Springifled, Illinois.

Hermann, B. P. and Riel, P. (1981). Interictal personality and behavioural traits in temporal lobe and generalized epilepsy. *Cortex* **17**, 125–8.

Hill, D. (1952). EEG in episodic psychotic and psychopathic behaviour. *Electroenceph. clin. Neurophysiol.* **4**, 419–42.

Hutt, S. J., Jackson, P. M., Belsham, A. N., and Higgins, G. (1968). Perceptual motor behaviour in relation to blood phenobarbitone level. *Develop. Med. child Neurol.* **10**, 626–32.

Ideström, C. M., Schalling, D., Carlquist, U., and Sjoquist, F. (1972). Behavioural and psychophysiological studies: acute effects of diphenylhydantoin in relation to plasma levels. *Psychol. Med.* **2**, 111–20.

Koella, W. P. and Trimble, M. R. (Eds.) (1982). *Temporal lobe epilepsy, mania, and schizophrenia and the limbic system,* Advances in Biological Psychiatry, No. 8. Karger, Basle.

Mungas, D. (1982). Interictal behavior abnormality in temporal lobe epilepsy. A specific syndrome or nonspecific psychopathology? *Arch. gen. Psychiat.* **39**, 108–11.

Parnas, J., Flachs, H., and Gram, L. (1979). Psychotropic effect of antiepileptic drugs. *Acta neurol. scand.* **16**, 329–43.

Pond, D. (1981). Psycho-social aspects of epilepsy – the family. In *Epilepsy and psychiatry* (ed. E. H. Reynolds and M. R. Trimble), pp. 291–5. Churchill Livingstone, Edinburgh.

Post, R. M. (1982). Use of the anticonvulsant carbamazepine in primary and secondary affective illness: clinical and theoretical implications. *Psychol. Med.* **12**, 701–4.

Pratt, J., Jenner, P., Johnson, A. L., Shorvon, S. D., and Reynolds, E. H. (1984). The influence of anticonvulsants on plasma tryptophan in epileptic patients: implications for antiepileptic action and mental function. *J. Neurol. Neurosurg. Psychiat.* [In press].

Reynolds, E. H. (1970). Iatrogenic disorders in epilepsy. In *Modern trends in neurology, No. 5* (ed. D. Williams), pp. 271–86. Butterworths, London.

—— (1975). Chronic antiepileptic toxicity: a review. *Epilepsia* **16**, 319–52.

—— (1981). Biological factors in psychological disorders associated with epilepsy. In *Epilepsy and psychiatry* (ed. E. H. Reynolds and M. R. Trimble), pp. 264–90. Churchill Livingstone, Edinburgh.

—— (1982). The pharmacological management of epilepsy associated with psychological disorders. *Br. J. Psychiat.* **141**, 549–57.

—— and Shorvon, S. D. (1981). Monotherapy or polytherapy for epilepsy? *Epilepsia* **22**, 1–10.

—— and Travers, R. D. (1974). Serum anticonvulsant concentrations in epileptic patients with mental symptoms. *Br. J. Psychiat.* **124**, 440–5.

—— and Trimble, M. R. (Eds.) (1981). *Epilepsy and psychiatry.* Churchill Livingstone, Edinburgh.

Robin, E. A. (1968). *The prognosis of patients with epilepsy.* Thomas, Springfield, Illinois.

——, Katz, M., and Lennox, K. (1976). Differences between patients with temporal lobe seizures and those with other forms of epileptic attacks. *Epilepsia* **17**, 313–20.

Shorvon, S. D. and Reynolds, E. H. (1979). Reduction of polypharmacy for epilepsy. *Br. med. J.* **2**, 1023–5.

—— and —— (1982). Early prognosis of epilepsy. *Br. med. J.* **285**, 1699–701.

Stevens, J. R. (1977). All that spikes is not fits. In *Psychopathology and brain dysfunction* (ed. C. Shagass, S. Gershon, and A. J. Friedhoff), pp. 183–98. Raven Press, New York.

—— (1982). Risk factors for psychopathology in individuals with epilepsy. In *Temporal lobe epilepsy, mania, and schizophrenia and the limbic system* (ed. W. P. Koella and M. R. Trimble), pp. 56–80. Karger, Basle.

Thompson, P. (1981). The effects of anticonvulsant drugs on the cognitive functioning of normal volunteers and patients with epilepsy. PhD thesis, London University.

—— and Trimble, M. R. (1981). Sodium valproate and cognitive functioning in normal volunteers. *Br. J. clin. Pharmacol.* **12**, 819–24.

Tomlinson, L., Andrewes, D. Merrifield, E., and Reynolds, E. H. (1982). The effects of antiepileptic drugs on cognitive and motor functions. *Br. J. clin. Practice* (Suppl. 18), 177–83.

Trimble, M. R. and Corbett, J. A. (1980). Behavioural and cognitive disturbances in epileptic children. *Irish med. J. (suppl.)* **73**, 21–8.

—— and Perez, M. M. (1982). The phenomenology of the chronic psychoses of epilepsy. In *Temporal lobe epilepsy, mania, and schizophrenia and the limbic system* (ed. W. P. Koells and M. R. Trimble) Advances in Biological Psychiatry, No. 8, pp. 98–105. Karger, Basle.

—— and Reynolds, E. H. (1976). Anticonvulsant drugs and mental symptoms. A review. *Psychol. Med.* **6**, 169–78.

——, Thompson, P. J., and Huppert, F. (1980). Anticonvulsant drugs and cognitive abilities. In *Advances in epileptology* (ed. R. Canger, F. Angeleri, and J. K. Penry), pp. 199–204. Raven Press, New York.

9

Carbamazepine in the treatment of mood and anxiety disorders: implications for limbic system mechanisms

ROBERT M. POST AND THOMAS W. UHDE

INTRODUCTION

Rationale for use of carbamazepine in affective illness

As reviewed earlier in this book, an extensive body of animal and human research has suggested that the limbic system may play a role in the modulation of affect and anxiety. On the basis of this analysis, we were particularly interested in exploring the utility of a drug which might have some relative selectivity for limbic-system effects. Many prior psychopharmacological approaches to the mood and anxiety disorders had attempted to utilize agents with relative selectivity for one neurotransmitter system or another. This approach might provide valuable evidence for the role of a given neurotransmitter system in the evolution and treatment of these disorders. However, since the limbic system appears to be influenced by a variety of classical and putative peptide neurotransmitters and modulators, it was thought that a practical alternative strategy might be to utilize a drug such as carbamazepine which has well documented action on limbic-system structures.

An additional rationale for our clinical trial of carbamazepine in affective illness was the consistent reports that carbamazepine was of use in the treatment of mood and behavioural disorders in patients with psychomotor seizures or complex partial seizures. Dalby (1975) reviewed the literature of 40 studies including 2500 patients with epilepsy treated with carbamazepine. Approximately 50 per cent of the patients included in these largely uncontrolled clinical trials were reported to show improvement in mood or behaviour. For example, in Dalby's own study (1971) of carbamazepine in psychomotor epileptics, he observed that 11 of 18 patients with periodic depressions 'symptomatically similar in many respects to endogenous depression' improved or showed complete remission of episodes. It is noteworthy that seven of the patients in his series improved psychiatrically in spite of continued seizures. This is of considerable import since many investigators have questioned whether the improvement on carbamazepine is related either to better seizure control or substitution of carbamazepine for more behaviourally toxic agents. While this argument continues in the literature, there is increasing clinical data indicating that carbamazepine may have some positive psychotropic effects in patients and normal volunteers. For example, in a study by Thompson *et al.* (1980), volunteers rated themselves nonsignificantly more active, less tired, and less depressed while taking carbamazepine compared with placebo, while taking phenytoin they were more depressed, less active, and significantly more fatigued. Furthermore, carbamazepine did not produce the significant and blood-level-related impairment in cognitive function in normal volunteers that phenytoin produced.

Although considerable controversy also remains regarding the relationship of psychomotor seizures to subsequent behavioural pathology and particularly the development of psychosis (Stevens 1982), it should be noted that prominent affective symptomatology has been reported to also develop in the ictal and interictal periods of patients with complex partial seizures (Bear *et al.* 1982; Hermann 1979; Bear and Fedio 1977; Currie *et al.* 1971; Flor-Henry 1969; Dongier 1959; Weil 1956). Several types of longlasting behavioural alterations have also been observed in experimental animals experiencing repeated seizures. In particular, altered learning, motor activity, social reactivity, and response to threat have been reported following repeated seizures kindled in the amygdala (see review in Post 1982). Since some of these long-term behavioural alterations appear to occur secondary to repeated seizure discharges of limbic system structures, were further impelled to initiate a clinical trial of a drug such as carbamazepine which might help stabilize limbic system excitability. The experimental model of kindling had been pursued as possibly relevant for exploring the relationship of repeated limbic seizure discharges to subsequent behaviour in animals and man. Specifically, we had postulated that a kindling-like effect could occur in man following repeated endogenous electrophysiological stimuli associated with complex partial seizures in a fashion similar to that observed in the kindling paradigm (Post and Kopanda 1976; Post 1977). Since carbamazepine has been reported to have potent effects on amygdala-kindled seizures (Babington 1977; Wada 1977; Albright and Burnham 1980), we were further attracted to the idea of studying this agent.

Several investigators in Japan had used carbamazepine with some success in open clinical trials of patients with manic-depressive illness (Takezaki and Hanaoka 1971; Okuma *et al.* 1973). Thus, there appeared to be several converging empirical and theoretical rationales for our controlled trial of carbamazepine in affective illness.

CARBAMAZEPINE'S EFFECTS ON LIMBIC SUBSTRATES

What is the evidence that carbamazepine has some relative selectivity for experimental and clinical seizures thought to arise from temporal-lobe and limbic system structures? Carbamazepine inhibits afterdischarges in seizures arising from a variety of limbic-system structures (Koella *et al.* 1976; Babington and Horovitz 1973) and inhibits the development and manifestation of amygdala-kindled seizures (Babington and Horovitz 1973; Wada *et al.* 1976; Wada 1977; Albright and Burnham 1980). Albright and Burnham (1980) reported that carbamazepine was the most effective of the anticonvulsants tested in inhibiting the amygdala-kindled, compared to the cortical-kindled, seizure focus. If these measures could be taken to assess the relative potency of anticonvulsants on the limbic system structures, carbamazepine was ranked first followed by sodium valproate, phenobarbital, phenytoin, methsuximide, clonazepam, ethosuximide, and, lastly, diazepam (Albright and Burnham 1980). The efficacy of carbamazepine in these animal models of limbic seizures is paralleled by the consensus that carbamazepine is one of the drugs of choice for the treatment of complex partial seizures and related psychomotor seizures that are thought to arise from foci in the temporal lobe or limbic system of man (see several chapters in Penry and Daly 1975).

CARBAMAZEPINE IN AFFECTIVE ILLNESS

Acute effects in mania and depression

Following the open clinical trials in Japan, Post *et al.* (1978, 1982*a*) and Ballenger and Post (1978, 1980) reported the first double-blind, placebo-controlled clinical trials of carbamazepine in manic and depressive illness. The time course and magnitude of antimanic effects of carbamazepine were roughly parallel to those observed in similarly diagnosed and rated patients receiving double-blind administration of the neuroleptic pimozide or lithium carbonate (see Fig. 9.1). Okuma *et al.* (1979) reported that carbamazepine was equally

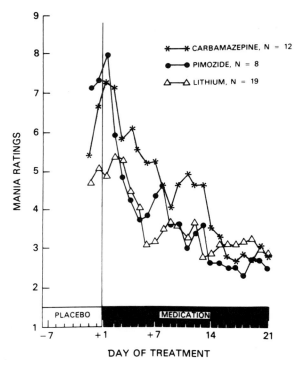

Fig. 9.1. Time course of antimanic effects of carbamazepine compared to lithium and pimozide.

effective when compared to chlorpromazine in a double-blind trial in mania. The drug was well tolerated and patients experienced fewer side-effects on carbamazepine than on chlorpromazine.

We observed mild-to-moderate antidepressant responses in 15 of our first 31 unipolar and bipolar depressed patients treated with carbamazepine in a double-blind fashion (Post *et al.* 1978, 1982*a*, 1984; Ballenger and Post 1978, 1980). CSF levels of carbamazepine and its active metabolite, carbamazepine-10,11-epoxide, were measured in 18 affectively ill patients. CSF levels of carbamazepine were 2.06 μg ml^{-1} or 31 per cent of those found in plasma (6.55 μg ml^{-1}). CSF carbamazepine-10, 11-epoxide concentrations averaged 0.9 μg ml^{-1} and were 63 per cent of those found in plasma. Levels of carbamazepine itself in either plasma or CSF were not significantly correlated with the degree of antidepressant response. However, in contrast, concentrations of the 10,11-epoxide were significantly correlated with the degree of antidepressant response ($r = 0.71$, $p < 0.01$) (see Fig. 9.2).

As illustrated in Fig. 9.3, in a subgroup of subjects who completed self-ratings and side-effect ratings during placebo and carbamazepine administration, there were no significant differences in the incidence or severity of these subjective changes during placebo compared to carbamazepine. These findings are of interest from several perspectives. Consistent with studies in the neurological literature, the drug does appear to have a high degree of patient acceptability and is well tolerated. From another perspective, the high incidence and severity of ratings of 'side-effects' often attributed to psychotropic drugs during

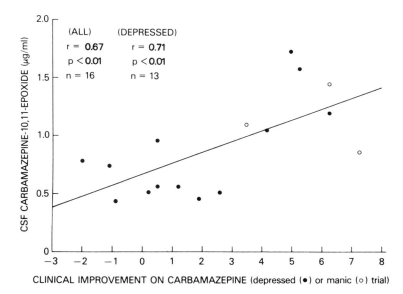

Fig. 9.2. Correlation of the 10,11-epoxide of carbamazepine in CSF with degree of clinical response in manic-depressive patients.

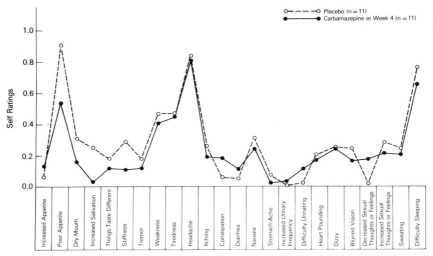

Fig. 9.3. Lack of subjective side-effects during carbamazepine treatment in depressed patients.

placebo administration, highlights the importance of this both as a methodological and theoretical issue. This high rate during placebo of symptoms usually considered to be drug-related needs to be assessed and corrected for in clinical trials of potential antidepressant treatments. In addition, the high incidence of these symptoms may be of theoretical interest in relationship to the possible influence of limbic system and other neuroanatomical substrates that might be involved in affective illness. Patients with a high degree of somatization are often viewed with a pejorative connotation that their complaints are somehow less real than the classical symptoms of depression, including sleep and appetite disturbance. Since various limbic structures appear to be highly involved in the modulation of a variety of somatic and autonomic responses as well as pain symptoms (Gloor 1972), this area may prove to be a useful one for further investigative analysis in relation to biological substrates involved in affective disorders, as well as the impact of treatment on these complaints.

Prophylactic effects of carbamazepine

A subgroup of patients with initial evidence of an acute response were continued on longer-term carbamazepine prophylaxis in either a double-blind fashion or in open clinical trials (Post *et al.* 1983*a*). Seven patients, all of whom were initially reported to be lithium-nonresponsive, were treated with carbamazepine for an average of 1.7 years (range 4 to 51 months). In this group of very rapidly cycling, treatment-resistance patients, the mean number of episodes per year decreased from 16.4 ± 5.7 in the year prior to carbamazepine administration to 5.7 ± 2.4 episodes per year during carbamazepine prophylaxis. There was a similar degree of approximately 65 per cent improvement in either the manic or the depressive phase of the illness. In addition to the decrease in episode frequency, affective episodes when they did occur during carbamazepine administration, were substantially less severe. Moreover, in five of six patients in whom there was opportunity to observe a period of carbamazepine discontinuation, relapses did occur during this interval, suggesting that carbamazepine was indeed contributing to the improvement in the course of illness rather that the patient's experiencing a spontaneous remission (Post *et al.* 1983*a*).

In a recent double-blind comparison of carbamazepine with placebo for a one-year prophylactic study, Okuma *et al.* (1981) reported a trend (*p*< 0.10) for the superiority of carbamazepine. Six of 12 patients responded to carbamazepine compared to two of nine placebo-treated patients. These figures are comparable with those of Okuma *et al.* 's earlier open study (1973) in which 34 of 51 patients (67 per cent) showed evidence of prophylaxis when carbamazepine was added to previously ineffective treatment regimes including lithium carbonate, neuroleptics, or antidepressant drugs.

In our studies, and in those of Okuma *et al.*, previous lithium nonresponders have been included in the group of patients who show adequate response to carbamazepine. Several open studies continue to support an antimanic and prophylactic effect of carbamazepine. Inoue *et al.* (1981) reported that carbamazepine and lithium carbonate, used in combination, were successful in the treatment of four out of five previously treatment-resistant manic-depressive patients. Lipinksi and Pope (1982) reported three patients who did not adequately respond to either treatment alone, but responded to carbamazepine and lithium in combination. Folks *et al.* (1982) reported responses to carbamazepine in eight out of 10 patients with affective illness including three with evidence of organic impairment.

Carbamazepine in anxiety disorders

Fear and anxiety are among the most common affects reported after exogenous stimulation of the temporal lobe's limbic system or associated with the onset of a psychomotor seizure (Gibbs 1956; Williams 1956; Feindel and Penfield 1954; Mulder and Daly 1952; Gloor 1972). Anxiety is also a common interictal affect (Hermann 1979; Standage and Fenton 1975; Currie *et al.* 1971; Roth and Harper 1962; Weil 1956). Compared to the earlier reports of carbamazepine's positive effects on mood and behaviour disorders in patients with psychomotor seizure disorders, there is relatively little evidence of carbamazepine's anti-anxiety effects in patients with epilepsy. However, Puente (1976) reported good-to-excellent effects of carbamazepine in 47 children with anxiety-related symptoms. Groh (1976) reported marked improvement in three of four patients presenting with predominant anxiety and neurotic symptoms. The effects of carbamazepine on anxiety components of depressive illness in patients with affective disorders have also been noteworthy. The anti-anxiety response closely parallels that of the antidepressant response in our study. Based on these theoretical and empirical observations Dr T. Uhde has initiated a double-blind investigation of carbamazepine in patients with primary panic-anxious and phobic-anxious disorders. The preliminary results are promising and suggest some effect of carbamazepine not only on the subjective sense of anxiety but also in objective measures of behaviour such as number of phone calls that the physicians and clinic staff receive from patients seeking help and reassurance. Further controlled clinical trials are required in order to assess the possible anti-anxiety effects of carbamazepine. Should significant clinical effects emerge, it will be of interest to ascertain whether improvement also occurs in response to classical tricyclic antidepressant drugs used in the treatment of anxiety disorders or whether any carbamazepine response is atypical or selective for a subgroup of patients.

Effects of carbamazepine on psychotic disorders

In our patients with manic-depressive disorders, the psychotic component of the symptomatology appears to improve in association with improvement in mood and motor aspects of the syndrome. However, like that observed in response to classical neuroleptic treatment, the antipsychotic response in mania often appears to lag behind that of the improvement in the mood and motor components (Post *et al.* 1984). Although Stevens *et al.* (1979), in open clinical observations, reported exacerbation of psychosis when carbamazepine was added to high-dose neuroleptic treatment, several subsequent reports have suggested that carbamazepine in combination with a neuroleptic may be useful in the treatment of some patients with schizoaffective and schizophrenic syndromes. Silberman and Post (1980) reported improvement in a patient with recurrent unipolar atypical psychosis and other schizoaffective patients appeared to respond well to carbamazepine (see Ballenger and Post 1978). Folks *et al.* (1982) reported improvement in two of three schizoaffective patients. Neppe (1982) compared six weeks of placebo with six weeks of carbamazepine treatment used in conjunction with neuroleptics in schizophrenics with abnormal EEGs. He reported that ratings in eight patients were definitely superior on carbamazepine compared to placebo, while three were largely unchanged. Parallel results were reported by Klein *et al.* (1982) who observed, in a double-blind study comparing carbamazepine to placebo when added to haloperidol treatment, that carbamazepine plus haloperidol was associated with a more significant improvement in schizophrenic symptomatology. Hakola and Laulumaa (1982), in an open clinical trial, reported carbamazepine to be of use in the treatment of aggressive schizophrenic patients.

Possible relationship of carbamazepine's efficacy in affective illness to the limbic system

Although carbamazepine is effective in the treatment of experimental models of limbic system epilepsy and in the treatment of patients with complex partial seizures in the clinic, its efficacy in affective illness does not imply an underlying seizure disorder. Carbamazepine is clearly also effective in a variety of paroxysmal pain syndromes which do not involve an ictal process (Post *et al.* 1982*a*, 1984). Thus, it remains an open question whether the effects of carbamazepine mediating its anticonvulsant properties are related to or different from those mediating its psychotropic effects in affective illness. One approach to the problem is the comparison of carbamazepine's psychotropic effects with other anticonvulsant agents thought to have a relatively less selective effect on limbic system dysfunction and excitability. In this light we initiated cross-over trials of carbamazepine to phenytoin and valproic acid in collaboration with T. Uhde and W. Berrettini. In our first patient to complete this double-blind clinical trial we observed notable antimanic and antipsychotic effects of carbamazepine but no effect of valproic acid or phenytoin (as illustrated in Fig. 9.4). These data suggest that individual patients may respond relatively selectively to one anticonvulsant agent but not to another. It will be of particular importance to ascertain whether a drug such as phenytoin has equal efficacy

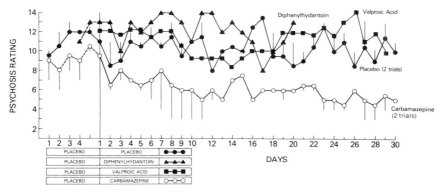

Fig. 9.4. Carbamazepine but not diphenylhydantoin or valproic acid decreases manic psychosis in a patient with manic-depressive illness.

compared to carbamazepine in systematic clinical trials in a large patient population. Both in animal models and in the clinic, phenytoin is thought to have relatively greater effects on seizures arising from areas other than the temporal-lobe and limbic systems. Thus, notable evidence of its efficacy in affective illness would vitiate against the hypothesis that carbamazepine's relative selectivity for limbic system substrates was directly related to its psychotropic properties.

In this regard it is of interest that Lambert *et al.* (1975) and Emrich *et al.* (1980) reported antimanic effects of valproic acid when used alone and prophylactic effects of valproic acid when it was added to lithium carbonate in previously lithium-resistant cycling patients. These data suggest that valproic acid in some treatment-resistant, rapidly cycling patients may, in addition to carbamazepine, have useful antimanic and prophylactic properties. As such it is of interest that valproic acid was second in potency in the study of Albright and Burnham (1980) which ranked drugs according to their relative efficacy in inhibiting the amygdala- compared to cortical-kindled focus. Chouinard *et al.* (1982), however, have recently recorded antimanic efficacy of clonazepam, an anticonvulsant with relatively less efficacy on the amygdala- compared to the cortical-kindled focus.

We have recently also initiated several other types of studies to indirectly attempt to assess whether limbic-system dysfunction is correlated with the positive antimanic and antidepressant effects of carbamazepine in affective illness. The local anaesthetic, procaine, is thought to have some selectivity for the limbic system based on a variety of animal studies. We are assessing the clinical, physiological, and EEG responses to infusions of procaine in order to assess possible relationships to subsequent carbamazepine response. In other psychological and physiological studies we are also investigating whether possible indirect clinical markers of limbic system dysfunction, such as the degree of psychosensory distortion experienced by a patient as rated on a systematic interview (in collaboration with E. Silberman) predict carbamazepine response. The assessment of patients' regional glucose utilization using PET-scan methodology also may provide important leads to the question of whether the limbic system dysfunction is associated with a carbamazepine response.

BIOCHEMICAL EFFECTS OF CARBAMAZEPINE: A SELECTIVE REVIEW

The biochemical profile of carbamazepine, particulary when compared with more routine psychotropic agents such as lithium carbonate and the classical neuroleptics, should provide valuable information not only about its mechanism of action but also about possible biochemical alterations underlying the disorders of affect regulation. The biochemical and physiological effects of carbamazepine are reviewed in detail elsewhere (Post *et al.* 1982*a, b,* 1983*b,* 1984; Post and Uhde 1983). Only several examples will be discussed here to illustrate differences from classical psychotropic agents such as neuroleptics or lithium carbonate. A variety of direct and indirect evidence supports the notion that carbamazepine does not exert its psychotropic effects in mania and psychosis by directly blocking dopamine receptors as has been postulated for classical neuroleptic agents. Our observations of affectively ill patients (Post *et al.* 1984) and those in the neurological literature (Reynolds 1975) indicate that carbamazepine does not produce parkinsonian side-effects or tardive dyskinesia, as has been reported with neuroleptics. Cerebrospinal fluid homovanillic acid (HVA), a dopamine metabolite, is not elevated in patients treated with carbamazepine as is the case with patients acutely treated with neuroleptics. In contrast, probenecid-induced accumulations of HVA were significantly reduced during carbamazepine administration compared to placebo treatment. This direction of HVA alteration is more consistent with a dopamine agonist rather than a dopamine antagonist profile. Carbamazepine also does not block cocaine- or amphetamine-induced hyperactivity in animals as do the neuroleptics (Koella *et al.* 1976; Post *et al.* 1983*a*). Similarly, while neuroleptics produce marked increases in plasma prolactin, only slight but statistically significant elevations in plasma prolactin are observed in our studies and those of London (1980). Drs Daniel Hommer and Paul Clarke (personal communication 1983) have recently observed that carbamazepine does not increase single-unit firing in dopaminergic neurones of the substantia nigra – an effect different from that observed with classical neuroleptics.

In summary, these several lines of evidence suggest that carbamazepine may be interacting with dopaminergic systems in an indirect fashion and at least by mechanisms different from those of classical neuroleptic agents. It is possible that this may be occurring secondarily to interactions with other neurotransmitter systems. A potential candidate is that of the gamma–amino–butyric acid (GABA) pathway. Although carbamazepine does not affect levels of GABA measured in rat brain or in human CSF (Post *et al.* 1980), Bernasconi and Martin (1979) reported that carbamazepine decreased GABA turnover. It is of interest though that valproic acid and electroconvulsive shock treatments, which have also been reported to be efficacious in affective illness, similarly decrease GABA turnover. The possible effects of carbamazepine on GABA systems as they may relate to either its psychotropic or anticonvulsive properties await further exploration, but are of interest in light of evidence linking alterations in GABA function to some seizure disorders and the recent reviews of Emrich *et al.* (1980) and Berrettini and Post (1982) suggesting possible GABA alterations in affective illness.

Emerging experimental data does not support a direct effect of carbamazepine on the benzodiazepine component of the benzodiazepine–GABA receptor complex. Preliminary data in collaboration with Marangos *et al.* 1983 does not indicate that carbamazepine displaces [^{3}H]-diazepam binding with high potency *in vitro*. Moreover, we have recently observed that treatment with the benzodiazepine receptor antagonists, RO15-1788 and CGS-8216, is not sufficient to reverse carbamazepine's anticonvulsant effects of amygdala kindled seizures (Weiss and Post 1982).

Carbamazepine does have an interesting profile of effects on other classical and peptide neurotransmiiter systems including norepinephrine, vasopressin, somatostatin, and the pituitary–adrenal axis regulating cortisol secretion (Post *et al.* 1982, 1984). Its vasopressin agonist-like properties may be of particular importance for some of its clinical effects and side-effects such as hyponatraemia (Uhde and Post 1982). Carbamazepine has been used to treat diabetes insipidus and is clearly an alternative treatment for affectively ill patients unable to tolerate lithium carbonate because of the diabetes insipidus syndrome.

ELECTROCONVULSIVE SEIZURES AS A LIMBIC-SYSTEM ANTICONVULSANT

The emerging evidence for the utility of carbamazepine in affective disorders raises the issue of why the apparently opposing treatments of electroconvulsive therapy and an anticonvulsant such as carbamazepine are both useful in the treatment of the affective disorders. One possible explanation of this paradox derives from recent studies in our laboratory where we have observed potent anticonvulsant and antikindling effects of electroconvulsive seizures (ECS) in the amygdala-kindling model. Pretreatment of animals with ECS compared to sham ECS six hours prior to kindling markedly inhibits development of amygdala-kindled seizures (Post *et al.* 1981). Moreover, seven once-daily ECS result in the suppression of amygdala-kindled seizures for up to five days following this course of chronic ECS. These data raise the possibility that the mechanism of action of ECS in affective illness could be related to physiological and biochemical mechanisms associated with this ability to stabilize limbic system excitability as revealed in the amygdala kindling model. In this fashion it is possible that electroconvulsive seizures themselves as well as treatment with anticonvulsants such as carbamazepine may share some common biochemical and physiological properties.

While it is possible that the efficacy of carbamazepine in affective illness is related to its action on limbic system substrates, direct confirmation of this hypothesis remains for further clinical and experimental work. Systematic description of the clinical and biological predictors of carbamazepine response, as well as elucidation of its biochemical and physiological mechanisms of action in affective illness, may ultimately provide further leverage in assessing the neural and biochemical substrates underlying the disorders of affective regulation.

REFERENCES

Albright, P. S. and Burnham, W. M. (1980). Development of a new pharmacological seizure model: effects of anticonvulsants on cortical- and amygdala-kindled seizures in the rat. *Epilepsia* **21**, 681–9.

Babington, R. G. (1977). The pharmacology of kindling. In *Animal models of psychiatry and neurology* (ed. I. Hanin and E. Usdin), pp. 141–9. Pergamon, Oxford/New York.

—— and Horovitz, Z. P. (1973). Neuropharmacology of SQ10,996, a compound with several therapeutic indications. *Arch. int. Pharmacodyn. Ther.* **202**, 106–18.

Ballenger, J. C. and Post, R. M. (1978). Therapeutic effects of carbamazepine in affective illness. A preliminary report. *Commun. Psychopharmacol.* **2**, 159–75.

—— and —— (1980). Carbamazepine (Tegretol) in manic-depressive illness. A new treatment. *Am. J. Psychiat.* **137**, 782–90.

Bear, D. M. and Fedio, P. (1977). Quantitative analysis of interictal behavior in temporal lobe epilepsy. *Arch. Neurol.* **34**, 454–67.

—— Levin, K., Blumer, D., Chetham, D., and Ryder, J. (1982). Interictal behavior in hospitalised temporal lobe epileptics: relationship to idiopathic psychiatric syndromes. *J. Neurol. Neurosurg. Psychiat.* **45**, 481–8.

Bernasconi, R. and Martin, P. (1979). Effects of antiepileptic drugs on the GABA turnover rate. *Arch. Pharmacol.* (suppl. 307), abstr. 251, R63.

Berrettini, W. H. and Post, R. M. (1984). GABA in affective illness. In *Neurobiology of mood disorders* (ed. R. M. Post and J. C. Ballenger), pp. 673–85. Williams and Wilkins, Baltimore, Maryland.

Chouinard, G., Young, S. Annable, L., and Bradwejn, J. (1983). Antimanic effects of clonazepam. *Biol. Psychiat.* **18**, 451–66.

Currie, S., Heathfield, K. W. G., Henson, R. A., and Scott, D. F. (1971). Clinical course and prognosis of temporal lobe epilepsy. *Brain* **94**, 173–90.

Dalby, M. A. (1971). Antiepileptic and psychomotor epilepsy. *Epilepsia* **12**, 325–34.

—— (1975). Behavioral effects of carbamazepine. In *Advances in neurology, Vol. 11: complex partial seizures and their treatment* (ed. J. K. Penry and D. D. Daly), pp. 331–43. Raven Press, New York.

Dongier, S. (1959). Statistical study of clinical and electroencephalographic manifestations of 536 psychotic episodes occurring in 516 epileptics between clinical seizures. *Epilepsia* **1**, 117–42.

Emrich, H. M., Zerssen, D. V., Kissling, W., Moller, H-J., and Windorfer, A. (1980). Effect of sodium valproate in mania. The GABA-hypothesis of affective disorders. *Arch. Psychiatr. Nervenkr.* **229**, 1–16.

Feindel, W. and Penfield, W. (1954). Localization of discharge in temporal lobe automatism. *Arch. Neurol. Psychiat.* **72**, 605–30.

Flor-Henry, P. (1969). Schizophrenic-like reactions and affective psychoses associated with temporal lobe epilepsy: etiological factors. *Am. J. Psychiat.* **126**, 148–51.

Folks, D. C., King, L. D., Dowdy, S. B., Petrie, W. M., Jack, R. A., Koomen, J. C., Swenson, B. R., and Edwards, P. (1982). Carbamazepine treatment of selected affectively disordered inpatients. *Am. J. Psychiat.* **139**, 115–17.

Gibbs, F. A. (1956). Abnormal electrical activity in the temporal regions and its relationship to abnormalities of behavior. *Res. Publ. Ass. res. nerv. ment. Dis.* **36**, 278–94.

Gloor, P. (1972). Temporal lobe epilepsy. Its possible contribution to the understanding of the functional significance of the amygdala and its interaction with neocortical-temporal mechanisms. In *The neurobiology of the amygdala* (ed. B. E. Eleftheriou), pp. 423–57. Plenum, New York.

Groh, C. (1976). The psychotropic effect of Tegretol in non-epileptic children, with particular reference to the drug's indications. In *Epileptic seizures—behavior—pain* (ed. W. Birkmayer), pp. 259–63. Hans Huber, Berne.

Hakola, H. P. A. and Laulumaa, V. A. O. (1982). Carbamazepine in treatment of violent schizophrenics. *Lancet* **i**, 1358.

Hermann, B. P. (1979). Psychopathology in epilepsy and learned helplessness. *Med. Hypotheses* **5**, 723–9.

Inoue, K., Arima, S., Tanaka, K., Fukui, Y., and Kato, N. (1981). A lithium and carbamazepine combination in the treatment of bipolar disorder – a preliminary report. *Folia Psychiat. Neurol. Jpn.* **35**, 465–76.

Klein, E., Bental, E., Lerer, B., and Belmaker, R. H. (1982). Combinations of carbamazepine and haloperidol versus placebo and haloperidol in excited psychoses: a controlled study. 13th Annual Congress, CNIP, Abstract.

Koella, W. P., Levin, P., and Baltzer, V. (1976). The pharmacology of carbamazepine and some other anti-epileptic drugs. In *Epileptic seizures—behavior—pain* (ed. W. Birkmayer), pp. 32–50. Hans Huber, Berne.

Lambert, P. A., Zarraz, G., Borselli, S., and Bouchardy, M. (1975). Le depropylacetamede dans le traitement de la psychose maniaco-depressive. *Encephale* **I**, 25–31.

Lipinksi, J. F. and Pope, H. G. (1982). Possible synergistic action between carbamazepine and lithium carbonate in the treatment of three acutely manic patients. *Am. J. Psychiat.* **139**, 948–9.

London, D. R. (1980). Hormonal effects of anticonvulsant drugs. In *Advances in epileptology* (ed. R. Carger, F. Angeleri, and J. K. Penry), pp. 399–405. Raven Press, New York.

Marangos, P. J., Post, R. M., Patel, J., Zander, C., Parma, A., and Weiss, S. (1983). Specific and potent interactions between carbamazepine and brain adenosine receptors. *Europ. J. Pharm.* **93**, 175–82.

Mulder, D. W. and Daly, D. (1952). Psychiatric symptoms associated with lesions of the temporal lobe. *J. Am. med. Ass.* **150**, 173–6.

Neppe, V. M. (1982). Carbamazepine in the psychiatric patient. *Lancet* **ii**, 334.

Okuma, T., Inanaga, K., Otsuki, S., Sarai, K., Takahashi, R., Hazama, H., Mori, A., and Watanabe, M. (1979). Comparison of the antimanic efficacy of carbamazepine and chlorpromazine. A double-blind controlled study. *Psychopharmacology* **66**, 211–17.

——, ——, ——, ——, ——, ——, ——, and —— (1981). A preliminary double-blind study of the efficacy of carbamazepine in prophylaxis of manic-depressive illness. *Psychopharmacology* **73**, 95–6.

——, Kishimoto, A., Inoue, K., Matsumoto, H., Ogura, A., Matsushita, T., Naklao, T., and Ogura, C. (1973). Anti-manic and prophylactic effects of carbamazepine on manic-depressive psychoses. *Folia Psychiat. Neurol. Jpn.* **27**, 283–97.

Penry, J. K. and Daly, D. D. (Eds.) (1975). *Advances in neurology, Vol. 11: complex partial seizures and their treatment.* Raven Press, New York.

Post, R. M. (1977). Clinical implications of a cocaine-kindling model of psychosis. In *Clinical neuropharmacology,* Vol. II (ed. H. L. Klawans), pp. 25–42. Raven Press, New York.

—— (1983). Behavioral effects of kindling. In *Advances in epileptology: XIVth International Symposium.* (Ed. M. Parsonage. *et al.*) pp. 173–80. Raven Press, New York.

—— and Uhde, T. W. (1982*b*). Biochemical and physiological mechanisms of action of carbamazepine in affective illness. In *Frontiers in neuropsychatric research* (ed. R. Usdin). MacMillan, London. In press.

——, Ballenger, J. C., Hare, T. A., and Bunney, W. E., Jr. (1980). Lack of effect of carbamazepine on gamma-aminobutyric acid levels in cerebrospinal fluid. *Neurology* **30,** 1008–11.

——, ——, Reus, V. I., Lake, C. R., Lerner, P., and Bunney, W. E., Jr. (1978). Effects of carbamazepine in mania and depression. Paper presented at Scientific Proc., 131st Annual Mtg, Am. Psychiatr. Assoc., *New Research Abstr. no. 7,* Atlanta, Georgia.

——, ——, Uhde, T. W., and Bunney, W. E., Jr. (1984). Carbamazepine in manic-depressive illness: implications for underlying mechanisms. In *Neurobiology of mood disorders* (ed. R. M. Post and J. C. Ballenger), pp. 777–816.

——, ——, ——, Smith, C., Rubinow, D. R. and Bunney, W. E., Jr. (1982*b*). Effect of carbamazepine on cyclic nucleotides in CSF of patients with affective illness. *Biol. Psychiat.* **17,** 1037–45.

—— and Kopanda, R. T. (1976). Cocaine, kindling, and psychosis. *Am. J. Psychiat.* **133,** 627–34.

——, Putnam, F. W., and Contel, N. R. (1981). Electroconvulsive shock inhibits amygdala kindling. *Abstracts of 11th Annual Meeting, Society for Neuroscience,* Los Angeles, October 1981, Vol. 7, p. 587, abstract no. 187.13.

——, Uhde, T. W., Ballenger, J. C., and Bunney, W. E. Jr. (1982*a*). Carbamazepine, temporal lobe epilepsy and manic-depressive illness. In *Advances in biological psychiatry, Vol. 8: Temporal lobe epilepsy, mania, schizophrenia, and the limbic system* (ed. M. Trimble and W. Koella), pp. 117–56. Karger, Basle.

——, ——, ——, and Squillace, K. M. (1983). Prophylactic efficacy of carbamazepine in manic-depressive illness. *Am. J. Psychiat.* **140,** 1602–4.

——, ——, Rubinon, D. R., Ballenger, J. C., Gold, P. W. (1983*b*). Biochemical effects of carbamazepine: relationship to its mechanisms of action in affective illness. *Prog. Neuropsychopharmacol. Biol. Psychiat.* **7,** 263–71.

Puente, R. M. (1976). The use of carbamazepine in the treatment of behavioural disorders in children. In *Epileptic seizures—behavior—pain* (ed. W. Birkmayer), pp. 243–7. Hans Huber, Berne.

Reynolds, E. H. (1975). Neurotoxicity of carbamazepine. In *Advances in neurology, Vol. 11: complex partial seizures and their treatment* (ed. J. K. Penry and D. D. Daly), pp. 345–53. Raven Press, New York.

Roth, M. and Harper, M. (1962). Temporal lobe epilepsy and the phobic anxiety-depersonalization syndrome. II. Practical and theoretical considerations. *Comp. Psychiat.* **3,** 215–26.

Silberman, E. K. and Post, R. M. (1980). The 'march' of symptoms in a psychotic decompensation: case report and theoretical implications. *J. nerv. ment. Dis.* **168,** 104–10.

Standage, K. F. and Fenton, G. W. (1975). Psychiatric symptom profiles of patients with epilepsy: a controlled investigation. *Psychol. Med.* **5,** 152–60.

Stevens, J. R. (1982). Risk factors for psychopathology in individuals with epilepsy. In *Advances in biological psychiatry: temporal lobe epilepsy, mania, and schizophrenia and the limbic system* (ed. W. P. Koella and M. R. Trimble), pp. 56–80. Karger, Basel.

——, Bigelow, L., Denney, D., Lipkin, J., Livermore, A., Rauscher, F., and Wyatt, R. J. (1979). Telemetered EEG–EOG during psychotic behaviors of schizophrenia. *Arch. gen. Psychiat.* **36,** 251–62.

Takezaki, H. and Hanaoka, M. (1971). The use of carbamazepine (Tegretol) in the control of manic-depressive psychoses and other manic, depressive states. *Clin. Psychiat.* **13,** 173–83.

Thompson, P., Huppert, F., and Trimble, M. (1980). Anticonvulsant drugs, cognitive function and memory. *Acta neurol. scand.* **63,** 75–81.

Uhde, T. W. and Post, R. M. (1983). Effects of carbamazepine on serum electrolytes: clinical and theoretical implications. *J. Clin. Psychopharmacol.* **3,** 103–6.

Wada, J. A. (1977). Pharmacological prophylaxis in the kindling model of epilepsy. *Arch. Neurol.* **34,** 389–95.

——, Sato, M., Wake, A., Green, J. R., and Troupin, A. S. (1976). Prophylactic effects of phenytoin, phenobarbital, and carbamazepine examined in kindled cat preparations. *Arch. Neurol.* **33,** 426–34.

Weil, A. (1956). Ictal depression and anxiety in temporal lobe disorders. *Am. J. Psychiat.* **113,** 149–57.

Weiss, S. R. B. and Post, R. M. (1983). Interactions of carbamazepine with benzodiazepine systems in amygdala kindling. *13th Annual Meeting, Society for Neuroscience Abst.* # *224.5,* Vol. 9, Pt. 2, p. 763, Boston, November, 1983.

Williams, D. (1956). The structure of emotions reflected in epileptic experiences. *Brain* **74,** 29–67.

10

Benzodiazepines and the limbic system

ROBERT DANTZER

Benzodiazepines have a variety of effects, including muscle-relaxant activity, anticonvulsant action, induction of sleep, and changes in behaviour. These compounds are among the most widely prescribed drugs in medicine and, within the 20 years after their introduction in psychopharmacology, an impressive number of papers, books, and symposia about their chemistry, pharmacology, metabolism, structure–activity relationships, and clinical properties have appeared.

The involvement of the limbic system in the effects of benzodiazepines was recognized very early on, first on intuitive grounds, based on the role of limbic structures in emotionality, and later on objective data gained in neuropharmacological experiments (e.g. effects on the electrical activity of the brain). The analysis of the relationships between benzodiazepines and specific brain areas is, however, complicated by the many different pharmacological properties of these compounds. At the same time, this task is made nearly insurmountable by the richness of experimental data and the large amount of theoretical speculation on the functional aspects of the limbic system.

This chapter is therefore deliberately limited to a re-examination of the behavioural data which point to an involvement of the limbic structures in the anti-anxiety effects of benzodiazepines.

DEFINITION OF THE BEHAVIOURAL SYNDROME INDUCED BY BENZODIAZEPINES

When an animal is rewarded for producing a particular response and is then simultaneously punished for that response, the occurrence of that response is reduced. Benzodiazepines and other anxiolytics typically reinstate responses suppressed by contingent punishment at doses which have little or no effect on food-reinforced responses, although higher doses of the anxiolytic agents do depress this last class of behaviour (Fig. 10.1). Since their demonstration by Geller *et al.* (1962), the antipunishment effects of anxiolytic agents have been observed in a number of species and on a variety of behavioural procedures generating conflict between simultaneous presentation of shocks and reward.

The release of punished responses produced by anxiolytic agents is highly specific and cannot be accounted for by an increase in food motivation, a reduced sensitivity to electric shock, or general stimulatory effects (McMillan 1975; Miczek 1973).

Many other behaviours are altered by benzodiazepines. Rats conditioned to

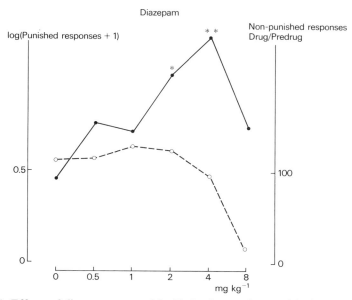

Fig. 10.1. Effects of diazepam on punished behaviour and unpunished behaviour in a multiple schedule of food reinforcement. Rats ($n = 6$) were conditioned to respond on a variable interval schedule (VI = 30 s) interrupted by two 2-min periods signalled by a light above the lever, during which every third response was simultaneously reinforced by a food pellet and punished by a weak electric shock (FR3). Diazepam was injected intraperitoneally at various doses 30 min before the session in well-trained animals. The solid line indicates effects on punished responses (after logarithmic transformation); the broken line indicates effects on non-punished responses (expressed as ratio of response rate after this treatment to preceding control response rate). Note that diazepam at 2 and 4 mg kg^{-1} increased the number of punished responses without altering non-punished responding while higher doses (8 mg kg^{-1}) disrupted non-punished responding.

respond for food and receiving no more reward usually stop responding within a few minutes (*extinction*). Extinction is slower in animals which have experienced partial reinforcement than in animals which have been reinforced after each response (*partial reinforcement effect*). Anti-anxiety agents increase resistance to extinction and disrupt the development of tolerance to extinction due to partial reward (Gray 1977).

Benzodiazepines reduce the inhibitory effects of novelty on locomotor activity or consummatory behaviour (Poschel 1971; Soubrié *et al.* 1976). In the same manner, these compounds increase the number of transitions between a darkened compartment and an illuminated compartment in a two-compartment cage (Crawley and Goodwin 1980) as well as the number of social interactions occurring between two rats when placed in a novel and brightly lit arena (File 1980).

Acquisition of a conditioned avoidance response can be facilitated by acute or chronic treatment with low doses of benzodiazepines in poor learners

(Bignami and de Acetis 1971; Dantzer 1977), while avoidance behaviour tends to be depressed by high doses of benzodiazepines in well-trained animals.

Facilitation as well as reduction of aggressive behaviour has been noted (Dimascio 1973). In general, defensive behaviour (e.g. shock-induced fighting or reactive biting in vicious monkeys) is more sensitive to the disruptive effects of benzodiazepines than attack behaviour (e.g. intermale fighting). In social groups, the effects of benzodiazepines depend on the social status of the treated subject. In pairs of rats, for example, low doses of chlordiazepoxide increase attack and threat when administered to the dominat animal, while high doses decrease submissiveness in the dominated rat (Miczek and Krsiak 1979).

During recent years, there have been a few attempts to construct theoretical models of the behavioural effects of benzodiazepines (Dantzer 1978; Gray 1981; Morris and Gebhart 1981). The main difficulty encountered is epistemological. Since benzodiazepines are clinically effective in reducing anxiety, it is tempting to relate the behavioural effects produced by these drugs in animals, to their anxiolytic activity. Such an attitude has been deliberately chosen for example by Gray who claims that the best way to understand the neuropsychology of anxiety is to consider the behavioural changes caused in animals by anxiolytic agents and their neural basis. This leads to the implicit assumption that behavioural tests sensitive to the effects of benzodiazepines are animal models of anxiety and that the changes in behaviour brought about by anxiolytic agents must reflect a reduction in anxiety. However, as pointed out by Carlton (1978), such a chain of logic is not a theory but a tautology. The problems with such a description become very clear when the effects of benzodiazepines which bear little or no relation to anxiety are considered.

In the case of the effects of benzodiazepines on feeding behaviour for example, there is evidence that a raised level of emotionality is not a necessary condition for the facilitation of the feeding response observed in benzodiazepine-treated rats (Cooper 1980). As a matter of fact, low doses of chlordiazepoxide increase consumption of familiar food and this effect is not altered by handling the subjects prior to the experiment or by administering electric shocks. Benzodiazepines may therefore enhance appetite for food by relatively direct means and specific screening for this pharmacological activity has led to the development of highly effective food-intake stimulants (Baile and McLaughlin 1979).

The role of anxiety is more difficult to assess in behavioural tests involving presentation of aversive events such as electric shock and non-reward or stimuli associated with those primary reinforcers. Benzodiazepines counteract the suppressing effects of a fear signal (i.e. a signal which has repeatedly been paired with electric shocks) and this action is consistent with an attenuation of fear. However, benzodiazepines are unable to block the facilitating effects that the same fear signal produces on avoidance responding (Dantzer *et al.* 1976; Morris *et al.* 1980) (Fig. 10.2). This result is incompatible with an interpretation in terms of fear. Recent research along these lines has led to the conclusion that benzodiazepines do not directly reduce emotionality but instead, produce their behavioural effects through some other mechanism.

It has been suggested that most of the behavioural changes induced by

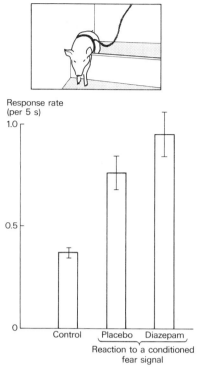

Fig. 10.2. Effects of diazepam on conditioned acceleration of responding produced by presentation of a fear signal. Pigs trained on a continuous avoidance procedure in a shuttle-box, were submitted to a fear-conditioning procedure during which a neutral tone was repeatedly paired with inescapable electric shock. The next day, they were put back in the shuttle-box and the fear signal was presented without shock while they were shuttling. Presentation of the fear signal induced a facilitation of avoidance responding which was increased by pretreatment with 1 mg kg^{-1} diazepam. Each result is the mean ($\pm$ SEM) of four pigs. (From Dantzer *et al.* 1976.)

benzodiazepines may be regarded as manifestations of a response persistence effect. Drug-treated animals persevere more with their most likely behaviour, on account of the situational characteristics and their previous experience (Dantzer 1978). Initiation and persistence of response are dependent on cues provided by the environment but also on response-associated cues (Fig. 10.3). There is evidence that benzodiazepines interfere mainly with the mechanisms controlling the processing of response-associated cues, especially in cases where the behaviour itself becomes a retrieval cue for the memory of past events to which it is correlated (Morris and Gebhart 1981). According to this analysis, anti-anxiety effects will be produced when aversive events are involved, while other behavioural changes such as increased resistance to satiation will appear with cues such as the appearance of familar food presentation.

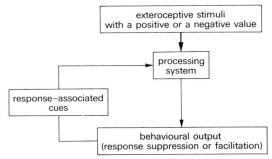

Fig. 10.3. A schematic model of the chain of events involved in the initiation and persistence of a conditioned response. The behavioural output is controlled by the affective value of the exteroceptive stimuli and by the consequence of the response. Benzodiazepines appear to selectively interfere with the response-associated cues. (From Dantzer 1978.)

INDIRECT EVIDENCE FOR THE INVOLVEMENT OF THE LIMBIC SYSTEM IN THE BEHAVIOURAL EFFECTS OF BENZODIAZEPINES

Apart from the commonly held idea that the limbic system plays a key role in the control of emotionality, there are at least two lines of evidence to suggest that benzodiazepines affect behaviour by acting on limbic structures

1. Specific high-affinity benzodiazepine-binding sites have been detected in various brain structures including the limbic system and the hypothalamus;

2. Specific lesions of limbic structures reproduce most of the behavioural effects of benzodiazepines.

Benzodiazepine binding to specific receptors in the brain was first demonstrated by Squires and Braestrup (1977) and Möhler and Okada (1977). These authors suggested a functional role for the receptors in the pharmacological potency of benzodiazepines by correlating their pharmacological effects with their ability to inhibit [³H]-diazepam binding to rat brain membranes *in vitro*. Such a correlation amounts to +0.78 in the case of the effects of benzodiazepine on behaviour suppressed by contingent punishment (Sepinwall and Cook 1980) (Fig. 10.4). The highest densities of benzodiazepine receptors are found in cortical and limbic forebrain areas (Speth *et al.* 1980). Data from binding assays have been confirmed by autoradiographic techniques emphasizing the importance of some limbic brain structures such as the amygdala and frontal cortex (Young and Kuhar 1979).

Rats selectively bred for a high defecation score (taken as an index of emotionality) in a brightly illuminated unfamiliar environment have a lower number of [³H]-diazepam-binding sites in the hypothalamus and hippocampus than animals selected for a low defecation score (Robertson *et al.* 1978). The same difference was found in an inbred strain of mice characterized as 'anxious' (Robertson 1979).

In view of the many pharmacological effects of benzodiazepines, there has been much effort devoted to finding subclasses of benzodiazepine receptors.

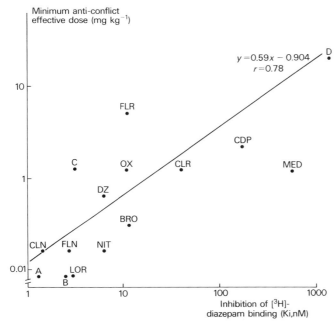

Fig. 10.4. Correlation of rat punishment potency (measured by the lowest effective dose per rat) with benzodiazepine receptor-binding (measured by inhibition of specific [³H]-diazepam binding in rat cortex). (A = 5-3027; B = 21-8324; LOR = lorazepam; CLN = clonazepam; FLN = flunitrazepam; NIT = nitrazepam; BRO = bromazepam; DZ = diazepam; C = 21-3981; OX = oxazepam; CLR = clorazepate; MED = medazepam; CDP = chlordiazepoxide; FLR = flurazepam; D = 5-5807. (Adapted from Sepinwall and Cook (1980).)

For example, Klepner *et al.* (1979) proposed the presence of two distinct binding sites of benzodiazepines according to their affinity for a new class of anti-anxiety compounds claimed to be devoid of sedative and ataxic effects. More recently, the presence of two benzodiazepine-binding sites in the hippocampus but not in other brain areas was postulated on the basis of a differential affinity for diazepam and flunitrazepam (Volicer and Biagioni 1982).

The pharmacological relevance of such a distinction and, *a fortiori*, its physiological significance must still await further experimentation. At the present time, the view that cortical benzodiazepine receptors may be involved in anticonvulsant activity while limbic receptors are involved in anxiolytic actions of benzodiazepines (Braestrup and Nielsen 1980) is no more than pure speculation.

Most of the available evidence in favour of the involvement of limbic brain structures in the action of benzodiazepines comes from comparisons between drug effects and behavioural consequences of lesions of specific neural pathways or structures belonging to the limbic system. An approach in terms of brain lesion rather than in terms of stimulation is justified on the basis of the depressing activity of benzodiazepines on spontaneous and evoked electrical activity of various brain areas after both systemic and local administration (for a recent review, see Haefely *et al.* 1981).

At an analytical level, close analogy can be found between behavioural effects produced by benzodiazepines and those observed after lesions of the septo–hippocampal system (Gray 1981). For example, benzodiazepines have been found to retard reversal learning, depress spontaneous alternation, disinhibit suppressed responding to a non-reward stimulus, facilitate acquisition of a two-way conditioned avoidance response in a shuttle-box, and disrupt successive but not simultaneous discrimination learning. These are all features which have been found in hippocampectomized animals.

On a more theoretical level, these effects have been interpreted in terms of perseverative behaviour arising from a reduced ability to delay or withhold responses (Kimble 1968). Other authors have suggested that behavioural deficits following hippocampal lesions are not due to response disinhibition, but rather are the result of a basic impairment in the processing of stimulus cues following a change in experimental conditions. As a typical example, disruption of passive avoidance or enhanced resistance to extinction shown by hippocampal rats in a straight runway was greatly reduced when additional cues facilitating the recognition of the goal box were introduced (Winocur and Bindra 1976). We have seen in the previous section that similar interpretations have been put forward to account for the behavioural effects of benzodiazepines.

Discussion of the behavioural functions of the hippocampus has also focused on memory (Iversen 1976), possibly in relation to spatial mapping (O'Keefe and Nadel 1979). Recognition memory (i.e. decision based on the past perception of a stimulus) can be assessed in animals by a delayed-matching-to-sample test in which a sample stimulus is presented followed by a delay interval, followed in turn by the presentation of a number of stimuli, one of which is the same as the one already presented; this last stimulus has to be selected by the subject in order to be rewarded. Transection of the fornix, which disconnects the hippocampus from the subcortical sites, produces a deficit in performance of this task in both monkeys and rats (Gaffan 1972, 1974). According to Olton *et al.* (1979), the critical procedural factor for hippocampal involvement in memory tasks is a working memory requirement, the sample stimulus to be remembered changing from trial to trial. When the subject does not have to remember what was the last stimulus presented, i.e. when the same information can be used for all trials, there is no deficit. Benzodiazepines fail to reproduce this key feature of the hippocampal syndrome since chlordiazepoxide does not disrupt the performance of monkeys in a delayed-matching-to-sample test, even at high doses (Saghal and Iversen 1980).

Another important difference between hippocampal lesions and benzodiazepines can be found in punishment experiments. When deprived rats with lesions are put into a box that contains a water cup and are punished for drinking from that cup, they inhibit the drinking response as readily as do normal rats (cf. Black *et al.* 1977). In contrast, benzodiazepine-treated rats continue to drink in spite of the shocks received.

Such negative data are difficult to reconcile with a preferential involvement of the hippocampus in behavioural effects of benzodiazepines. As a matter of fact, it is unlikely that benzodiazepines depress only one limbic area and leave other areas unaffected. Septum lesions lead to several features of the benzo-diazepine syndrome, such as response persistence under shock and response persistence under non-reward (Dickinson 1974; Fried 1972). However, several qualitative differences are apparent. For example, septal lesions have been found to decrease the suppressive effect of contingent shock on an intermit-tently food-reinforced lever press. The effect is the same, however, when non-contingent shock is used, in sharp contrast with the differential activity of benzodiazepines on contingent and noncontingent shock (Huppert and Iversen 1975). Other differences between the effects of benzodiazepines and those of septal damage can be found in aggressive behaviour. Footshock-induced fighting is increased by lesions of all or selected nuclei of the septum (e.g. Blanchard and Blanchard 1968), but is decreased by benzodiazepines. In paired encounters, behaviours characteristic of attack and threat are disrupted while defensive and submissive reactions are not changed by septal lesions (Lau and Miczek 1977). This is again at variance with the effects of benzodiazepines (see the preceding section).

Another key limbic structure may be the amygdala. Amygdaloid lesions attenuate emotional/motivational reactions in a number of situations and interfere with response suppression (Isaacson 1974). Once again, many of the behavioural alterations observed in animals with amygdaloid damage are found in benzodiazepine-treated animals (e.g. impaired learning of successive reversals, perseveration in drinking a water solution adulterated with quinine, disinhibition of suppressed behaviour under conditions of non-reward) but noticeable exceptions are found (e.g. impairment of conditioned avoidance responding).

It is therefore difficult to conclude that the limbic system is the critical target for the behavioural effects of benzodiazepines from a comparison of the behavioural alterations induced by benzodiazepines with those resulting from limbic lesions. Although in many cases the same interpretations have been put forward to account for both effects, such a comparison is of little significance when it is recognized that the nature and intensity of the behavioural deficits after either a lesion or a drug are subjected to so many treatment and procedural factors. In addition, even if the benzodiazepine drugs constitute a relatively homogeneous pharmacological class, every limbic structure discussed so far can hardly be conceived of as an entity, but must rather be considered as functionally heterogeneous. The only way therefore to build a credible account of the neural structures involved in benzodiazepine action is to bring together carefully selected data from many disciplines, e.g. electrophysiology, neuro-psychology, neurochemistry, and behavioural pharmacology. This has been done in a very elegant way by Gray, but his model of a behavioural inhibition system including the septo–hippocampal system, the Papez circuit, and the prefrontal cortex as well as the ascending monoaminergic and cholinergic path-ways which innervate these forebrain structures (Gray 1981), is still awaiting more direct experimental confirmation.

It is still true that a theoretical account based mainly on indirect evidence is like a chain which is as weak as the weakest link. A typical example of the shaky grounds on which theories are sometimes built can be found in the field of the mesocorticolimbic dopaminergic pathway. Thus, it has been suggested that the mesocorticofrontal dopaminergic neurones originating in the ventral tegmental area are involved in motor inhibition under conditions of stress (Tassin *et al.* 1978; Simon 1981). In addition, their stress-induced activation has been shown to be prevented by benzodiazepines (Fadda *et al.* 1978; Lavielle *et al.* 1978). From these data, it has been proposed that this dopaminergic tract might be one route by which the anti-anxiety drugs disinhibit suppressed motor behaviour (Gray 1981). If this interpretation is true, it means that benzodiazepines should be highly effective in a situation in which the mesocorticofrontal dopaminergic pathway is preferentially activated. Such a situation has been identified recently (Herman *et al.* 1982). Rats having been submitted to inescapable electric shocks typically display a decreased locomotor activity when replaced in the same situation without shock. Exposure to environmental stimuli paired with footshock markedly increased the activity of the anteromedial frontal dopaminergic terminals but not the activity of other cortical and limbic structures (Sulcal frontal cortex, nucleus accumbens, amygdaloid complex, olfactory tubercle) (Fig. 10.5). In contrast repeated exposure to footshock produced an undifferentiated increase of dopamine turnover in all of these areas, although more marked in the anteromedial frontal cortex and the amygdala.

These data can be interpreted either in terms of response suppression or in

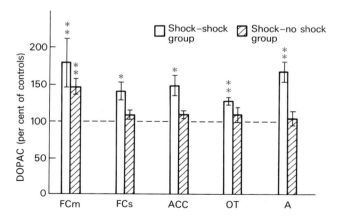

Fig. 10.5. Effects of repeated electric shocks and exposure to an environment previously paired with shock upon DOPAC content of different brain areas. Results are mean (± SEM) of data obtained in six rats and are expressed as percentages of respective control value (FCM = anteromedial frontal cortex; FC$_s$ = sulcal frontal cortex; ACC = nucleus accumbens; OT = olfactory tubercle; A = amygdaloid complex; *p<0.05; **p<0.01).

terms of emotionality. In either case, before further speculating on the possible role of the anteromedial frontal cortex in the activity of anti-anxiety agents, the minimum caution is to make sure that the behavioural paradigm used is sensitive to the effects of benzodiazepines. Unfortunately, this is not the case, since rats treated with low or medium doses of diazepam before being placed into the environment in which they have been previously shocked still show suppressed behaviour. Diazepam is only effective when it is given before administration of footshock (Fig. 10.6).

DIRECT EVIDENCE FOR THE INVOLVEMENT OF THE LIMBIC SYSTEM IN THE BEHAVIOURAL EFFECTS OF BENZODIAZEPINES

If a structure or pathway plays a critical role in the appearance of the behavioural effects of benzodiazepines, several predictions can be made.

1. There should be a temporal correlation between changes in neuronal activity and behavioural effects;

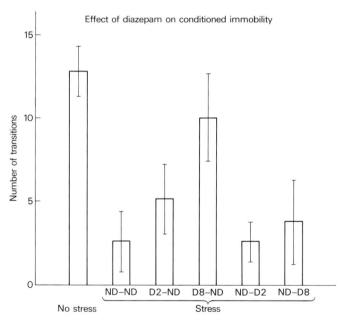

Fig. 10.6. Effects of diazepam on conditioned immobility in rats. Rats were given 60 shocks (1 mA; 3-s duration; 15-s intershock interval) followed 48 hours later by placement in a two-compartment cage in which the number of transitions from one compartment to the other was counted for five min. Each result is the mean ($\pm$ SEM) of six rats. Each group is identified by two terms, the first indicating the treatment given before the stress session, the second the treatment given before the test session (ND = no drug, placebo; D2 = diazepam 2 mg kg^{-1}; D8 = diazepam 8 mg kg^{-1}). All injections were given intraperitoneally in 1 per cent carboxymethylcellulose, 30 min before the experimental session. Diazepam released suppressed behaviour in a dose-related way only when injected before the stress session.

2. Local injection of drug should reproduce the effect of systemic administration;

3. Ablation of the critical structure should abolish the effects of benzo-diazepines.

Instead of examining successively these three different lines of evidence, I will discuss the various systems which have been proposed as the neural basis of the anti-anxiety effects of benzodizepines.

The dosal noradrenergic bundle (DNB)

The DNB originates from the locus ceruleus and innervates the cortex, thalamus, and hippocampus. Depletion of forebrain noradrenaline mimics some of the effects of chlordiazepoxide on extinction of appetitive tasks and on hippocampal theta rhythm (Gray *et al.* 1975; Mason and Iversen 1979). In addition, benzodiazepines reduce the activity of forebrain noradrenergic neurones, especially under conditions of stress (for a recent review see Haefely *et al.* 1981). Using rats trained to run a straight alley for food reward, Morris *et al.* (1979) showed that the increased resistance to extinction produced by chlor-diazepoxide was blocked in animals with DNB radiofrequency lesions.

A critical role for the DNB in the effects of benzodiazepines is however elimi-nated by the results of two recently published reports. Koob *et al.* (1980) found that 6-hydroxydopamine lesions of the DNB in adult rats failed to alter the release of responding by chlordiazepoxide in an approach–avoidance conflict. In the same manner, rats depleted of forebrain noradrenaline by neonatal systemic administration of 6-hydroxydopamine and submitted at 70 days of age to a water extinction test or to punishment of the licking response by electric shock, were still responsive to disinhibiting effects of chlordiazepoxide (Bialik *et al.* 1982). There is therefore at least some direct evidence for the absence of a critical involvement of forebrain noradrenaline in the effects of benzodiazepines in suppressed behaviour.

Forebrain serotonin

Based on release of punishment-suppressed responding after systemic admini-stration of serotoninergic antagonists (e.g. methysergide, cianserin) and sero-tonin-synthesis inhibitors (parachlorophenylalanine) or intracerebral admini-stration of 5,7-dihydroxytryptamine (5,7-DHT), a specific neurotoxin for sero-toninergic neurones, several authors have suggested that serotoninergic neurones originating from the raphe and innervating limbic structures are involved in the anticonflict effects of benzodiazepines (Stein *et al.* 1973; Tye *et al.* 1979). Local injection of chlordiazepoxide into the raphe dorsalis reversed behavioural suppression in rats submitted to an approach–avoidance conflict (Thiebot *et al.* 1980*a, b*). Chlordiazepoxide was however still effective in blocking conflict when systemically administered to 5,7-DHT-treated rats, indicating that intact serotonin neurones are not essential for this drug effect (Tye *et al.* 1977; see also Chapter 11, this volume).

Specific limbic structures

The first attempt to locate the main sites of action of benzodiazepines by intra-cerebral injection was initiated by Nagy and Decsi (1973). Diazepam was injected into various parts of the limbic system through cannulae implanted in freely-moving rats, and its effectiveness in blocking the rage reaction elicited by chemical stimulation of the hypothalamus with carbachol was studied (Nagy and Desci 1973). Application of diazepam into the amygdaloid complex elicited the same taming effect as systemic administration of the drug. There was no simultaneous impairment of the motor co-ordination. Levomepromazine and phenobarbital injected into the same area had no effect at all. Diazepam was ineffective when it was administered into the hippocampus. In a follow-up of this study, the same authors showed that the intra-amygdaloid application of diazepam in rats blocked response suppression induced by conflict and they were therefore able to replicate one of the most specific effects of benzodiaze-pines (Nagy *et al.* 1979). The site of action was found to be the central nucleus of the amygdala; injection into the medial and the basolateral portions of the amygdala had no effect (Shibata *et al.* 1982). Since morphine injected into the same area had partial anxiolytic activity in the social interaction test (File and Rodgers 1979), it would certainly be of interest to determine whether the effects of intra-amygdaloid applications of benzodiazepines are naloxone-sensitive.

Another limbic structure found to be involved in benzodiazepine activity is the mamillary body (MB). Muricide behaviour elicited in rats by olfactory bulbectomy is inhibited by microinjections of chlordiazepoxide into the MB, but not into various nuclei of the hypothalamus and midbrain (Hara *et al.* 1975). The same local injections markedly increased lever pressing suppressed by contingent punishment (Kataoka *et al.* 1982).

These results are still too fragmentary to be integrated in a coherent account of the neural basis of the behavioural effects of benzodiazepines. They are, however, sufficient to suggest that further research along these lines, could be extremely promising.

CONCLUSION

This chapter has purposely been limited to behavioural effects of benzodiaze-pines. The role of the limbic system in those effects is substantiated by both indirect and direct evidence. However the exact sites of action still need to be determined. Neurophysiological data have pointed out a selective interference with the hippocampus by benzodiazepines (e.g. Olds and Olds 1969). There has been much speculation about the possible involvement of this structure in the anti-anxiety effect of benzodiazepines. Available behavioural data do not however permit the main site of benzodiazepine action to be ascribed solely to the hippocampus. Other areas likely to be involved include at least the septum, the amygdala, and possibly the frontal cortex. Theoretical interpretations of the behavioural effect produced by lesions in the frontal cortical–limbic system have close similarities to those which have been put forward to account for the behavioural effects of benzodiazepines (e.g. Gray 1981; Numan 1978). But when examined in detail, several pieces of the puzzle do not fit together. There is an obvious difference between an approach in terms of brain lesioning and one by pharmacological treatment with a psychotropic drug. It would be naïve of course to expect that benzodiazepines when injected systematically affect only one or a few brain structures. The comparison between the two approaches is therefore inevitably biased.

This is less of a problem when effects of drugs are investigated in animals bearing specific brain lesions. At the neurochemical level, there is clear evidence that neither the dorsal noradrenergic bundle nor the ascending serotoninergic projections from the raphe dorsalis are critically involved in the effects of benzodiazepines. This does not discard the possibility that one of these pathways could substitute for the other in the effects of benzodiazepines, since it could easily be hypothesized that benzodiazepines release suppressed behaviour indifferently by release of inhibition or by facilitation of activation. The only answer to this alternative would be to study animals bearing lesions of both the DNB and ascending 5-HT pathways. Other possibilities of study include the combination of intracerebral injections of drug with specific brain lesions.

Such a direct approach is not easy to put into practice, due to possible alterations in the behavioural baseline with consecutive ceiling or floor effects. In addition, brain lesions may disturb the brain distribution of drugs and therefore lead to pure pharmacokinetic differences. Much more effort is therefore needed to elucidate precisely the relationships between the limbic system and the anti-anxiety effects of benzodiazepines.

REFERENCES

Baile, C. A. and McLaughlin, C. L. (1979). A review of the behavioral and physiological response to elfazepam, a chemical feed intake stimulant. *J. animal. Sci.* **49**, 1371–95.

Bialik, R. J., Pappas, B. A., and Pusztay, W. (1982). Chlordiazepoxide-induced released responding in extinction and punishment–conflict procedures is not altered by neonetal forebrain norepinephrine depletion. *Pharmacol. Biochem. Behav.* **16**, 279–83.

Bignami, G. and de Acetis, L. (1971). Facilitation and impairment of avoidance responding by phenobarbital sodium, chlordiazepoxide and diazepam. The role of performance base lines. *J. Pharmacol. exp. Ther.* **176**, 725–32.

Black, A. H., Nadel, L., and O'Keefe, J. (1977). Hippocampal function in avoidance learning and punishment. *Psychol. Bull.* **84**, 1107–29.

Blanchard, R. J. and Blanchard, D. C. (1968). Limbic lesions and reflexive fighting. *J. comp. Physiol. Psychol.* **66**, 603–5.

Braestrup, C. and Nielsen, M. (1980). Benzodiazepine receptors. *Arzmein. Forsch. Drug Res.* **30**, 852–7.

Carlton, P. L. (1978). Theories and models in psychopharmacology. In *Psychopharmacology: a generation of progress* (ed. M. A. Lipton, A. Dimascio, and K. F. Killam), pp. 553–61. Raven Press, New York.

Cooper, S. J. (1980). Benzodiazepines as appetite-enhancing compounds. *Appetite* **1**, 7–19.

Crawley, J. and Goodwin, F. K. (1980). Preliminary report of a simple animal behavior model for the anxiolytic effects of benzodiazepines. *Pharmacol. Biochem. Behav.* **13**, 167–70.

Dantzer, R. (1977). Etude des effets du diazepam sur le comportement d'évitement continu chez le porc. *J. Pharmacol., Paris* **8**, 415–26.

—— (1978). Dissociation between suppressive and facilitating effects of aversive stimuli on behavior by benzodiazepines. A review and a reinterpretation. *Prog. Neuropsychopharmacol.* **2**, 559–67.

——, Mormède, P., and Favre, B. (1976). Fear-dependent variations in continuous avoidance behavior of pigs. II. Effects of diazepam on acquisition and performance of Pavlovian fear conditioning and plasma corticosteroid levels. *Psychopharmacology* **49**, 75–78.

Dickinson, A. (1974). Response suppression and facilitation by aversive stimuli following septal lesions in rats: a review and model. *Physiol. Psychol.* **2**, 444–56.

Dimascio, A. (1973). The effects of benzodiazepines on aggression: reduced or increased? *Psychopharmacologia* **30**, 95–102.

Fadda, F., Argiolas, A., Melis, M. R., Tissari, A. H., and Onali, P. L. (1978). Stress-induced increase in 3,4-dihydroxyphenylacetic acid (DOPAC) levels in the cerebral cortex and in n. accumbens: reversal by diazepam. *Life Sci.* **23**, 2219–24.

File, S. E. (1980). The use of social interaction as a method for detecting anxiolytic activity of chlordiazepoxide-like drugs. *J. Neurosci. Meth.* **2**, 219–38.

—— and Rodgers, R. J. (1979). Partial anxiolytic action of morphine sulphate following microinjection into the central nucleus of the amygdala in rats. *Pharmacol. Biochem. Behav.* **11**, 313–18.

Fried, P. A. (1972). Septum and behavior – a review. *Psychol. Bull.* **78**, 292–310.

Gaffan, D. (1982). Loss of recognition memory in rats with lesions of the fornix. *Neuropsychologia* **10**, 327–41.

—— (1974). Recognition impaired and association intact in the memory of monkeys after transection of the fornix. *J. comp. Physiol. Psychol.* **86**, 1100–9.

Geller, I., Kulak, J.T., and Seifter, J. (1962). The effects of chlordiazepoxide and chlorpromazine on a punishment discrimination. *Psychopharmacologia* **3**, 374–85.

Gray, J. A. (1977). Drug effects on fear and frustration: possible limbic site of action of minor tranquilizers. In *Handbook of psychopharmacology* (ed. L. Iversen, S. Iversen, and S. Snyder), Vol. 8, pp. 433–529. Plenum, New York.

—— (1981). *The neuropsychology of anxiety: an enquiry into the functions of the septo–hippocampal system.* Oxford University Press, Oxford.

——, McNaughton, N., James, D. T. D., and Kelly, P. H. (1975). Effect of minor tranquilizers on hippocampal theta rhythm mimicked by depletion of forebrain noradrenaline. *Nature* **258**, 424–5.

Hara, C., Watanabe, S., and Ueki, S. (1975). Effects of drugs injected into the hypothalamus on muricide and EEG in the olfactory bulbectomized rats. *Jap. J. Pharmacol. (Suppl.)* **25**, 452.

Haefely, W., Pieri, L., Polc, R., and Schaffner, R. (1981). General pharmacology and neuropharmacology of benzodiazepine derivatives. In *Handbook of experimental pharmacology,* Vol. 55 (ed. F. Hoffmeister and G. Stille), pp. 13–262. Springer-Verlag, Berlin.

Herman, J. P., Guilloneau, D., Dantzer, R., Scatton, B., Semerdjian-Rouquier, L., and Le Moal, M. (1982). Differential effects of inescapable footshock and of stimuli previously paired with inescapable footshocks on dopamine turnover in cortical and limbic areas of the rat. *Life Sci.* **30**, 2207–14.

Huppert, F. A. and Iversen, S. D. (1975). Response suppression in rats: a comparison of response contingent and non contingent punishment and the effect of the minor tranquilizer, chlordiazepoxide. *Psychopharmacologia* **44**, 67–75.

Isaacson, R. L. (1974). *The limbic system.* Plenum Press, New York.

Iversen, S. D. (1976). Do hippocampal lesions produce amnesia in animals? In *International review of neurobiology,* Vol. 19 (ed. C. C. Pleiffer and J. R. Smythies), pp. 1–49. Academic Press, New York.

Kataoka, Y., Shibata, K., Gomita, Y., and Ueki, S. (1982). The mamillary body is a potential site of anti anxiety action of benzodiazepines. *Brain Res.* **241**, 374–7.

Kimble, D. P. (1968). Hippocampus and internal inhibition. *Psychol. Bull.* **70**, 285–95.

Klepner, C. A., Lippa, A. S., Benson, D. I., Sano, M. C., and Beer, B. (1979). Resolution of two biochemically and pharmacologically distinct benzodiazepine receptors. *Pharmacol. Biochem. Behav.* **11**, 457–62.

Koob, G., Strecker, R., Roberts, D. C. S., and Bloom, F. (1980). Failure to alter anxiety or the anxiolytic properties of chlordiazepoxide and ethanol by destruction of the dorsal noradrenergic system. *Soc. Neurosci. Abstr.* 1980.

Lau, P. and Miczek, K. A. (1977). Differential effects of septal lesions on attack and defensive–submissive reactions during intra-species aggression in rats. *Physiol. Behav.* **18**, 479–85.

Lavielle, S., Tassin, J. P., Thierry, H. M., Blanc, G., Hervé, D., Barthelemy, C., and Glowinski, J. (1978). Blockade by benzodiazepines of the selective high increase of dopamine turnover induced by stress in mesocortical dopaminergic neurons in rats. *Brain Res.* **168**, 585–94.

Mason, S. T. and Iversen, S. D. (1979). Theories of the dorsal bundle extinction effect. *Brain Res. Rev.* **1**, 107–37.

McMillan, D. E. (1975). Determinants of drug effects on punished responding. *Fed. Proc.* **34**, 1870–9.

Miczek, K. A. (1973). Effects of scopolamine, amphetamine and benzodiazepines on conditioned suppression. *Pharmacol. Biochem. Behav.* **1**, 401–11.

—— and Krsiak, M. (1979). Drug effects on agonistic behavior. In *Advances in behavioral pharmacology,* Vol. 2 (ed. T. Thompson and P. B. Dews), pp. 87–162. Academic Press, new York.

Möhler, H. and Okada, T. (1977). Benzodiazepine receptors: demonstration in the central nervous system. *Science* **198**, 849–51.

Morris, M. D. and Gebhart, G. F. (1981). Antianxiety agents and emotional behavior: an information processing analysis. *Prog. Neuropsychopharmacol.* **5**, 219–40.

——, Berger, A. B., and Gebhart, C. (1980). Effect of chlordiazepoxide on conditioned and unconditioned fear in rats. *Prog. Neuropsychopharmacol.* **4**, 153–60.

——, Trenamel, F., and Gebhart, G. (1979). Forebrain noradrenaline depletion blocks the release by chlordiazepoxide of behavioural extinction in rats. *Neurosci. Lett.* **12**, 343–8.

Nagy, J. and Decsi, L. (1973). Location of the site of tranquillizing action of diazepam by intralimbic application. *Neuropharmacology* **12**, 757–68.

——, Zambo, K., and Decsi, L. (1979). Anti-anxiety action of diazepam after intra-amygdaloid application in rats. *Neuropharmacology* **18**, 573–6.

Numan, R. (1978). Cortical-limbic mechanisms and response control: a theoretical review. *Physiol. Psychol.* **6**, 445–70.

O'Keefe, J. and Nadel, L. (1979). The hippocampus as a cognitive map. *Behav. Brain Sci.* **2**, 487–533.

Olds, M. E. and Olds, J. (1969). Effects of anxiety-relieving drugs on unit discharges in hippocampus, reticular mid-brain and pre-optic area in the freely moving rat. *Int. J. Neuropharmacol.* **7**, 231–9.

Olton, D. S., Becker, J. T., and Handelmann, G. E. (1979). Hippocampus, space and memory. *Behav. Brain Sci.* **2**, 313–65.

Poschel, B. H. P. (1971). A simple and specific screen for benzodiazepine-like drugs. *Psychopharmacologia* **19**, 193–8.

Robertson, H. A. (1979). Benzodiazepine receptors in "emotional" and non "emotional" mice: comparison of four strains. *Eur. J. Pharmacol.* **56**, 163–6.

——, Martin, I. L., and Candy, J. M. (1978). Differences in benzodiazepine receptor binding in Maudsley reactive and Maudsley non-reactive rats. *Eur. J. Pharmacol.* **50**, 455–7.

Saghal, A. and Iversen, S. D. (1980). Recognition memory, chlordiazepoxide and rhesus monkey: some problems and results. *Behav. Brain Res.* **1**, 227–43.

Sepinwall, J. and Cook, L. (1980). Mechanism of action of the benzodiazepines: behavioural aspect. *Fed. Proc.* **39**, 3021–31.

Shibata, K., Kataoka, Y., Comita, Y., and Ueki, S. (1982). Localization of the site of the anticonflict action of benzodiazepines in the amygdaloid nucleus of rats. *Brain Res.* **234**, 442–6.

Simon, H. (1981). Neurones dopaminergiques A10 et système frontal. *J. Physiol., Paris* **77**, 81–95.

Soubrié, P., Kulkarni, S., Simon, P., and Boissier, J. R. (1976). Effect des anxiolytiques sur la prise de boisson en situation nouvelle et familière. *Psychopharmacologia* **45**, 203–10.

Speth, R. C., Johnson, R. W., Regan, J., Reisine, T., Kobayashi, R. M., Bresolin, N., Roeske, W. R., and Yamamura, H. I. (1980). The benzodiazepine receptor of mammalian brain. *Fed. Proc.* **39**, 3032–8.

Squires, R. F. and Braestrup, C. (1977). Benzodiazepine receptors in rat brain. *Nature* **266**, 732–4.

Stein, L., Wise, C. D., and Berger, B. D. (1973). Anti-anxiety action of benzodiazepines: decrease in activity of serotonin neurons in the punishment system. In *The benzodiazepines* (ed. S. Garattini), pp. 299–326. Raven Press, New York.

Tassin, J. P., Stinus, L., Simon, H., Blanc, G., Thierry, A. M., Le Moal, M., Cardo, B., and Glowinski, J. (1978). Relationship between the locomotor hyperactivity induced by A10 lesions and the destruction of the fronto-cortical dopaminergic innervation in the rat. *Brain Res.* **141**, 267–81.

Thiébot, M. H., Jobert, A., and Soubrié, P. (1980a). Conditioned suppression of behaviour: its reversal by intra-raphe microinjection of chlordiazepoxide and GABA. *Neurosci. Lett.* **16**, 213–17.

——, ——, —— (1980b). Chlordiazepoxide and GABA injected into raphe dorsalis release the conditioned behavioural suppression induced in rats by a conflict procedure without nociceptive component. *Neuropharmacology* **19**, 633–42.

Tye, N. C., Everitt, B. J., and Iversen, S. D. (1977). 5-hydroxytryptamine and punishment. *Nature* **268**, 741–3.

Volicer, L. and Biagioni, T. M. (1982). Presence of two benzodiazepine binding sites in the rat hippocampus. *J. Neurochem.* **38**, 591–3.

Winocur, G. and Bindra, D. (1976). Effects of additional cues on passive avoidance learning and extinction in rats with hippocampal lesions. *Physiol. Behav.* **17**, 915–20.

Young, W. S. and Kuhar, M. J. (1979). Autoradiographic localization of benzodiazepine receptors in the brain of humans and animals. *Nature* **280**, 395–7.

11

Serotonergic neurones and anxiety-related behaviour in rats

MARIE-HÉLÈNE THIÉBOT, MICHEL HAMON, AND
PHILIPPE SOUBRIÉ

The serotonin (5-hydroxytryptamine, 5-HT)-containing pathways originating from the raphe nuclei (Bobiller *et al.* 1976; Azmitia and Segal 1978; Dray *et al.* 1978; Jacobs *et al.* 1978; Parent *et al.* 1981) have been tentatively implicated in a variety of functions including pain sensitivity, sleep,aggressiveness, and locomotor activity. In particular, the ascending 5-HT systems seem to play a crucial role in controlling experimentally-induced anxiety in animals (e.g. the behavioural suppression caused by punishment). Thus, like benzodiazepines, 5-HT antagonists (Geller and Blum 1970; Hartmann and Geller 1971; Schoenfeld 1976) or specific lesions of serotonergic neurones (Tye *et al.* 1977, 1979) apparently release behaviour from inhibitory influences. In contrast, increasing 5-HT transmission exacerbates behavioural suppression (Stein *et al.* 1973; Schoenfeld 1976; Graeff and Silveira-Filho 1978). Moreover, there is some evidence to suggest that 5-HT neurones are involved in the attenuation by benzodiazepines of experimentally-induced anxiety. The attenuation of 5-HT neuronal activity and the decreased 5-HT turnover caused by these drugs (Wise *et al.* 1972; Jenner *et al.* 1975; Saner and Pletscher 1979; Hunkeler *et al.* 1981) have been tentatively related to their antipunishment activity (Wise *et al.* 1972). Furthermore, the destruction of ascending 5-HT pathways prevents the release of behaviour elicited by benzodiazepines (Tye *et al.* 1977, 1979).

However, the target 5-HT processes involved in the control of punished behaviour and/or the antipunishment activity of benzodiazepines are not fully elucidated. For instance, the relative contribution of the pathways originating from the dorsal versus the median raphe, where these nuclei innervate (although to a different magnitude) the same brain areas, remains to be specified. Likewise, the respective role of the various brain structures receiving such 5-HT innervation is not fully understood. The following experiments were performed in order to provide information regarding these two main points.

INVOLVEMENT OF SEROTONERGIC NEURONES OF THE NUCLEUS RAPHE DORSALIS IN THE ANTIPUNISHMENT ACTIVITY OF BENZODIAZEPINES

In rats with chronically implanted cannulae, the inhibition of responding for food initiated by a signal previously associated with punishment could be reversed by application of chlordiazepoxide (5×10^{-9} M to 5×10^{-5} M) to the dorsal raphe. The magnitude of this effect was similar to that generally obtained after i.p. injection of chlordiazepoxide (8 mg kg^{-1}) or diazepam (2 mg kg^{-1}) (Thiébot *et al.* 1980; Fig. 11.1). Thus, one classical experimental correlate of the anxiolytic activity of benzodiazepines, an increase of responding for food that is depressed by punishment (or signal of punishment), can be obtained after intra-raphe dorsalis application of chlordiazepoxide. This effect seems to be dependent on 5-HT processes since chlordiazepoxide failed to release responding for food during the signal of punishment, in rats whose 5-

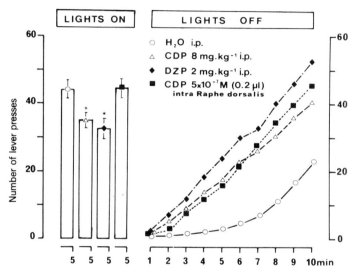

Fig. 11.1. Effects of parenteral administration of chlordiazepoxide (CDP) diazepam (DZP), and of microinjection of CDP into the dorsal raphe, on lever-pressing for food during the delivery of a signal (lights-off) previously paired with punishment. The rats were trained to press a lever for food reward according to a continuous reinforced schedule (CRF = 1 pellet for each lever press), in an illuminated Skinner box. They were then subjected to CRF sessions with alternating signalled periods of punishment (lights-off, one electric footshock, 1.5 mA, 45 ms), following every second press and nonpunished (lights-on) responding for food. At the end of these learning sessions, rats exhibited a strong inhibition of pressing for food in presence of the lights-off signal 4 ± 1 presses/10 min, as compared to 70 ± 5 presses during the 10-min nonpunished periods. On the test day, rats were given either CDP infused (0.2 μl in 3 min) into the raphe dorsalis via a chronically implanted cannula (stereotaxic co-ordinates according to König and Klippel 1963: *A* = 0.16; *L* = 0; *H* = − 0.8) or CDP or DZP i.p. They were subjected to the experimental session which started with a 5-min (lights-on) nonpunished period followed by a continuous 10-min lights-off period during which *responding was no longer punished* (conditioned conflictual response).

The values are the mean ± SEM of the presses effected by separate groups of rats (*n* = 7 to 12) during the initial 5-min lights-on period and the mean of the presses per minute during the subsequent 10-min lights-off period. *$p < 0.05$ as compared to controls (Student's *t*-test). During the lights-off period the three groups of treated rats did not differ from each other ($F(2, 367) = 2.047$, NS) but differ significantly from the control group ($F(1, 367) = 42.315$; $p < 0.01$) as calculated by two-way (time and subjects) analysis of variance.

HT neurones of the dorsal raphe were substantially destroyed by prior local administration of 5,7-dihydroxytryptamine (5,7-DHT) (Thiébot *et al.* 1982). Moreover, punishment-induced suppression of ongoing behaviour seems also to be dependent on the integrity of 5-HT projections of the raphe dorsalis. Indeed, rats with a 5,7-DHT lesion of the raphe dorsalis exhibited no

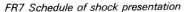

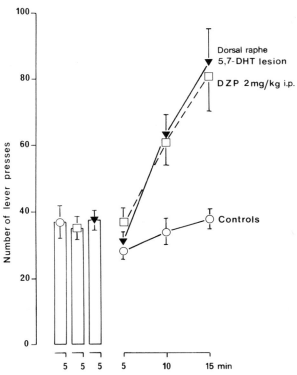

Fig. 11.2. Effects of intra-raphe application of 5,7-DHT on responding for food during a fixed ratio 7 (FR7) schedule of shock presentation. Rats were trained to press a lever for food reward in a Skinner box according to a CRF schedule. Afterwards, 5,7-DHT (1 μg free base in 0.4 μl) was infused into the nucleus raphe dorsalis ($A = 0.16$; $L = 0$; $H = -0.8$) in one group of rats ($n = 6$) pretreated with desipramine (25 mg kg^{-1} i.p.). The rats of a second group ($n = 7$) were sham-operated (controls). 15 days post-surgery, the animals were subjected to one experimental session divided into two components: a 5-min CRF period followed by a 15-min period of FR7 schedule of shock presentation (each seventh lever press for food was paired with the delivery of one electric footshock, 0.5 mA, 45 ms). One group of non-operated rats ($n = 7$) received diazepam (DZP) i.p. 30 min before the experimental session.

The values are the mean number of lever presses ± SEM effected during the initial 5-min CRF period and the mean cumulated number of presses effected per 5 min during the subsequent 15 min of FR7 schedule of shock presentation.

During the 15-min FR7 period, the 5,7-DHT-lesioned and the DZP-treated rats did not differ from each other ($F(1,55) < 1$, NS) but they differ significantly from the control group ($F(1,55) = 25.772$, $p < 0.001$) as calculated by a two-way (time and subjects) analysis of variance.

behavioural suppression in a paradigm in which each seventh level press for food-reward was paired with the delivery of an electric footshock (FR7 schedule of shock presentation) (Fig. 11.2). The 5-HT neurones of these rats were substantially lesioned as revealed by a significant loss of the tryptophan hydroxylase activity ranging from 75 to 80 per cent in the dorsal raphe, 55 to 60 per cent in the median raphe, and 70 to 80 per cent in several forebrain structures receiving 5-HT projections (Thiébot *et al.* 1982).

Consonant with these results, we found intra-raphe dorsalis applications of 5-HT were able to mimic (10^{-7} M) the effects of the benzodiazepine or to act synergistically (10^{-8} M) with chlordiazepoxide (5×10^{-9} M) (Thiébot *et al.* 1982). Microiontophoretic injections of 5-HT into the dorsal raphe have been found to depress the firing rate of neurones of this nucleus, perhaps through an autoreceptor-mediated feedback (Bramwell and Gönye 1976; Haigler and Aghajanian 1977). Taken together, these findings extend previous pharmacological studies (Wise *et al.* 1972; Tye *et al.* 1977, 1979) concerning the role of central serotonergic processes in the antipunishment activity of benzodiazepines and also give more direct support to the hypothesis of a decreased 5-HT neuronal activity as the relevant mechanism for the attenuation of behavioural suppression.

There is considerable evidence to suggest that benzodiazepines exert a facilitatory role on GABAergic transmission (Costa *et al.* 1978; Haefely 1978). Such a facilitation – perhaps through an interaction with benzodiazepine-binding sites within the raphe dorsalis (Thiébot *et al.* 1982) – could be one of the mechanisms by which intra-raphe dorsalis chlordiazepoxide may depress the activity of 5-HT neurones. Indeed, cells accumulating GABA have been described in the dorsal raphe (Belin *et al.* 1979) and microiontophoretic applications of GABA exerted an inhibitory control (amplified by benzodiazepines) upon 5-HT neurones in this nucleus (Gallager 1978). In addition, *in vitro* experiments (Kerwin and Pycock 1979) revealed that this amino acid may facilitate [³H]-5-HT release from rat midbrain slices. In keeping with these results, we found that intra-raphe applications of GABA enhanced responding for food during signal of punishment (Thiébot *et al.* 1980).

Anatomical and electrophysiological studies reveal that 5-HT(?) neurones originating from the dorsal raphe project to various forebrain structures and, in particular, to areas belonging to the limbic system such as the amygdala, the septum, and the hippocampus. However, these same structures also receive a major innervation from the median raphe (Azmitia and Segal 1978; Parent *et al.* 1981; Steinbusch 1981).

SEROTONERGIC PROJECTIONS WITHIN VARIOUS LIMBIC STRUCTURES AND THE CONTROL OF PUNISHED BEHAVIOUR

The 5-HT innervation of either the amygdala, the nucleus accumbens, or the hippocampus was bilaterally destroyed in separate groups of rats with infusions of 5,7-DHT. Two weeks post-surgery, the rats were subjected to punishment-induced inhibition (FR7 schedule of shock presentation) and were killed for tryptophan hydroxylase activity assay. Although a significant local

TABLE 11.1. *Effects of bilateral applications of 5,7-DHT into various brain regions on local tryptophan hydroxylase activity and on lever-pressing for food during a fixed ratio (FR7) schedule of shock presentation*

	Amygdala			N. accumbens			Hippocampus		
	Sham	5,7-DHT	Per cent	Sham	5,7-DHT	Per cent	Sham	5,7-DHT	Per cent
Tryptophan hydroxylase activity (mean ±SEM)*	1.151 ±0.061 $p < 0.001$	0.627 ±0.037	54	1.835 ±0.070 $p < 0.001$	0.762 ±0.110	42	0.430 ±0.022 $p < 0.001$	0.007 ±0.003	<5
FR7 schedule of shock presentation (mean ±SEM)†	47.57 ±10.77 NS	40.50 ±8.51	85	48.00 ±6.05 NS	38.50 ±5.41	80	40.60 ±7.94 NS	35.59 ±4.52	88

5,7-DHT (2 µg free base in 0.5 µl) was bilaterally infused under stereotaxic procedure either into the amygdala ($A = 4.5$; $L = 3.8$; $H = -3.4$), into the nucleus accumbens ($A = 8.9$; $L = 1.0$; $H = -1.4$), or at the level of ascending 5-HT fibres to the hippocampus ($A = 1.7$; $L = 0.4$; $H = -2.6$) of separate groups of adult male rats ($n = 8$ to 10) pretreated with desipramine (25 mg kg^{-1} ip.). Behavioural testing was performed 15 days later as described in the legend of Fig. 11.2 and the rats were sacrified 24 hours later for tryptophan hydroxylase assay. Tryptophan hydroxylase activity was measured according to Hamon *et al.* (1978) in the 35 000 × g supernatant of tissues homogenates with 0.15 mM tryptophan, 0.16 mM 6-MPH4, and 0.01 per cent sodium dodecyl sulphate.
  *n mol of 5-hydroxytryptophan synthesized by mg of protein in 15 min.
  †Number of lever presses effected in 15 min.

destruction of 5-HT terminals was observed in each case, no significant change in behavioural suppression could be detected in each of these three groups of lesioned animals (Table 11.1).

These results do not imply that these structures are neither involved in the control of punishment-induced suppression, nor in the anti-anxiety activity of benzodiazepines, especially as Nagy *et al.* (1979) reported that injections of diazepam into the amygdaloid complex were able to induce anticonflict effects. However, our findings suggest that the 5-HT innervation of these structures plays a minimal role in such behavioural responses.

Since the dorsal raphe sends projections to nonlimbic areas such as extra-pyramidal structures (caudate nucleus, globus pallidus, and substantia nigra), we decided to investigate the possible influence of the nigral 5-HT innervation in the control of punished behaviour.

INVOLVEMENT OF THE SEROTONERGIC INNERVATION OF THE SUBSTANTIA NIGRA IN PUNISHMENT-INDUCED BEHAVIOURAL SUPPRESSION

Following a bilateral infusion of 5,7-DHT into the substantia nigra, a significant correlation ($r = 0.62$; $p < 0.01$) has been found between the decrease of nigral tryptophan hydroxylase activity and the attenuation of punishment-induced behavioural suppression. Indeed, when nigral 5-HT innervation was significantly destroyed (loss of nigral tryptophan hydroxylase activity greater than 55 per cent), a significant release of punished behaviour was observed in rats subjected to a FR7 schedule of shock presentation (Fig. 11.3(A)) or to a schedule of shock-induced suppression of drinking (Fig. 11.3(B)). Interestingly, in the FR7 schedule of shock presentation, lever pressing for food during an initial 5-min nonpunished period did not differ between control and lesioned rats (42 ± 5 versus 39 ± 12). In rats with a severe damage of the nigral 5-HT innervation, the attenuation of punished behaviour was similar to that displayed by nonlesioned rats treated with diazepam (2 mg kg^{-1}): 49 ± 7 lever presses for food in the FR7 schedule of shock presentation or 25 ± 5 shocks received during the drinking test. Conversely, when nigral 5-HT innervation was not markedly damaged (loss of tryptophan hydroxylase activity less than 40 per cent), punished behaviour in the two experimental situations did not statistically differ from controls (Fig. 11.3(A), (B)).

Serotonergic neurones arising from raphe nuclei and innervating limbic structures have been found to pass within the vicinity of the substantia nigra (Parent *et al.* 1981; Steinbusch 1981). As a matter of fact, a decrease of tryptophan hydroxylase activity (-37 ± 5 per cent) within the hippocampus was obtained following intra-nigral 5,7-DHT. This change correlated with the reduction of nigral tryptophan hydroxylase activity ($r = 0.47$; $p < 0.02$). However, in accordance with our previous results (*vide supra*) this damage seemed not to play an important role in the animal's response to aversive events. Indeed there was no direct significant correlation between hippocampal tryptophan hydroxylase activity and punished behaviour since this correlation (0.30) fell to 0.10 (non-significant) when the decrease of nigral tryptophan

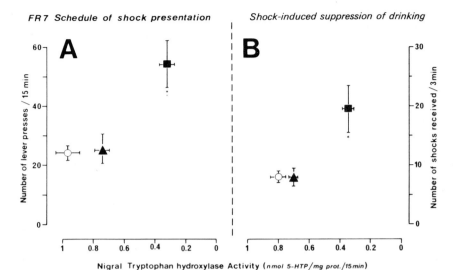

Fig. 11.3. Relationship between the severity of the lesion of nigral 5-HT innervation and punished behaviour. Punished behaviour was assessed in: (○) Control rats; Rats with a bilateral 5,7-DHT (2 μg free base in 0.5 μl) infusion into the substantia nigra ($A = 2.2$; $L = 1.8$; $H = -2.8$): (▲) loss of nigral tryptophan hydroxylase activity less than 40 per cent when compared to control values; (■) loss of nigral tryptophan hydroxylase activity more than 55 per cent when compared to control values. (A) Suppression of lever pressing for food in a FR7 schedule of shock presentation. Rats were trained, lesioned, and tested as described in the legend of Fig. 11.2. The results are the mean number of lever presses ± SEM (vertical bars) effected during the 15 min of FR7 schedule of shock presentation. (○) $n = 12$; (▲) $n = 5$; (■) $n = 10$. (B) Shock-induced suppression of drinking. After daily 15-min sessions of habituation to the test apparatus, rats were subjected to a test slightly modified from the punishment procedure used by Vogel *et al.* (1971): during a 3-min test session, the animals, deprived of water, were given one electric foot shock (0.40 mA; 45 ms) every time a 3-s period of licking water was completed. The results are the mean number of shocks ± SEM (vertical bars) received during 3 min. (○) $n = 15$; (▲) $n = 8$; (■) $n = 8$. Nigral tryptophan hydroxylase activity, mean ± SEM (horizontal bars), was measured as described in the legend of Table 11.1. *$p < 0.05$; **$p < 0.01$ as compared to respective control values (Student's *t*-test).

hydroxylase activity was kept constant by calculating the partial coefficient of correlation.

Taken together, all these results confirm the involvement of 5-HT neurones in the control of punishment-induced behavioural suppression (Stein *et al.* 1975, 1977; Tye *et al.* 1977, 1979; Thiébot *et al.* 1982) and indicate a significant contribution of 5-HT innervation to the substantia nigra. These findings suggest additional functional roles for the substantia nigra. Indeed, this structure has never been claimed to be crucially involved in the control of emotional responses, although a nigro–raphe dorsalis pathway, which can be the anatomical substratum for such an activity, has been described (Beckstead *et al.* 1979).

The role of the nigral 5-HT innervation in controlling punishment-induced suppression raises questions as to whether an attenuation of punished behaviour following the damage of 5-HT neurones, especially those innervating the substantia nigra, is really associated with lessened anxiety. As an alternative hypothesis, one can suggest that a 5-HT lesion prevents the animals from expressing their anxiety through an inhibition of ongoing behaviour. Indeed, the destruction of 5-HT neurones of the raphe nuclei has been reported to facilitate motor behaviour (Kostowski *et al.* 1972) and to reduce the ability of animals to adopt passive behaviour such as neuroleptic-induced catalepsy (Costall *et al.* 1975). Interestingly, nigral 5-HT innervation seems to play a crucial role in these behavioural changes: the infusion of 5,7-DHT into the substantia nigra caused hyperactivity, enhanced amphetamine-induced excitation, and reduced the potency of neuroleptics to elicit catalepsy (Carter and Pycock 1978, 1979).

A large amount of data support the contention that nigral 5-HT innervation plays an inhibitory role upon nigro-striatal dopaminergic neurones (Carter and Pycock 1978, 1979; Giambalvo and Snodgrass 1978). One can therefore assume that such a 5-HT influence on dopaminergic neurones is essential for response suppression, such as occurs during punishment procedures, to be achieved.

REFERENCES

Azmitia, E. and Segal, M. (1978). Autoradiographic analysis of differential ascending projections of the dorsal and median raphé nuclei in the rat. *J. comp. Neurol.* **179**, 641–51.

Beckstead, E. M., Domesick, V. B., and Nauta, W. J. H. (1979). Efferent connections of substantia nigra and ventral tegmental area in the rat. *Brain Res.* **175**, 191–9.

Belin, M. F., Aguera, M., Tappaz, M., McRae-Degueurce, A., Bobiller, P., and Pujol, J. F. (1979). GABA-accumulating neurons in the nucleus raphé dorsalis and periaqueductal gray in the rat: a biochemical and autoradiographic study. *Brain Res.* **170**, 279–97.

Bobiller, P., Seguin, S., Petitjean, P., Salvert, M., Touret, M., and Jouvet, M. (1976). The raphé nuclei of the cat brain stem, a topographical atlas of their efferent projections as revealed by autoradiography. *Brain Res.* **113**, 449–86.

Bramwell, G. J. and Gönye, T. (1976). Response of midbrain neurons to micro-iontophoretically applied 5-hydroxytryptamine: comparison with the response to intravenously administered lysergic acid diethylamide. *Neuropharmacology* **15**, 457–64.

Carter, C. J. and Pycock, C. J. (1978). A study of the sites of interaction between dopamine and 5-hydroxytryptamine for the production of fluphenazine-induced catalepsy. *Naunyn–Schmiedeberg's Arch. Pharmacol.* **304**, 135–9.

—— and —— (1979). The effects of 5,7-dihydroxytryptamine lesions of extrapyramidal and mesolimbic sites on spontaneous motor behaviour and amphetamine-induced stereotypy. *Naunyn–Schmiedeberg's Arch. Pharmacol.* **308**, 51–4.

Costa, E., Guidotti, A., and Toffano, G. (1978). Molecular mechanisms mediating the action of diazepam on GABA receptors. *Br. J. Psychiat.* **133**, 239–48.

Costall, B., Fortune, D., Naylor, R. J., Marsden, C. D., and Pycock, C. J. (1975). Serotonergic involvement with neuroleptic catalepsy. *Neuropharmacology* **14**, 859–68.

Dray, A., Davies, J., Oakley, N. R., Tongroach, P., and Vellucci, S. (1978). The dorsal and medial projections to the substantia nigra in the rat: electrophysiological, biochemical and behavioural observations. *Brain Res.* **151**, 431–42.

Gallager, D. W. (1978). Benzodiazepines potentiation of a GABA inhibitory response in the dorsal raphé nucleus. *Eur. J. Pharmacol.* **49**, 133–43.

Geller, I. and Blum, K. (1970). The effects of 5-HTP on parachlorophenylalanine (pCPA) attenuation of "conflict" behavior in the rat. *Eur. J. Pharmacol.* **9**, 319–24.

Giambalvo, C. T. and Snodgrass, S. R. (1978). Biochemical and behavioral effects of serotonin neurotoxins on the nigrostriatal dopamine system: comparison of injection sites. *Brain Res.* **152**, 555–6.

Graeff, F. G. and Silveria-Filho, N. G. (1978). Behavioral inhibition induced by electrical stimulation of the median raphé nucleus of the rat. *Physiol. Behav.* **21**, 477–84.

Haefely, W. E. (1978). Behavioral and neurochemical aspects of drugs used in anxiety and related states. In *Psychopharmacology: a generation of progress* (ed. A. Lipton, A. Di Mascio, and K. F. Killam), pp. 1359–74. Raven Press, New York.

Haigler, H. J. and Aghajanian, G. K. (1977). Serotonin receptors in the brain. *Fed. Proc.* **36**, 2159–64.

Hamon, M., Bourgoin, S., Hery, F., and Simonnet, G. (1978). Phospholipid-induced activation of tryptophan hydroxylase from the rat brainstem. *Biochem. Pharmacol.* **27**, 915–22.

Hartmann, R. J. and Geller, I. (1971). p-Chlorophenylalanine effects on a conditioned emotional response in rats. *Life Sci.* **10**, 927–33.

Hunkeler, W., Möhler, H., Pieri, L., Polc, P., Bonetti, E. P., Cumin, R., Schaffner, R., and Haefely, W. (1981). Selective antagonists of benzodiazepines. *Nature* **290**, 514–16.

Jacobs, B. L., Foote, S. L., and Bloom, F. (1978). Differential projections of neurons within the dorsal raphé nucleus of the rat: a horseradish peroxidase (HRP) study. *Brain Res.* **147**, 149–53.

Jenner, P., Chadwick, D., Reynolds, E. H., and Marsden, C. D. (1975). Altered 5-HT metabolism with clonazepam, diazepam and diphenylhydantoin. *J. Pharm. Pharmacol.* **27**, 707–10.

Kerwin, R. W. and Pycock, C. J. (1979). The effect of some putative neurotransmitters on the release of 5-hydroxytryptamine and γ-aminobutyric acid from slices of the rat midbrain raphé area. *Neuroscience* **4**, 1359–65.

König, J. F. R. and Klippel, R. A. (1963). *The rat brain.* Williams and Wilkins, Baltimore, Maryland.

Kostowski, W., Gumulka, W., and Czlonkowski, A. (1972). Reduced cataleptogenic effects of some neuroleptics in rats with lesioned midbrain raphé and pretreated with parachlorophenylalanine. *Brain Res.* **48**, 443–6.

Nagy, J., Zambo, K., and Decsi, L. (1979). Anti-anxiety action of diazepam after intra-amygdaloid application in the rat. *Neuropharmacology* **18**, 573–6.

Parent, A., Descarries, L., and Beaudet, A. (1981). Organization of ascending serotonin systems in the adult rat brain. A radioautographic study after intraventricular administration of ^{3}H-5-hydroxytryptamine. *Neuroscience* **6**, 115–38.

Saner, A. and Pletscher, A. (1979). Effects of diazepam on cerebral 5-hydroxytryptamine synthesis. *Eur. J. Pharmacol.* **55**, 315–18.

Schoenfeld, R. I. (1976). Lysergic acid diethylamide and mescaline-induced attenuation of the effect of punishment in the rat. *Science* **192**, 801–3.

Stein, L., Belluzzi, J. D., and Wise, C. D. (1977). Benzodiazepines: behavioral and neurochemical mechanisms. *Amer. J. Psychiat.* **177**, 665–9.

——, Wise, C. D., and Berger, B. D. (1973). Anti-anxiety action of benzodiazepines: decrease in activity of serotonin neurons in the punishment system. In *The benzodiazepines* (ed. E. Costa and P. Greengard), pp. 299–326. Raven Press, New York.

——, ——, and Belluzzi, J. D. (1975). Effects of benzodiazepines on cerebral serotoninergic mechanisms. In *Mechanisms of action of benzodiazepines* (ed. E. Costa and P. Greengard), pp. 29–44. Raven Press, New York.

Steinbusch, H. W. M. (1981). Distribution of serotonin-immunoreactivity in the central nervous system of the rat-cell bodies and terminals. *Neuroscience* 6, 557–618.

Thiébot, M. H., Hamon, M., and Soubrié, P. (1982). Attenutation of induced anxiety in rats by chlordiazepoxide: role of raphé dorsalis benzodiazepine binding sites and serotonergic neurons. *Neuroscience* 7, 2287–94.

——, Jobert, A., and Soubrié, P. (1980). Chlordiazepoxide and GABA injected into raphé dorsalis release the conditioned behavioural suppression induced in rats by a conflict procedure without nociceptive component. *Neuropharmacology* 19, 633–41.

Tye, N. C., Everitt, B. J., and Iversen, S. D. (1977). 5-hydroxytryptamine and punishment. *Nature, Lond.* 268, 741–43.

——, Iversen, S. D., and Green, A. R. (1979). The effects of benzodiazepines and serotoninergic manipulations on punished responding. *Neuropharmacology* 18, 689–95.

Vogel, J. R., Beer, B., and Clody, D. E. (1971). A simple and reliable conflict procedure for testing antianxiety agents. *Psychopharmacologia, Berlin* 21, 1–7.

Wise, C. D., Berger, B. D., and Stein, L. (1972). Benzodiazepines: anxiety-reducing activity by reduction of serotonin turnover in the brain. *Science* 177, 180–3.

12

Neuroleptics and the limbic system

BERNARD SCATTON AND BRANIMIR ZIVKOVIC

INTRODUCTION

Neuroleptics are clinically effective against the core symptoms of schizophrenia and other major psychoses. It is now widely accepted that the antipsychotic action of neuroleptics is related to the effects of these drugs on specific brain neurones. Since the pioneering work of Papez (1937), the limbic system has been recognized as the neuroanatomical substrate of emotion. Drugs that alter emotionality or which are used in clinical states where an emotional abnormality exists are therefore capable of altering limbic function, and consequently the limbic system has been proposed as the anatomical substrate of the antipsychotic action of neuroleptics (Matthysse 1973; Stevens 1973; Snyder *et al.* 1974).

The neurochemical anatomy of the limbic system has considerably progressed during the past two decades. Most of the areas constituting the limbic system are now known to be innervated by a variety of neurotransmitter systems including monoaminergie, amino-acidergic, and peptidergic neurones. By acting on these different neuronal pathways, neuroleptic drugs are therefore capable of altering the limbic system function.

In this chapter we will briefly consider the influence of neuroleptics on limbic neurones which utilize dopamine (DA), noradrenaline (NA), GABA, and peptides as neurotransmitters. The relationship between the biochemical alterations induced by neuroleptics and their therapeutic benefits and/or side-effects will also be considered.

LIMBIC DOPAMINERGIC NEURONES AND NEUROLEPTICS

Dopaminergic innervation of the limbic system

Histofluorescence and biochemical studies have provided evidence for the existence of a dopaminergic innervation of most of the limbic areas in a variety of animal species. The limbic dopaminergic neurones originate in cell bodies mostly located in the ventral tegmental area (area of Tsaï) but also partly in the substantia nigra and project to the olfactory tubercle, amygdala, septum, nucleus accumbens, limbic cortex (frontal, cingulate, and entorhinal divisions) (for review see Lindvall and Björklund 1978; Chapter 1, this volume), and hippocampus (Scatton *et al.* 1980). Other structures classically connected to the limbic system, e.g. the habenula and hypothalamus, also receive a dopaminergic innervation (Lindvall and Björklund 1978; Simon *et al.* 1979). The hypothalamus contains dopaminergic neurones of both intrinsic (tubero-infundibular system) and extrinsic (nigro-hypothalamic system) origins. Neuro-chemical studies performed on post-mortem tissue support the existence of a dopaminergic innervation of the limbic system in the human brain as well. Thus, appreciable amounts of DA and its major metabolites, homovanillic acid (HVA) and dihydroxyphenylacetic acid (DOPAC), have been detected in a variety of human limbic areas including the nucleus accumbens, septum, amygdala, hippocampus, and cerebral cortex (Farley *et al.* 1977; Scatton *et al.* 1982*b*). In these areas, DA and its metabolites are present in lower amounts than in the caudate–putamen suggesting a lower density of dopaminergic innervation of the limbic as compared to the extrapyramidal system. However, the turnover rate of DA (as measured by the HVA/DA ratio) is much higher in the limbic areas than in the caudate nucleus indicating that DA, although present in low amounts in limbic areas, might subserve an important functional role. The existence of a dopaminergic innervation of the human limbic system is also supported by histochemical studies which revealed the presence of dopaminergic terminals in biopsies of the cerebral cortex (Berger 1977). A fall of DA and its metabolites has been observed in the limbic areas of parkinson-ian patients (Price *et al.* 1978; Scatton *et al.* 1982*a, b*). This, together with the loss of DA-containing cells in the ventral tegmental area observed in this disease (Taquet *et al.* 1982), supports the existence in the normal human brain of a mesocorticolimbic dopaminergic system analogous to that described in other mammals.

Further evidence for the existence of a limbic dopaminergic innervation has been obtained by binding studies using different DA agonist radioligands. Thus, DA, apomorphine, N-propylapomorphine, and bromocriptine binding sites have been found in the nucleus accumbens, olfactory tubercle, and septum of a variety of mammalian species including man (for a review see Seeman 1980).

Influence of neuroleptic drugs on limbic dopaminergic neurones

The hypothesis that neuroleptic drugs block central DA receptors, originally postulated by Carlsson and Lindqvist in 1963, has received extensive support. All clinically active neuroleptics have been shown to antagonize, to varying degrees, some of the physiological actions of DA and the pharmacological effects of DA agonists (for a review see Scatton 1981*b*). The striking correlation between the capacity of neuroleptics of different chemical families to block brain DA receptors (as evaluated by the displacement of [^{3}H]-neuroleptic binding) and their average clinical dose for controlling schizophrenia (Creese *et al.* 1976; Seeman 1977) supports the view that the therapeutic action of neuroleptics is connected to DA-receptor blockade.

A blockade by neuroleptics of DA receptors has been demonstrated without ambiguity in the striatum (for a review see Scatton 1981*b*). There is also exten-sive pharmacological, electrophysiological, and biochemical evidence supporting a blockade by these drugs of limbic DA receptors. Thus neuro-leptics given systemically antagonize the increase in locomotor activity (a behaviour of limbic origin) induced by a systemic administration of apomorphine or by a local injection of DA or apomorphine in the nucleus accumbens of the rat (Pijnenburg and Van Rossum 1973; Andén 1975). Moreover, intra-accumbens infusion of neuroleptics blocks the motor hyperactivity elicited by a local infusion of amphetamine (Andén 1976) or a systemic administration of apomorphine (Pijnenburg *et al.* 1975) in reserpinized rats. Electrophysiological studies have also revealed that iontophoretic application of neuroleptics in the nucleus accumbens or in the DA-rich layers of the frontal cortex antagonizes the depression of cell activity induced by the local iontophoretic application of DA (Bunney and Aghajanian 1976). Similarly, systemic administration of neuroleptics blocks the inhibition of the spontaneous firing rate of neurones in the prefrontal cortex of the rat induced by local injections of direct or indirect DA agonists (Mora *et al.* 1976).

Further evidence indicating that neuroleptics act at DA receptors in the limbic system has been obtained from binding studies using a variety of neuroleptics (e.g. [^{3}H]-haloperidol, [^{3}H]-sulpiride, [^{3}H]-spiperone, [^{3}H]-flupenthixol) as radioligands. These have been shown to bind stereospecifically to membranes prepared from olfactory tubercle, nucleus accumbens, septum (Seeman 1980), and ventral tegmental area (Gundlach *et al.* 1982). The potency of a given neuroleptic drug for displacement of [^{3}H]-haloperidol binding is very similar in the nucleus accumbens or olfactory tubercle and in the striatum (Seeman 1980). Haloperidol binding has also been observed *in vitro* in amygdala (Creese *et al.* 1975) and cortex (Howlett and Nahorski 1978) and *in vivo* in the nucleus accumbens, olfactory tubercle, frontal cortex, septum, and hippocampus (Table 12.1) (Scatton *et al.* 1982*a*). In most limbic areas, the neuroleptic binding is likely to occur at the DA receptor as the bound neuroleptic is displaced by DA and DA agonists as well as by other neuroleptics of the same or a different chemical family (Seeman 1980). Some neuroleptics, e.g. spiperone have also been found to mainly bind *in vitro* to 5-HT receptors in some limbic areas, e.g. the frontal cortex and hippocampus (Creese and Snyder 1978). However, spiperone mainly labels DA receptors in these areas under *in vivo* binding assay conditions (Murrin and Kuhar 1979; Bischoff *et al.* 1980).

Finally, evidence for a blockade by neuroleptics of DA receptors in limbic areas has been obtained in DA turnover studies. It is well known that neuroleptics increase the turnover of DA in the striatum and that this effect is connected to a blockade of DA receptors with a subsequent feedback activation of nigro-striatal DA neurones (Carlsson and Lindqvist 1963). Neuroleptics also increase DA turnover in limbic regions. Thus, a variety of neuroleptics of different chemical families have been shown to increase HVA and DOPAC levels (Scatton 1977; Westerink *et al.* 1977), tyrosine hydroxylase activity (Zivkovic *et al.* 1975), DA synthesis (Scatton *et al.* 1976), release (Bartholini *et al.* 1976), and utilization (Andén 1972; Waldmeier and Maitre 1976; Scatton 1977) in the nucleus accumbens, olfactory tubercle, and frontal cortex of rat, rabbit, or cat (see also Fig. 12.1). Neuroleptics also elevate DOPAC levels in the rat ventral tegmental area (Scatton 1979), septum, and hippocampus (Scatton

TABLE 12.1. *In vivo [³H]-haloperidol binding in rat brain regions: effect of neuroleptics*

Pre-treatment	Dose (mg kg⁻¹, i.p.)	[³H]-haloperidol (dpm mg⁻¹ wet weight)					
		Striatum	Nucleus accumbens	Olfactory tubercle	Frontal cortex	Hippocampus	Septum
Saline		50.4± 5.0	39.2± 3.0	48.3± 4.0	32.0± 3.0	35.4± 2.0	20.6± 3.0
Haloperidol	5	13.4± 2.0*	16.3± 2.0*	14.7± 2.0*	13.5± 3.0*	22.2± 1.0*	11.4± 1.0*
Metoclopramide	50	20.5± 2.0*	25.6± 2.0*	25.3± 3.0*	24.5± 1.7*	25.0± 1.0*	13.9± 1.0*
Butaclamol	5	23.9± 2.0*	26.5± 3.0*	36.3± 2.0*	–	–	–

Neuroleptics were administered 30 min before intracarotid pulse injection of 0.1 μCi of [³H]-haloperidol (D'Ambrosio *et al.* 1982). Rats were killed 20 min after injection of [³H]-ligand.
*$p < 0.01$ vs. controls.

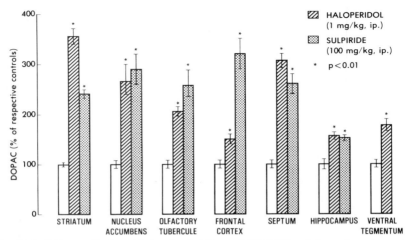

Fig. 12.1. Effects of haloperidol and sulpiride on DOPAC levels in various dopamine-rich brain areas of the rat. Rats were sacrificed 2 hours after drug injection. Results are means ± SEM of data obtained on 8–16 rats and are expressed as a per cent of respective controls. *$p < 0.01$ vs. related controls.

1981a) (Fig. 12.1). As has been proposed for the striatum, the neuroleptic-induced increase in limbic DA turnover might be connected to the blockade of DA receptors with a subsequent activation of mesocorticolimbic dopaminergic neurones. This view is supported by the fact that neuroleptics increase the firing rate of DA cells located in the ventral tegmental area and antagonize the inhibition of the electrical activity of these cells induced by systemic administration of amphetamine (Bunney and Aghajanian 1975).

Taken together, these results support the notion that neuroleptic drugs block limbic DA-receptor function. Recently, attention has been focused on the possibility that DA receptors do not constitute a single and homogeneous population but rather are heterogeneous and can be differentiated into distinct classes according to anatomical localization, pharmacological properties, and functional significance (Spano *et al.* 1978; Kebabian and Calne 1979). Among the various criteria for DA-receptor classification, the one most widely used distinguishes between those receptors which are coupled to adenylate cyclase (D_1 receptors) and those which are not (D_2 receptors). As in the striatum, both types of DA receptors appear to be present in the limbic system, although the relative distribution of D_2/D_1 sites is somewhat different (Seeman 1980). Most of the neuroleptics block, to varying extent, the two types of DA receptors. There are however some neuroleptics which are relatively selective for DA-receptor subtypes. Thus, substituted benzamides specifically block the D_2 receptor as they fail to antagonize DA-sensitive adenylate cyclase but displace [^{3}H]-spiperone binding in both striatum and limbic areas (Spano *et al.* 1978; Seeman 1980). In contrast, thioxanthenes (flupenthixol, piflutixol) act preferentially on D_1 receptors. The question arises as to the relative importance of DA-receptor categories in the impairment of central dopaminergic transmission caused by neuroleptics and, consequently, in their therapeutic effects. Recent studies have indicated that, irrespective of their ability to antagonize DA-sensitive adenylate cyclase, neuroleptics produce qualitatively similar effects on dopaminergic neurones (striatal tyrosine hydroxylase activity and HVA levels) and postsynaptic events of a functional nature (e.g. apomorphine-induced climbing) (Zivkovic *et al.* 1982). Moreover, the potencies of neuroleptics on these various parameters are strikingly correlated. The qualitative similarities between the effects induced by those neuroleptics which antagonize and those which do not block D_1 receptors indicate that the blockade of this DA-receptor category does not play an important role in the changes of striatal DA transmission induced by these drugs. Inasmuch as the effects of neuroleptics on DA-sensitive adenylate cyclase in the striatum are similar to those in limbic areas, it can be inferred that D_1 receptors do not play a major role in the neuroleptic-induced blockade of DA transmission in limbic areas. Therefore, according to the above classification of DA-receptor subtypes, the changes in striatal and limbic DA transmission induced by neuroleptics and consequently the therapeutic action of these drugs appear to be related to the D_2-receptor blockade.

Supporting evidence for this view is found in the striking correlation between the ability of neuroleptics of different chemical structure to block D_2 receptors (displacement of [^{3}H]-haloperidol binding) and their average clinical dose for controlling schizophrenia (Creese *et al.* 1976; Seeman 1977). In contrast, the clinical doses of neuroleptics do not correlate with their capacity to block the D_1 receptor (Seeman 1977).

The limbic dopaminergic system as the neuroanatomical substrate of antipsychotic action

Neuroleptics have been shown to block dopaminergic transmission in all brain areas innervated by dopaminergic neurones. The question naturally arises as to the relative involvement of the different dopaminergic systems in the therapeutic benefits and/or side-effects of these drugs. It is now widely accepted that the neuroendocrine side-effects of neuroleptics (e.g. galactorrhoea, amenorrhoea) are linked to the blockade of DA receptors located in the anterior pituitary (see Muller 1979). Since DA exerts an inhibitory influence on the release of prolactin from the pituitary, the blockade by neuroleptics of DA receptors leads to enhanced secretion of the hormone and to the subsequent neuroendocrine side-effects.

It is also widely believed that impairment of striatal DA transmission is responsible for the extrapyramidal side-effects (Parkinson-like state) of neuroleptics. Neuroleptic-induced parkinsonism is indeed indistinguishable from Parkinson's disease in which a degeneration of the nigro-striatal DA system is known to occur.

The anatomical substrate of the antipsychotic action of neuroleptics is still a matter of debate, but, in view of its involvement in the regulation of emotional behaviour, the limbic system appears to be the most likely candidate. There are some biochemical studies which support this contention.

The earlier (but controversial) indication for a preferential involvement of the limbic dopaminergic system in the antipsychotic action of neuroleptics has been obtained by Andén (1972) who studied the effects of anticholinergic drugs on the neuroleptic-induced alterations of DA turnover in the striatum and limbic areas. Anticholinergic drugs are used clinically in combination with neuroleptics to overcome the Parkinson-like syndrome induced by the latter drugs. Although, the influence of anticholinergics on the antipsychotic effects of neuroleptics is still a matter of controversy, it is generally agreed that these drugs do not reverse the antipsychotic effect of neuroleptics or, if they do so, the reversal of the antipsychotic effect occurs to a lesser degree than the amelioration of the pseudoparkinsonism (see Lloyd 1978*b*). Biochemical studies performed in rabbit have indicated that anticholinergic drugs block the neuroleptic-induced elevation of HVA (a reflection of the blockade of dopaminergic transmission) to a much greater extent in the striatum than in the limbic areas. These initial data were therefore taken as evidence for the hypothesis that neuroleptics exert their antipsychotic action via blockade of limbic dopaminergic receptors. However, more recent studies performed in other animal species failed to confirm these data (Westerink and Korf 1975; Waldmeier and Maitre 1979).

A much stronger argument in favour of the involvement of limbic DA neurones in the antipsychotic action of neuroleptics has been obtained in studies comparing the changes in DA turnover in the striatum and various limbic areas after acute treatment with 'classical' (drugs possessing antipsychotic action but inducing Parkinson-like states in man and inducing catalepsy in animals) and 'atypical' (drugs exerting antipsychotic action with minimal extrapyramidal side-effects in man and devoid of cataleptic effect in animals) neuroleptics. As discussed above, neuroleptics increase DA turnover in both limbic areas and striatum of various laboratory animals. However, those drugs (e.g. haloperidol, chlorpromazine) which can be classified as 'classical' neuroleptics consistently produce a greater increase of DA turnover in the striatum than in the limbic areas of the rat, cat, or rabbit, whereas 'atypical' neuroleptics (e.g. thioridazine, clozapine, sulpiride) increase DA turnover preferentially in the limbic regions (Andén and Stock 1973; Zivkovic *et al.* 1975; Bartholini 1976; Scatton *et al.* 1977*a*; Westerink *et al.* 1977). A similar differential regional effect of 'classical' as compared to 'atypical' neuroleptics has also been obtained by measuring other biochemical parameters, e.g. DA release (Bartholini *et al.* 1976) or tyrosine hydroxylase activity (Zivkovic *et al.* 1975).

Several hypotheses have been put forward to explain the preferential effect of 'atypical' neuroleptics on limbic DA turnover. Since some 'atypical' neuroleptics like clozapine and thioridazine are also potent anticholinergic agents (Snyder *et al.* 1974) and since a link between dopaminergic and cholinergic neurones exists in striatum but not in limbic areas (see Lloyd 1978*a*), it has been proposed that the association of anticholinergic with DA-receptor blocking properties in the same molecule would abolish the biochemical and functional consequences of DA-receptor blockade in the extrapyramidal system without affecting the expression of DA receptor blockade in limbic structures (Andén 1972; Bartholini *et al.* 1975). However, while this hypothesis might be valid for clozapine and thioridazine, it could hardly be tenable for sulpiride which has negligible anticholinergic effects. It has also been postulated that 'atypical' neuroleptics, by blocking preferentially presynaptic inhibitory DA receptors in the striatum, may enhance DA release from the nigro-striatal dopaminergic neurones and consequently overcome the blockade of postsynaptic DA receptors (Bartholini *et al.* 1972). This suggestion is however not supported by our recent findings showing that 'atypical' as well as 'classical' neuroleptics possess a similar affinity for pre- and postsynaptic DA receptors in the rat striatum (Zivkovic *et al.* 1982).

The most likely explanation for the differential regional effects of 'atypical' versus 'classical' neuroleptics on DA turnover is that these drugs block to a different degree DA receptors in the limbic areas and striatum. This view is supported by recent *in vivo* binding studies performed in our laboratory. After pulse injection of [³H]-N-propylnorapomorphine ([³H]-NPA) into the internal carotid artery of the rat, the tritiated ligand selectively labels DA (D₂) receptors. As shown in Fig. 12.2, both haloperidol and sulpiride antagonize in a dose-dependent manner [³H]-NPA binding in both the striatum and the nucleus accumbens of the rat. However, the 'classical' neuroleptic haloperidol appears to be more effective in antagonizing [³H]-NPA retention in the striatum as compared to the nucleus accumbens, whereas the reverse holds true for the 'atypical' neuroleptic sulpiride. Interestingly, both the efficacy and the potency of sulpiride in displacing [³H]-NPA were greater in the nucleus accumbens than in striatum. Similar results have been obtained by others using [³H]-spiperone to label DA receptors *in vivo*; in low doses sulpiride was found to displace the

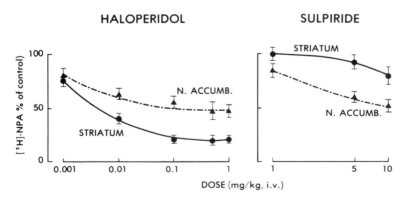

Fig. 12.2. Displacement by haloperidol and sulpiride of [³H]-NPA binding to striatum (•) and nucleus accumbens (▲). Rats received intravenous injection of various doses of the neuroleptics 15 min before intracarotid pulse injection (D'Ambrosio *et al.* (1982) of [³H]-NPA (5 ng in 10 μl; 1 μCi). Rats were sacrificed 20 min after the ligand application. Each point with vertical bars represents the mean with SEM of five experimental values. Saline-injected controls (100 per cent).

binding of the [³H]-ligand in the olfactory tubercle, frontal cortex, and septum but not in striatum, whereas haloperidol displaced the radioligand to a similar extent in all areas (Köhler *et al.* 1979).

A different sensitivity of the DA effector cells in limbic and striatal structures to 'classical' and 'atypical' neuroleptics is also suggested by electrophysiological studies. Thus, clozapine blocks more markedly the depression of cell activity induced by micro-iontophoretic application of DA in the nucleus accumbens than in striatum, whereas haloperidol depressed cell firing similarly in both areas (Bunney and Aghajanian 1976). Moreover, clozapine but not haloperidol reversed the depression of firing rate of amygdaloid neurones produced by amphetamine (Rebec *et al.* 1981). Therefore, these data indicate unequivocally that 'atypical' neuroleptics block DA receptors preferentially (exclusively in low doses) in the limbic areas, whereas 'classical' neuroleptics block DA receptors in both striatum and limbic areas. This supports the hypothesis that impairment of limbic DA transmission is mainly responsible for the antipsychotic action of neuroleptics whereas blockade of striatal DA transmission is connected to extrapyramidal side-effects.

The latter hypothesis has also been substantiated by chronic studies with neuroleptics. Neuroleptic drugs are administered clinically over prolonged periods. Under these conditions, the antipsychotic effects of these drugs are maintained throughout the treatment period, in contrast to many of the side-effects (e.g. the Parkinson-like symptoms), which are attenuated during chronic administration (cf. Hollister 1972). If the blockade by neuroleptics of limbic and striatal DA receptors is responsible for the antipsychotic action and extrapyramidal side-effects, respectively, then the impairment of dopaminergic transmission must be maintained in the limbic areas but attenuated in striatum during chronic administration. Biochemical studies performed in rats or monkeys receiving neuroleptics subacutely, or long-acting injectable neuroleptics, have revealed that this is indeed the case. Thus, after repeated treatment with moderate doses (comparable to those used clinically) of haloperidol, chlorpromazine, thioproperazine, or sulpiride, tolerance to the increase in HVA and DA synthesis seen after a single drug injection is observed in striatum but not in the nucleus accumbens, olfactory, tubercle, frontal cortex, septum, or ventral tegmental area (Bowers and Rozitis 1974; Scatton *et al.* 1975, 1976; Scatton 1977, 1979, 1981*b*; Bacopoulos *et al.* 1978). Moreover, administration of a depot neuroleptic (pipotiazine palmitate) has been shown to induce a long-lasting and sustained increase in DA synthesis in the frontal cortex whereas abolition of the early increase in DA synthesis is observed in the striatum within a few days following the ester injection (Scatton *et al.* 1977*b*).

Post-mortem studies in schizophrenic patients who had been under neuroleptic medication for several years have also revealed that the levels of HVA were unchanged in the caudate–putamen of these patients but were markedly elevated in the nucleus accumbens, the cingulate, and frontal cortex (Bacopoulos *et al.* 1979). These data indicate a persistence of the augmentation of DA metabolism (and consequently of DA-receptor blockade) in limbic areas of schizophrenic patients after prolonged neuroleptic treatment and are therefore compatible with the involvement of the limbic dopaminergic system in the antipsychotic action of neuroleptics. In contrast, the tolerance to the neuroleptic-induced increase in striatal DA metabolism observed after chronic administration (generally attributed to the development of DA target-cell supersensitivity linked to an increased number of DA receptors and to a subsequent wearing off of DA-receptor blockade) implicates the striatum as being responsible for the Parkinson-like side-effects of neuroleptics.

In the rat, after repeated treatment of long duration with doses of neuroleptics far higher than those used clinically, tolerance to the increase in DA metabolism also occurs in some limbic areas, e.g. hippocampus, septum (Scatton 1981*a*), nucleus accumbens, and olfactory tubercle (Bowers and Rozitis 1976; Scatton 1977). Increased DA-receptor density is also observed in these areas under these conditions of treatment (Clow *et al.* 1980). These findings, which suggest a disappearance of neuroleptic-induced DA-receptor blockade in certain limbic areas after chronic neuroleptic treatment, may have some relation to the loss of therapeutic efficacy of neuroleptics which is known to occur in some patients treated for several years with these drugs (Chouinard and Jones 1980).

In conclusion, the above data are supportive of the hypothesis that limbic dopaminergic neurones are associated with the long-term antipsychotic effect of neuroleptics. This is compatible with the clinical description of the schizophrenic process which suggests an impairment of prefrontal cortex function (defect in attention, concept formation, abstraction). A deficit in frontal cortex function in schizophrenia has also been demonstrated recently by measuring cerebral blood flow in various cortical regions in awake schizophrenics: under resting conditions, the cerebral blood flow is high in the frontal cortex of control patients but reduced in schizophrenic patients (Ingvar 1976).

It should be underlined that, even if the biochemical data discussed above are strongly suggestive of an involvement of limbic dopaminergic neurones in the antipsychotic action of neuroleptics, this does not exclude the possibility that transmitters other than DA are also involved in the therapeutic action of these drugs (see below). The underlying dysfunction in schizophrenia is far from being understood. Based on the facts that

1. Neuroleptics block central DA transmission;
2. DA agonists like amphetamine induce paranoid psychosis in nonschizophrenic subjects and exacerbate schizophrenia (Snyder 1973);
3. Apomorphine given in 'autoreceptor' doses alleviates some schizophrenic symptoms (Tamminga *et al.* 1978*b*);

the hypothesis has been put forward that schizophrenia may be due to excessive central dopaminergic transmission. However neurochemical investigations of post-mortem brain material from schizophrenic patients have failed to show convincing and reproducible alterations in dopaminergic systems. Those few changes that have been consistently demonstrated, notably supersensitivity of DA receptors (Lee *et al.* 1978) have yet to be proven independent of long-term neuroleptic administration. It is therefore possible that the primary alteration in schizophrenia does not occur in dopaminergic neurones but rather occurs in a neuronal system functionally connected with the dopaminergic system.

LIMBIC NORADRENERGIC NEURONES AND NEUROLEPTICS

Noradrenergic innervation of the limbic system

Noradrenergic nerve terminals are distributed widely in the limbic system of mammals. Histofluorescence studies as well as the analysis of the effects of brain lesions on noradrenaline (NA) levels and DA-β-hydroxylase activity have revealed the existence of both ventral and dorsal noradrenergic paths projecting from the brainstem to the forebrain (Saavedra and Zivin 1976; Chapter 1, this volume). In the rat, the ventral noradrenergic pathway originates in cell bodies located in the pons–medulla and projects mainly to the hypothalamus and nucleus interstitialis striae terminalis. The dorsal noradrenergic pathway origi-nates in the locus ceruleus and sends fibres to cortical areas, amygdala, septum, hippocampus, and olfactory tubercle. Noradrenergic fibres are characterized by an extreme divergence and a high degree of plasticity. In general there is no coincidence between the distribution of noradrenergic and dopaminergic terminals in cortical and subcortical limbic regions. For instance, in the prefrontal cortex, dopaminergic terminals are found in layers V and VI whereas noradrenergic terminals are mainly distributed in the superficial layers.

As suggested by neurochemical studies performed on post-mortem brain tissue, the pattern of distribution of NA in the human limbic system grossly corresponds to that found in the rat, with the exception of the cortical areas where proportionally less NA is present (Farley and Hornykiewicz 1975; Scatton *et al.* 1983). In Parkinson's disease there is a drop of NA content in several limbic regions (e.g. nucleus accumbens, hippocampus, cerebral cortex) as well as in the locus ceruleus (Farley and Hornykiewicz 1975; Scatton *et al.* 1983) suggesting the existence of a dorsal noradrenergic pathway in the human brain as well.

In addition to presynaptic markers for the presence of noradrenaline-containing neurones, there is biochemical evidence for the presence of adrenergic receptors in mammalian limbic structures. Radioligand binding studies have demonstrated the occurrence of α (α_2 and α_1) and β receptors in a variety of limbic areas (U'Prichard *et al.* 1976; Alexander *et al.* 1976; Weinreich and Seeman 1981). Moreover, the presence of an NA- and isoprenaline-sensitive adenylate cyclase in limbic forebrain regions has been reported (Blumberg *et al.* 1976). This NA- sensitive adenylate cyclase appears to be regu-lated by both α and β components, although the exact nature of the receptor involved is as yet unclear.

Influence of neuroleptic drugs on limbic noradrenergic neurones

The available data support the hypothesis that neuroleptic drugs exert their antipsychotic action via blockade of limbic DA receptors. It is however hardly conceivable that a similar alteration of limbic DA transmission might be responsible for the therapeutic action of neuroleptics in both hebephrenia and paranoid schizophrenia, which exhibit opposite symptomatologies. This suggests that, apart from the limbic dopaminergic system, other neuronal systems might be involved in the therapeutic effects of neuroleptics. In this respect the limbic noradrenergic system has received much attention and has been proposed as a possible target for antipsychotic action (see Robinson *et al.* 1979).

There is indeed some evidence that neuroleptics may affect limbic noradrenergic transmission. Thus, the concentrations of MOPEG, the major NA metabolite, NA synthesis as well as the rate of NA utilization are increased by some neuroleptics in individual limbic regions presumably via a feedback response mediated by receptor blockade (Andén *et al.* 1970; Keller *et al.* 1973; Scatton *et al.* 1976). Moreover, the release of NA into the perfusate of nucleus accumbens is increased by clozapine and chlorpromazine (Bartholini *et al.* 1976). In general the extent of increased NA turnover is considerably smaller than that observed for DA. Moreover only those neuroleptics which possess adrenolytic properties increase NA turnover. This effect, as well as the α-adrenoceptor blocking properties of neuroleptics (as measured by displacement of [^{3}H]-clonidine binding), do not correlate well with antipsychotic activity (Keller *et al.* 1973; Peroutka and Snyder 1980) and are more likely connected to the sedative and/or neurovegetative effects of neuroleptics.

Neuroleptics also inhibit NA-sensitive adenylate cyclase in the limbic system (Blumberg *et al.* 1976). This effect is apparently independent from their α-adrenoceptor antagonistic activity as it is also induced by sulpiride and pimozide which are virtually devoid of α-adrenoceptor blocking properties (Robinson *et al.* 1979; Zivkovic *et al.* 1982). It is noteworthy that sulpiride or metoclopramide, which do not block the DA-sensitive, potently inhibit NA-sensitive adenylate cyclase in the limbic system. Although the pharmacological implication of the blockade by neuroleptics of NA-sensitive adenylate cyclase is still uncertain, a relationship to the antipsychotic action of these drugs (possibly the disinhibitory component) has been suggested (Robinson *et al.* 1979). It is worth mentioning in this respect that, carpipramine, a drug acting essentially on the negative symptoms of schizophrenia, also blocks NA-sensitive adenylate cyclase (Zivkovic *et al.*, unpublished results). Moreover, the β-antagonist propranolol which also antagonizes limbic NA-sensitive adenylate cyclase has been reported to possess antipsychotic action when administered in high doses (Middlemiss *et al.* 1981). Finally, drugs that increase central noradrenergic activity exacerbate the schizophrenia condition (see Lloyd 1978*c*).

An involvement of the limbic noradrenergic system in the action of neuroleptics is also suggested by chronic studies. If the blockade by neuroleptics of the NA receptor coupled to adenylate cyclase has some functional significance, then repeated treatment with these drugs would be expected to induce adaptive changes in the NA-regulated cyclic AMP system. Indeed, daily administration of pimozide, haloperidol, or sulpiride for three weeks produces an increased response to NA of septal adenylate cyclase (Table 12.2 and Zivkovic *et al.* 1982). The development of the supersensitivity of noradrenergic receptors coupled to adenylate cyclase appears to have behavioural and biochemical

TABLE 12.2. *Effect of repeated neuroleptic treatment on the sensitivity to noradrenaline of rat septal adenylate cyclase*

Treatment	Dose	Cyclic AMP (per cent of controls)
Saline	–	100± 5
Pimozide	5 mg kg^{-1}, i.p.	141± 14*
Haloperidol	4 mg kg^{-1}, p.o.	129± 3‡
Sulpiride	100 mg kg^{-1}, i.p.	141± 10†

Rats were treated once daily for 18 days and sacrificed 48 h after the last dose. Slices (0.5 mm) of the rat septal regions were washed and pre-incubated in Krebs–Ringer solution for 45 min. The slices were then transferred into fresh medium (5 ml) and after a 10-min pre-incubation period, noradrenaline was added (final concentration 10 μM). The incubation was continued for another 10 min after which the slices were separated from the medium and homogenized. Cyclic AMP was isolated by column chromatography and quantified by a protein binding assay. Each value is the mean with S.E.M. of data obtained on six rats.

*p 0.05; †$p < 0.01$; ‡$p < 0.001$ as compared to saline-treated rats.

correlates. Thus, in mice a marked increase in locomotor activity (which can be antagonized by the NA-synthesis inhibitor FLA 63 but not by the DA-receptor blocker haloperidol) has been shown to occur during the withdrawal phase following a repeated treatment with haloperidol (Jackson *et al.* 1979). The neuroleptic-induced changes in cyclase-linked NA receptors are also accompanied by alterations of brain NA metabolism. Chronic treatment with thioproperazine or haloperidol increases NA synthesis in the limbic cortex (Scatton *et al.* 1976) and enhances tyrosine hydroxylase activity in the locus ceruleus of the rat (Guidotti *et al.* 1978). Therefore, it does not seem unreasonable to believe that adaptive changes in limbic noradrenergic systems induced by repeated administration of neuroleptics may play a part in the overall therapeutic action of these drugs. This view is supported by recent post-mortem studies which have reported supra-normal levels of NA in limbic forebrain areas in chronic paranoid schizophrenics treated with neuroleptics (Farley *et al.* 1978). Elevated concentrations of NA in the cerebrospinal fluid have also been found in chronic schizophrenic patients under neuroleptic treatment (Gomes *et al.* 1980). These results would indicate that a relative hyperactivity of limbic noradrenergic neurones may be of aetiological importance in certain subgroups of schizophrenics and that neuroleptics may exert their curative effect in these patients by blocking limbic NA transmission.

LIMBIC GABAERGIC NEURONES AND NEUROLEPTICS

In addition to DA and NA, recent evidence indicates that other putative neurotransmitters may also play an important role in the neuronal function of limbic circuitry and as such may be involved in the actions of neuroleptics. One of the candidates is the inhibitory neurotransmitter γ-aminobutyric acid (GABA). GABA and its synthetic enzyme, glutamic acid decarboxylase (GAD), as well as GABA receptors are present in high amounts in all limbic structures, particularly the medial part of the nucleus accumbens, the olfactory areas, amygdala, and septum in various animal species (Walaas and Fonnum 1978; Palacios *et al.* 1981). The development of immunocytochemical techniques for GAD has allowed the visualization of GABAergic terminals in these areas (Perez de la Mora *et al.* 1981). In most limbic regions (e.g. nucleus accumbens, septum, hypothalamus, olfactory tubercle) GABA is present in interneurones. The nucleus accumbens has also been found to send GABA-containing fibres to the globus pallidus and to the rostral ventral tegmental area (Walaas and Fonnum 1980). The latter pathway may take part in the feedback regulation of the activity of the mesolimbic DA neurones. A link between GABAergic and dopaminergic neurones in the limbic system is also compatible with the decrease of limbic DA turnover observed after administration of GABA agonist agents (Scatton *et al.* 1982c) and the decrease of GABA turnover in mesolimbic cell body areas induced by apomorphine (Perez de la Mora *et al.* 1975). Injection of GABA antagonists into the ventral tegmental area in cats has been shown to produce behavioural and motor abnormalities not unlike those seen in schizophrenia (Stevens *et al.* 1974). This is consistent with GABAergic neurones acting to inhibit the mesolimbic dopaminergic system and the overactivity of this system in schizophrenia.

There is evidence that neuroleptics affect GABAergic transmission in several brain areas including the limbic regions. Thus, acutely administered neuroleptics increase GABA turnover in the nucleus accumbens and globus pallidus (Mao *et al.* 1977b) and decrease GABA content in several brain regions (Lloyd *et al.* 1977). After repeated neuroleptic treatment, the former effect persists in nucleus accumbens but is attenuated in the globus pallidus. Chronic adminis-tration of chlorpromazine or haloperidol does not alter GABA content in nucleus accumbens and olfactory tubercle (Perry *et al.* 1978). Acute admini-stration of clozapine results in lowered GAD activity in several brain regions (including limbic areas) but this effect is not maintained on chronic administration (Lloyd *et al.* 1977). Finally, in schizophrenic patients, neuroleptic administration has been shown to elevate GABA levels in the cerebrospinal fluid (Zarifian *et al.* 1982). The alteration by neuroleptics of GABAergic activity in the limbic structures may be connected to the blockage of DA transmission, as GABA neurones are target cells for DA neurones in a number of limbic regions (*vide supra*). Another possibility is that neuroleptic drugs have a direct effect on GABA receptors. Thus, acute or chronic neuroleptics significantly decrease the specific [^{3}H]-GABA binding in membranes prepared from various brain areas (Lloyd *et al.* 1977; Trabucchi *et al.* 1978). Moreover, there is some structural analogy between haloperidol and the GABA molecule (Janssen 1965) and neuroleptics have been shown to alter GABA-induced depolarization (Higashi *et al.* 1981).

These data raise the possibility that limbic GABA neurones may be involved in the therapeutic action of neuroleptics. A hypothesis suggesting that there is a defect in the GABA system in schizophrenia has been put forward on theoretical considerations (Roberts 1976) and also on post-mortem studies indicating a decrease of GAD activity in the putamen, nucleus accumbens, amygdala, and hippocampus in the brains of patients with schizophrenia (Bird *et al.* 1977;

Perry *et al.* 1979). The latter finding was subsequently characterized as arte-factual, being dependent primarily on the mode of death (Iversen *et al.* 1979). Nevertheless, GABA content appears to be slightly reduced in the nucleus accumbens and in the amygdala in schizophrenic patients irrespective of the mode of death (Perry *et al.* 1979); whether this effect is connected to the schizo-phrenic process itself or subsequent to the neuroleptic treatment remains to be elucidated. Cerebrospinal fluid GABA concentrations have been reported as normal in schizophrenia (Bowers *et al.* 1980; Gerner and Hare 1981). Recent clinical studies also strongly argue against GABA involvement in schizophrenia. Thus, either given as monotherapy or in combination with neuroleptic drugs, the GABA agonist agents (progabide, muscimol, valproate) were found to be devoid of any antipsychotic action (Morselli *et al.* 1980; Tamminga *et al.* 1978*a*; Lautin *et al.* 1980). Muscimol even produced a definite worsening of the schizo-phrenic syndrome with an increase in thought disorder. Nonetheless, although devoid of antipsychotic activity, GABA agonists diminish tardive dyskinesia (Morselli *et al.* 1980).

From the available data, it appears therefore that the involvement of limbic GABA neurones in the antipsychotic action of neuroleptics remains an open question.

LIMBIC PEPTIDERGIC NEURONES AND NEUROLEPTICS

A major advance in neuroscience during the past decade has been the discovery of endogenous opiate-like peptides, the endorphins and enkephalins in the central nervous system. A variety of endogenous opiate-like peptides (β-endorphin, met-enkephalin, leu-enkephalin) as well as other peptides (e.g. substance P, neurotensin) are present throughout the brain especially in limbic structures. They thus seem to be well positioned for a role in the regulation of mood and affect. Structures, such as the lateral septal nucleus, the nucleus accumbens, several amygdaloid nuclei, the stria terminalis, and the hypothal-amus, have major opiate peptide contributions. The fact that, in the limbic areas, peptide-containing neurones are in close connection with monoaminergic systems together with the recent discovery that polypeptides often coexist in the same axons with monoamine neurotransmitters (Hökfelt *et al.* 1980), brings up the question of whether putative peptide neurotransmitters are involved in (at least some of) the therapeutic effects of neuroleptics.

Recent neurochemical studies in the rat indicate that neuroleptics modify the levels of some of these neuropeptides in the limbic system. Neurotensin is a tridecapeptide which exhibits a widespread distribution in the mammalian CNS. Highest concentrations in the human brain have been found in the hypothalamus, the periventricular grey area, and certain limbic regions, including the nucleus accumbens, septum, and amygdala (Emson *et al.* 1981). In the nucleus accumbens, neurotensin is associated with intrinsic short-axon neurones and is also located in a polysynaptic chain of neurones originating from the stria terminalis. A significant increase of neurotensin content has been found in the nucleus accumbens after both acute and subacute treatments with haloperidol (Govoni *et al.* 1980). Interestingly, the modification of accumbens neurotensin content is elicited only by those neuroleptics which possess antipsychotic action. This effect is thought to reflect an overproduction of neurotensin in response to the blockade of dopaminergic synapses. This, together with the fact that intra-cerebral injection of neurotensin elicits neuroleptic-like behavioural effects in rats (e.g. antagonism of amphetamine-induced locomotor activity, inhibition of conditioned avoidance response and self-stimulation from the ventral tegmental area (Nemeroff 1980)) and, like neuroleptics, increases DA turnover in various brain areas (see Widerlöv *et al.* 1982), has led to the suggestion that an increased production of neurotensin may be involved in the mediation of certain actions of antipsychotics (Govoni *et al.* 1980; Costa 1980). Of particular interest in this respect is the recent finding that the CSF neurotensin levels are subnormal in a subgroup of schizophrenic patients (exhibiting the more severe symptoms) and are increased to the normal range during neuroleptic treatment (Widerlöv *et al.* 1982; see also Chapter 16, this volume). Patients with the hebephrenic type of schizophrenia had normal levels of neurotensin in CSF. These findings are consistent with the view that at least part of the beneficial effects of neuroleptics may be caused by increased levels of neurotensin, an 'endogenous neuroleptic-like' peptide.

Neuroleptics also affect endogenous opiate-like peptides in the limbic system. Thus, an increase in the level of total opiate activity in the brain of rats has been observed after chronic chlorpromazine treatment. Moreover, chronic treatment with haloperidol elevates β-endorphin levels in the hypothalamus and the septum (Höllt and Bergmann 1982) and diminishes the peptide content in the nucleus accumbens (Kato *et al.* 1981). Repeated but not acute administration of haloperidol, chlorpromazine, or pimozide has also been found to increase the met-enkephalin and leu-enkephalin contents in the nucleus accumbens and striatum but not in the hypothalamus, septum, and medulla oblongata (Hong *et al.* 1978). Evidence has been provided that the increased met-enkephalin levels are due to an accelerated biosynthesis of the peptide (Costa 1980).

These data raise the possibility that the therapeutic action of neuroleptics in schizophrenic patients might be associated with changes in peptidergic activity. However, there is controversial evidence as to the role of endorphins in the pathogenesis of schizophrenia. A role for endorphins in schizophrenia was first suggested by Terenius *et al.* (1976) who reported that schizophrenic patients exhibited elevated levels of endorphins in the cerebrospinal fluid which returned to normal after neuroleptic treatment. The reported improvement in schizo-phrenic symptoms following haemodialysis, proposed to be secondary to the removal of leucine-5-β-endorphin (Palmour *et al.* 1979), was also compatible with an endorphin excess in schizophrenia. However, the haemodialysis trial was not double-blind and more recent haemodialysis studies have failed to confirm these initial data (see Berger *et al.* 1980 for a review). A series of experiments using naloxone in schizophrenic patients has yielded either negative results or, particularly with large doses of naloxone, a tendency to improve some symptoms, notably auditory hallucinations (for a review, See Berger *et al.* 1980; Malek-Ahmadi and Callen 1980).

Evidence for an endorphin deficiency in schizophrenia has also been provided. Thus, the met-enkephalin analogue FK 33-824 has been reported to decrease psychotic symptoms in an open study in a limited number of patients (Jorgensson *et al.* 1979). The endorphin-like peptide Des-tyr-γ-endorphin, a peptide structurally related to β-endorphin but devoid of opiate activity, has been reported to ameliorate schizophrenic symptoms (Verhoeven *et al.* 1979) but this finding was not confirmed (Tamminga *et al.* 1981). Finally, β-endorphin was reported to exert a beneficial effect (particularly on hallucinations) in schizophrenic patients but these results were not replicated (cf. Berger *et al.* 1980).

From the above data it appears that a link between neuropeptides and psychosis is currently very weak or absent and that more careful studies must be performed before an overall hypothesis integrating these findings can be proposed.

CONCLUSION

At present, it appears that neuroleptics affect a variety of neurotransmitter systems in the limbic system, and it is probable that their clinical efficacy cannot be ascribed to their effect on a single neurotransmitter. There is a host of evidence that neuroleptic agents produce a consistent impairment of limbic DA synaptic function which is very closely related to their antipsychotic action. However, other limbic neuronal pathways using noradrenaline, GABA, or polypeptides as neurotransmitters might also play a part in the overall therapeutic action of these drugs. As it is likely that different subclasses of psychotic illnesses may be related to different pathogenesis (neurochemical-neuroanatomical), the differential effect of neuroleptics on these limbic neuronal systems will probably determine their clinical specificities. The development of neuroleptic agents acting selectively on a particular limbic transmitter system should lead to a better understanding of the relative roles of these systems in determining the clinical specificity of neuroleptics.

REFERENCES

Alexander, R. W., Davis, J. N., and Lefkowitz, R. J. (1976). Direct identification and characterization of β-adrenergic receptors in rat brain. *Nature* **258**, 437–40.

Andén, N. E. (1972). Dopamine turnover in the corpus striatum and the limbic system after treatment with neuroleptic and anticholinergic drugs. *J. Pharm. Pharmacol.* **24**, 905–6.

—— (1975). Animal models of brain dopamine function. In *Advances in parkinsonism* (ed. W. Birkmayer and O. Hornykiewicz), pp. 169–77. Editions Roche, Basle.

—— (1976). Effects of drugs on the dopamine mechanisms in the corpus striatum and in the limbic system and on the noradrenaline and 5-hydroxytryptamine mechanisms in the spinal cord. In *Drugs and central synaptic transmission* (ed. P. B. Bradley and B. N. Dhawan), pp. 49–62. MacMillan Press Ltd, London.

——, Butcher, S. G., Corrodi, H., Fuxe, K., and Ungerstedt, U. (1970). Receptor activity and turnover of dopamine and noradrenaline after neuroleptics. *Eur. J. Pharmacol.* **11**, 303–14.

——, and Stock, G. (1973). Effects of clozapine on the turnover of dopamine in the corpus striatum and in the limbic system. *J. Pharm. Pharmacol.* **25**, 346–8.

Bacopoulos, N. G., Bustos, G., Redmond, D. E., Baulu, J., and Roth, R. H. (1978).

Regional sensitivity of primate brain dopaminergic neurons to haloperidol: alterations following chronic treatment. *Brain Res.* **157**, 396–401.

——, Spokes, E. G., Bird, E. D., and Roth, R. H. (1979). Cortical dopamine metabolism after long term treatment. *Science* **205**, 1405–7.

Bartholini, G. (1976). Differential effect of neuroleptic drugs on dopamine turnover in the extrapyramidal and limbic system. *J. Pharm. Pharmacol.* **28**, 429–33.

——, Haefely, W., Jalfre, M., Keller, H. H., and Pletscher, A. (1972). Effect of clozapine on cerebral catecholaminergic neuron systems. *Br. J. Pharmacol.* **46**, 736–40.

——, Keller, H., and Pletscher, A. (1975). Drug-induced changes of dopamine turnover in striatum and limbic system of the rat. *J. Pharm. Pharmacol.* **27**, 439–41.

——, Stadler, H., Gadea-Ciria, M., and Lloyd, K. G. (1976). The effect of antipsychotic drugs on the release of neurotransmitters in various brain areas. In *Antipsychotic Drugs, pharmacodynamics and pharmacokinetics* (ed. G. Sedvall, B. Uvnäs, and I. Zotterman), pp. 105–16. Pergamon Press, Oxford.

Berger, B. (1977). Histochemical identification and localization of dopaminergic axons in rat and human cerebral cortex. In *Non striatal dopaminergic neurons* (ed. E. Costa and G. L. Gessa), Advances in biochemical psychopharmacology Vol. 16, pp. 13–20. Raven Press, New York.

Berger, P., Watson, S., Akil, H., Elliott, G., Rubin, R., Pfefferbaum, A., Davis, K., Barchas, J., and Li, C. (1980). β-endorphin and schizophrenia. *Arch. gen. Psychiat.* **37**, 635–40.

Bird, E. D., Spokes, E. G., Barnes, J., McKay, A. V. P., Iversen, L. L., and Shepherd, M. (1977). Increased brain dopamine and reduced glutamic acid decarboxylase and choline acetyltransferase activity in schizophrenia and related psychoses. *Lancet* ii, 1247–9.

Bischoff, S., Bittiger, H., and Krauss, J. (1980). In vivo ^{3}H-spiperone binding to the rat hippocampal formation: involvement of dopamine receptors. *Europ. J. Pharmacol.* **68**, 305–15.

Blumberg, J. B., Vetulani, J., Stawarz, R., and Sulser, F. (1976). The noradrenergic cyclic AMP generating system in the limbic forebrain: pharmacological characterization in vitro and possible role of limbic noradrenergic mechanisms in the mode of action of antipsychotics. *Eur. J. Pharmacol.* **37**, 357–6.

Bowers, M. B., Gold, B. I., and Roth, R. H. (1980). CSF GABA in psychotic disorders. *Psychopharmacology* **70**, 279–82.

—— and Rozitis, A. (1974). Regional differences in homovanillic acid concentrations after acute and chronic administration of antipsychotic drugs. *J. Pharm. Pharmacol.* **26**, 743–6.

—— and —— (1976). Brain homovanillic acid: regional changes over time with antipsychotic drugs. *Eur. J. Pharmacol.* **39**, 109–14.

Bunney, B. S. and Aghajanian, G. K. (1975). The effect of antipsychotic drugs on the firing of dopaminergic neurons: a reappraisal. In *Antipsychotic drugs, pharmacodynamics and pharmacokinetics* (ed. G. Sedvall, B. Uvnäs, and Y. Zotterman), pp. 305–18. Pergamon Press, Oxford.

—— and —— (1976). Dopamine and norepinephrine innervated cells in the rat prefrontal cortex: pharmacological differentiation using microiontophoretic techniques. *Life Sci.* **19**, 1783–93.

Carlsson, A. and Lindqvist, M. (1963). Effect of chlorpromazine or haloperidol on formation of 3-methoxytyramine and normetanephrine in mouse brain. *Acta Pharmacol. Toxicol.* **20**, 140–4.

Chouinard, G. and Jones, B. D. (1980). Neuroleptic-induced supersensitivity psychosis: clinical and pharmacological characteristics. *Am. J. Psychiat.* **137**, 16–21.

Clow, A., Theodorou, A., Jenner, P., and Marsden, C. D. (1980). A comparison of striatal and mesolimbic function in the rat during 6-month trifluoperazine administration. *Psychopharmacology* **69**, 227–33.

Costa, E. (1980). Mode of action of antipsychotics: the peptide connection. In *Synaptic constituents in health and disease* (ed. M. Brzin, D. Sket, and H. Bachelard), pp. 687–9. Pergamon Press, Oxford.

Creese, I., Burt, D. R., and Snyder, S. (1975). Dopamine receptor binding: differentiation of agonist and antagonist states with ³H-dopamine and ³H-haloperidol. *Life Sci.* **17**, 993–1002.

——, ——, and —— (1976). Dopamine receptor binding predicts clinical and pharmacological potencies of antischizophrenic drugs. *Science* **192**, 481–4.

—— and Snyder, S. H. (1978). ³H-spiroperidol labels serotonin receptors in rat cerebral cortex and hippocampus. *Eur. J. Pharmacol.* **49**, 201–2.

D'Ambrosio, A., Zivkovic, B., and Bartholini, G. (1982). ³H-haloperidol labels brain dopamine receptors after its injection into the internal carotid artery of the rat. *Brain Res.* **238**, 470–4.

Emson, P. C., Rossor, M., and Lee, C. M. (1981). The regional distribution and chromatographic behaviour of somatostatin in human brain. *Neurosci. Lett.* **22**, 319–24.

Farley, I. J. and Hornykiewicz, O. (1975). Noradrenaline in subcortical brain region of patients with Parkinson's disease and control subjects. In *Advances in parkinsonism* (ed. W. Birkmayer and O. Hornykiewicz), pp. 178–84, Editions Roche, Basle.

——, Price, K. S., and Hornykiewicz, O. (1977). Dopamine in the limbic regions of the human brain: normal and abnormal. In *Non striatal dopaminergic neurons* (ed. E. Costa and G. L. Gessa), Advances in biochemical psychopharmacology, Vol. 16, pp. 57–64. Raven Press, New York.

——, ——, McCullough, E., Deck, J. H. N., Hordynski, W., and Hornykiewicz, O. (1978). Norepinephrine in chronic paranoid schizophrenia: above-normal levels in limbic forebrain. *Science* **200**, 456–8.

Gerner, R. H. and Hare, T. A. (1981). CSF GABA in normal subjects and patients with depression, schizophrenia, mania and anorexia nervosa. *Am. J. Psychiat.* **138**, 1098–101.

Gomes, U. C. R., Shanley, B. C., Potgieter, L., and Roux, J. T. (1980). Noradrenergic overactivity in chronic schizophrenia: evidence based on cerebrospinal fluid noradrenaline and cyclic nucleotide concentrations. *Br. J. Psychiat.* **137**, 346–51.

Govoni, S., Hong, J. S., Yang, H., and Costa, E. (1980). Increase of neurotensin content elicited by neuroleptics in nucleus accumbens. *J. Pharmacol. exp. Ther.* **215**, 413–17.

Guidotti, A., Gale, K., Toffano, G., and Vargas, F. M. (1978). Tolerance to tyrosine hydroxylase activation in n. accumbens and c. striatum after repeated injections of "classical" and "atypical" antischizophrenic drugs. *Life Sci.* **33**, 501–6.

Gundlach, A. L., McDonald, D., and Beart, P. M. (1982). ³H-spiperone labels non-cyclase-linked dopamine receptors in the vental tegmental area of the rat. *J. Neurochem.* **39**, 890–4.

Higashi, H., Inokuchi, H., Nishi, S., Inanaga, K., and Gallagher, J. P. (1981). The effects of neuroleptics on the GABA receptor of cat primary afferent neurons. *Brain Res.* **222**, 103–17.

Hökfelt, T., Lundberg, J. M., Schultzberg, M., Johansson, O., Ljungdahl, A., and Rehfeld, J. (1980). Coexistence of peptides and putative transmitters in neurons. In *Neural peptides and neuronal communication* (ed. E. Costa and M. Trabucchi), pp. 1–23. Raven Press, New York.

Hollister, L. E. (1972). Mental disorders, antipsychotic and antimanic drugs. *New Engl. J. Med.* **286**, 984–7.

Höllt, V. and Bergmann, M. (1982). Effects of acute and chronic haloperidol treatment on the concentrations of immunoreactive β-endorphin in plasma pituitary and brain of rats. *Neuropharmacology* **21**, 147–54.

Hong, J. S., Yang, M., Fratta, W., and Costa, E. (1978). Rat striatal methionine-enkephalin content after chronic treatment with cataleptogenic and non cataleptogenic antischizophrenic drugs. *J. Pharmacol. exp. Ther.* **205**, 141–7.

Howlett, D. R. and Nahorski, S. R. (1978). A comparative study of ³H-haloperidol and ³H-spiroperidol binding to receptors on rat cerebral membranes. *FEBS Lett.* **87**, 152–6.

Ingvar, D. H. (1976). Functional landscapes of the dominant hemisphere. *Brain Res.* **107**, 181–97.

Iversen, L. L., Bird, E., Spokes, E., Nicholson, S. H., and Suckling, C. J. (1979). Agonist specificity of GABA binding sites in human brain and GABA in Huntington's disease and schizophrenia. In *GABA-neurotransmitters* (ed. P. Krogsgaard-Larsen, J. Scheel-Krüger, and H. Kofod), pp. 179–90. Munksgaard, Copenhagen.

Jackson, D. M., Dunstan, R., and Perrington, A. (1979). The hyperkinetic syndrome following long-term haloperidol treatment: involvement of dopamine and noradrenaline. *J. neural Transmission* **44**, 175–8.

Janssen, P. A. (1965). Pharmacological aspects. In *Neuropsychopharmacology* (ed. D. Bente and P. B. Bradley), Vol. 4, pp. 151–9. Elsevier, Amsterdam.

ˈJorgensson, A., Fog, R., and Veili, S. B. (1979). Synthetic enkephalin analogue in treatment in schizophrenia. *Lancet* i, 935.

Kato, N., Shah, K. R., Friesen, H. G., and Havlicek, V. (1981). Effect of chronic treatment with haloperidol on serum prolactin, striatal opiate receptors and β-endorphin content in rat brain and pituitary. *Prog. Neuropsychopharmacol.* **5**, 549–52.

Kebabian, J. W. and Calne, D. B. (1979). Multiple receptors for dopamine. *Nature* **277**, 93–6.

Keller, H. H., Bartholini, G., and Pletscher, A. (1973). Increase of 3-methoxy-4-hydroxyphenylethyleneglycol in rat brain by neuroleptic drugs. *Eur. J. Pharmacol.* **23**, 183–6.

Köhler, C., Ogren, S. O., Haglund, L., and Angeby, T. (1979). Regional displacement by sulpiride of ³H-spiperone binding in vivo. Biochemical and behavioural evidence for a preferential action on limbic and nigral dopamine receptors. *Neurosci. Lett.* **13**, 51–5.

Lautin, A., Angrist, B., Stanley, M., Gershon, S., Heckl, K., and Karobath, M. (1980): Sodium valproate in schizophrenia: some biochemical correlates. *Br. J. Psychiat.* **137**, 240–4.

Lee, T., Seeman, P., Tourtelotte, W. W., Farley, I. J., and Hornykiewicz, O. (1978). Binding of ³H-neuroleptics and ³H-apomorphine in schizophrenic brains. *Nature* **274**, 897–900.

Lindvall, O. and Björklund, A. (1978). Anatomy of the dopaminergic neuron systems in the rat brain. In *Dopamine* (ed. P. J. Roberts, G. N. Woodruff, and L. L. Iversen), Advances in biochemical psychopharmacology, Vol. 19, pp. 1–24. Raven Press, New York.

Lloyd, K. G. (1978a). Neurotransmitter interactions related to central dopamine neurons. In *Essays in neurochemistry and neuropharmacology* (ed. M. B. H. Youdim, W. Lovenberg, D. F. Sharman, and J. R. Lagnado), Vol. 3, pp. 131–207. Wiley New York.

—— (1978b). Observations concerning neurotransmitter interaction in schizophrenia. In *Cholinergic–monoaminergic interactions in the brain* (ed. L. L. Butcher), pp. 363–92. Academic Press, New York.

—— (1978c). The biochemical pharmacology of the limbic system: neuroleptic drugs. In *Limbic mechanisms* (ed. K. E. Livingston and O. Hornykiewicz), pp. 263–305. Plenum, New York.

——, Shibuya, M., Davidson, L., and Hornykiewicz, O. (1977). Chronic neuroleptic therapy: tolerance and GABA systems. In *Non striatal dopaminergic neurons* (ed. E Costa and G. L. Gessa), Advances in biochemical psychopharmacology, Vol. 16, pp. 409–15. Raven Press, New York.

Malek-Ahmadi, P. and Callen, K. E. (1980). Endorphins and schizophrenia – narcotic antagonists in the treatment of chronic schizophrenia. *Gen. Pharmacol.* **11**, 149–51.

Mao, C. C., Cheney, D. L., Marco, E., Revuelta, A., and Costa, A. (1977a). Turnover times of gamma-aminobutyric acid and acetylcholine in nucleus accumbens, globus pallidus and substantia nigra: effects of repeated administration of haloperidol. *Brain Res.* **132**, 375–9.

——, Marco, E., Revuelta, A., Bertilsson, L., and Costa, E. (1977*b*). The turnover rate of γ-aminobutyric acid in the nuclei of telencephalon: implications in the pharmacology of antipsychotics and of a minor tranquilizer. *Biol. Psychiat.* **12**, 359–71.

Matthysse, S. (1973). Antipsychotic drug actions: a clue to the neuropathology of schizophrenia? *Fed. Proc.* **32**, 200–5.

Middlemiss, D. N., Buxton, D. A., and Greenwood, D. T. (1981). Beta-adrenoceptor antagonists in psychiatry and neurology. *Pharmacol. Ther.* **12**, 419–37.

Mora, F., Sweeney, K. F., Rolls, E. T., and Sanguinetti, A. M. (1976). Spontaneous firing rate of neurones in the prefrontal cortex of the rat: evidence for a dopaminergic inhibition. *Brain Res.* **116**, 516–26.

Morselli, P. L., Bossi, L., Henry, J. F., Zarifian, E., and Bartholini, G. (1980). On the therapeutic action of SL 76 002, a new GABA-mimetic agent: preliminary observations in neuropsychiatric disorders. *Brain Res. Bull.* **5** (suppl. 2), 411–14.

Muller, E. E. (1979). Dopaminergic receptors and the secretion of anterior pituitary hormones. In *Sulpiride and other benzamides* (ed. P. F. Spano, M. Trabucchi, G. U. Corsini and G. L. Gessa), pp. 149–56. Italian Brain Research Foundation Press, Milan.

Murrin, L. C. and Kuhar, M. J. (1979). Dopamine receptors in the rat frontal cortex: an autoradiographic study. *Brain Res.* **177**, 279–85.

Nemeroff, C. B. (1980). Neurotensin: perchance an endogenous neuroleptic. *Biol. Psychiat.* **15**, 283–302.

Palacios, J. M., Wamsley, J. K., and Kuhar, M. J. (1981). High affinity GABA receptors-autoradiographic localization. *Brain Res.* **222**, 285–307.

Palmour, R., Ervin, F., and Wagemaker, H. (1979). Characterization of a peptide from the serum of psychotic patients. In *Endorphins in mental health research* (ed. E. Usdin, W. E. Jr. Bunney, and N. S. Kline), pp. 581–93. MacMillan Press, London.

Papez, J. W. (1937). A proposed mechanism of emotion. *Arch. Neurol. Psychiat.* **38**, 725–43.

Perez De La Mora, M., Fuxe, K. Hökfelt, T., and Ljungdahl, A. (1975). Effect of apomorphine on the GABA turnover in the DA cell group rich area of the mesencephalon. Evidence for the involvement of an inhibitory GABAergic feedback control of the ascending DA neurons. *Neurosci. Lett.* **1**, 109–14.

——, Possani, L. D., Tapia, R., Teran, L., Palacios, R., Fuxe, K., Hökfelt, T., and Ljungdahl, A. (1981). Demonstration of central γ-aminobutyrate-containing nerve terminals by means of antibodies against glutamate decarboxylase. *Neuroscience* **6**, 875–95.

Peroutka, S. J. and Snyder, S. H. (1980). Relationship of neuroleptic drug effects at brain dopamine, serotonin, α-adrenergic and histamine receptors to clinical potency. *Am. J. Psychiat.* **137**, 1518–22.

Perry, T. L., Buchenan, J., Kish, S. J., and Hansen (1979). Gamma-aminobutyric acid deficiency in brain of schizophrenic patients. *Lancet* **i**, 237–9.

——, Hansen, S., and Kish, S. J. (1978). Effects of chronic administration of antipsychotic drugs on GABA and other amino acids in the mesolimbic area of rat brain. *Life Sci.* **24**, 283–8.

Pijnenburg, A. J., Hönig, W. M., and Van Rossum, J. M. (1975). Effects of antagonists upon locomotor stimulation induced by injection of dopamine and noradrenaline into the nucleus accumbens of nialamide-pretreated rats. *Psychopharmacology* **41**, 175–85.

——, and Van Rossum, J. M. (1973). Stimulation of locomotor activity following injection of dopamine into the nucleus accumbens. *J. Pharm. Pharmacol.* **25**, 1003–13.

Price, K. S., Farley, I. J., and Hornykiewicz, O. (1978). Neurochemistry of Parkinson's disease: relation between striatal and limbic dopamine. In *Dopamine* (ed. P. J.

Roberts, G. N. Woodruff, and L. L. Iversen), Advances in biochemical psychopharmacology, Vol. 18, pp. 293–300. Raven Press, New York.

Rebec, G. V., Alloway, K. D., and Bashore, T. R. (1981). Differential actions of classical and atypical antipsychotic drugs on spontaneous neuronal activity in the amygdaloid complex. *Pharmacol. Biochem. Behav.* **14**, 49–56.

Roberts, E. (1976). An hypothesis suggesting that there is a defect in the GABA system in schizophrenia. *Neurosci. Res. Prog. Bull.* **10**, 468–82.

Robinson, S. E., Berney, S., Mishra, R., and Sulser, F. (1979). The relative role of dopamine and norepinephrine receptor blockade in the action of antipsychotic drugs: metoclopramide, thiethylperazine and molindone as pharmacological tools. *Psychopharmacology* **64**, 141–7.

Saavedra, J. M. and Zivin, J. (1976). Tyrosine hydroxylase and dopamine-β-hydroxylase: distribution in discrete areas of the rat limbic system. *Brain Res.* **105**, 517–24.

Scatton, B. (1977). Differential regional development of tolerance to increase in dopamine turnover upon repeated neuroleptic administration. *Eur. J. Pharmacol.* **46**, 363–9.

—— (1979). Acute and subacute effects of haloperidol on DOPAC levels in substantia nigra and ventral tegmental area in the rat brain. *Eur. J. Pharmacol.* **56**, 183–4.

—— (1981*a*). Differential changes in DOPAC levels in the hippocampal formation, septum and striatum of the rat induced by acute and repeated neuroleptic treatment. *Eur. J. Pharmacol.* **71**, 499–503.

Scatton, B. (1981*b*). Modes d'action biochimiques des neuroleptiques. *L'Encéphale* **7**, 201–14.

——, Bischoff, S., Dedek, J., and Korf, J. (1977*a*). Regional effects of neuroleptics on dopamine metabolism and dopamine-sensitive adenylate cyclase activity. *Eur. J. Pharmacol.* **44**, 287–92.

——, Boireau, A., Garret, C., Glowinski, J., and Julou, L. (1977*b*). Action of the palmitic ester of pipotiazine on dopamine metabolism in the nigrostriatal, mesolimbic and mesocortical systems. *Naunyn Schmiedeberg's Arch. Pharmacol.* **296**, 169–75.

——, D'Ambrosio, A., Javoy-Agid, F., Agid, Y., Bischoff, S., Simon, H., and Le Moal, M. (1982*a*). Evidence for the existence of a dopaminergic innervation of the rat and human hippocampal formation. In *Advances in dopamine research* (ed. M. Kohsaka, T. Shohmori, Y. Tsukada, and G. N. Woodruff), Advances in the biosciences, Vol. 37, pp. 377–82. Pergamon Press, Oxford.

——, Garret, C., and Julou, L. (1975). Acute and subacute effects of neuroleptics on dopamine synthesis and release in the rat striatum. *Naunyn Schmiedeberg's Arch. Pharmacol.* **289**, 419–34.

——, Glowinski, J., and Julou, L. (1976)., Dopamine metabolism in the mesolimbic and mesocortical dopaminergic systems after single or repeated administration of neuroleptics. *Brain Res.* **109**, 184–9.

——, Javoy-Agid, F., Rouquier, L. Dubois, B., and Agid, Y. (1983). Reduction of cortical dopamine, noradrenaline, serotonin and their metabolites in Parkinson's disease. *Brain Res.* **275**, 321–8.

——, Rouquier, L., Javoy-Agid, F., and Agid, Y. (1982*b*). Dopamine deficiency in the cerebral cortex in Parkinson disease. *Neurology* **32**, 1039–40.

——, Simon, H., Le Moal, M., and Bischoff, S. (1980). Origin of dopaminergic innervation of the rat hippocampal formation. *Neurosci. Lett.* **18**, 125–131,

——, Zivkovic, B., Dedek, J., Lloyd, K. G., Constantinidis, J., Tissot, R., and Bartholini, G. (1982*c*). γ-aminobutyric acid (GABA) receptor stimulation. III. Effect of progabide (SL 76 002) on norepinephrine, dopamine and 5-hydroxytryptamine turnover in rat brain areas. *J. Pharmacol. exp. Ther.* **220**, 678–88.

Seeman, P. (1977). Antischizophrenic drugs. Membrane receptor sites of action. *Biochem. Pharmacol.* **26**, 1741–48.

—— (1980). Brain dopamine receptors. *Pharmacol. Rev.* **32**, 229–313.

Simon, H., Le Moal, M., and Calas, A. (1979). Efferents and afferents of the ventral tegmental-A10 region studied after local injection of ³H-leucine and horseradish peroxidase. *Brain Res.* **178**, 17–40.

Snyder, S. H. (1973). Amphetamine psychosis: a "model" schizophrenia mediated by catecholamines. *Am. J. Psychiat.* **130**, 61–7.

——, Banerjee, S. P., Yamamura, H. I., and Greenberg, D. (1974). Drugs, neurotransmitters and schizophrenia. *Science* **184**, 1243–53.

——, Greenberg, D., and Yamamura, H. I. (1974). Antischizophrenic drugs and brain cholinergic receptors. *Arch. gen. Psychiat.* **31**, 58–61.

Spano, P. F., Govoni, S., and Trabucchi, M. (1978). Studies on the pharmacological properties of dopamine receptors in various areas of the central nervous system. In *Dopamine* (ed. P. J. Roberts, G. N. Woodruff, and L. L. Iversen), Advances in biochemical psychopharmacology, Vol. 19, pp. 155–65. Raven Press, New York.

Stevens, J. R. (1973). An anatomy of schizophrenia? *Arch. gen. Psychiat.* **29**, 177–89.

——, Wilson, R., and Foote, W. (1974). GABA blockade, dopamine and schizophrenia: experimental studies in the cat. *Psychopharmacologia* **39**, 105–19.

Tamminga, C. A., Crayton, J. W., and Chase, T. N. (1978a). Muscimol: GABA agonist therapy in schizophrenia. *Am. J. Psychiat.* **135**, 746–7.

——, Schaffer, M. H., Smith, R. C., and Davis, J. M. (1978b). Schizophrenic symptoms improve with apomorphine. *Science* **200**, 567–8.

——, Tighe, P., Chase, T., Defraites, E. G., and Schaffer, M. (1981). Des-tyrosine-γ-endorphin administration in chronic schizophrenics. *Arch. gen. Psychiat.* **38**, 167–8.

Taquet, H., Javoy-Agid, F., Cesselin, F., Hamon, M., Legrand, J. C., and Agid, Y. (1982). Microtopography of methionine–enkephalin, dopamine and noradrenaline in the ventral mesencephalon of human control and parkinsonian brains. *Brain Res.* **235**, 303–14.

Terenius, L., Wahlström, A., Lindstrom, L., and Widerlöv, E. (1976). Increased CSF levels of endorphines and chronic psychoses. *Neurosci. Lett.* **3**, 157–62.

Trabucchi, M., Govoni, S., Tonon, G. C., and Spano, P. F. (1978). Functional interaction between receptors for dopamine antagonists and GABA central receptors. *Life Sci.* **23**, 1751–6.

U'Prichard, D. C., Greenberg, D. A., and Snyder, S. H. (1976). Binding characteristics of a radiolabeled agonist and antagonist at central nervous system α-noradrenergic receptors. *Mol. Pharmacol.* **13**, 454–73.

Verhoeven, W., Van Praag, H., Van Ree, J., and De Wied, D. (1979). Improvement of schizophrenic patients treated with [Des-Tyr¹]-γ-endorphin (DTγE). *Arch. gen. Psychiat.* **36**, 294–8.

Walaas, I. and Fonnum, F. (1978). The distribution of putative monoamine, GABA acetylcholine and glutamate fibers in the mesolimbic system. In *GABA-neurotransmitters* (ed. P. Krogsgaard-Larsen, J. Scheel-Kruger, and H. Kofod), pp. 60–73. Munksgaard, Copenhagen.

—— and —— (1980). Biochemical evidence for γ-aminobutyrate containing fibers from the nucleus accumbens to the substantia nigra and ventral tegmental area in the rat. *Neuroscience* **5**, 63–72.

Waldmeier, P. C. and Maitre, L. (1976). On the relevance of preferential increases of mesolimbic versus striatal dopamine turnover for the prediction of antipsychotic activity of psychotropic drugs. *J. Neurochem.* **27**, 589–97.

—— and —— (1979). The use of scopolamine for the estimation of the central antiacetylcholine properties of neuroleptics. *J. Pharm. Pharmacol.* **31**, 553–5.

Weinreich, P. and Seeman, P. (1981). Binding of adrenergic ligands (³H-clonidine and ³H-WB 4101) to multiple sites in human brain. *Biochem. Pharmacol.* **30**, 3115–20.

Westerink, B. H. and Korf, J. (1975). Influence of drugs on striatal and limbic homovanillic acid concentrations in the rat brain. *Eur. J. Pharmacol.* **33**, 31–40.

——, Lejeune, B., Korf, J., and Van Praag, H. M. (1977). On the significance of regional dopamine metabolism in the rat brain for the classification of centrally acting drugs. *Eur. J. Pharmacol.* **42**, 179–89.

Widerlöv, E., Lindström, L. H., Beseve, G., Manberg, P. J., Nemeroff, C. B., Breese, G. R., Kizer, J., and Prange, A. J. (1982). Subnormal CSF levels of neurotensin in a subgroup of schizophrenic patients: normalisation after neuroleptic treatment. *Am. J. Psychiat.* **139**, 1122–6.

Zarifian, E., Scatton, B., Bianchetti, G., Cuche, H., Loo, H., and Morselli, P. L. (1982). High doses of haloperidol in schizophrenia: clinical, biochemical and pharmacokinetic study. *Arch. gen. Psychiat.* **39**, 212–15.

Zivkovic, B., Guidotti, A., Revuelta, A., and Costa, E. (1975). Effect of thioridazine, clozapine and other antipsychotics on the kinetic state of tyrosine hydrolase and on the turnover rats of dopamine in striatum and nucleus accumbens. *J. Pharmacol. exp. Ther.* **194**, 37–47.

——, Worms, P., Scatton, B., Dedek, J., Oblin, A., Lloyd, K. G., and Bartholini, G. (1983). Functional similarities between benzamides and other neuroleptics. In *Receptors as supramolecular entities* (ed. G. Biggio, E. Costa, G. L. Gessa, and P. F. Spano), pp. 155–70. Pergamon Press, Oxford.

13

Antidepressants and the limbic system

MAURICE JALFRE AND ROGER D. PORSOLT

INTRODUCTION

Although there exist a large number of publications treating effects of anti-depressants on the central nervous system as a whole, relatively few publications have been devoted to differential effects of antidepressants on the various anatomical structures of the brain, in particular those of the limbic system. The paucity of data in this field contrasts with the relative wealth of information about the effects of anti-epileptics, anxiolytics, and neuroleptics in these brain regions.

This chapter reviews the experimental literature concerning the effects of antidepressants on the limbic system and covers primarily behavioural but also electrophysiological and biochemical aspects. Among the behavioural aspects we shall deal with the effects of systemic or local administration of antidepres-sants on spontaneous behaviour or on behaviour during tests for antidepressants in animals where various limbic structures have been either lesioned or subjected to electrical stimulation. Under electrophysiological aspects we shall discuss the effects of antidepressants on electrical activity in different brain regions after stimulation of limbic structures, and on intralimbic evoked potentials. Finally we shall attempt to determine whether antidepressants show differential effects between intralimbic structures and between intra- and extralimbic structures on biochemical parameters such as neurotransmitter turnover or tricyclic binding.

BEHAVIOURAL ASPECTS

Effects of antidepressants on behaviour induced by stimulation or lesion of limbic structures

To our knowledge this kind of effect has only been studied in the amygdala. Electrical stimulation of the amygdala in cats produces a characteristic 'rage' reaction which can be suppressed by peripheral administration of imipramine (Penaloza-Rojas *et al.* 1961). This inhibitory action of antidepressants on the amygdala was confirmed by Allikmets and Delgado (1968) who showed that amitriptyline or imipramine injected directly into the amygdala in monkeys suppressed the 'attention response' (ear movements, licking, and searching) induced by electrical stimulation of the same structure. This suppressive effect of imipramine on amygdaloid excitability does not appear to be related to the known local anaesthetic effect of tricyclics since novocaine was inactive. In . contrast to electrical stimulation, bilateral destruction of the amygdala lowers emotional reactivity in rats. This effect can be at least partially reversed by chronic treatment with imipramine (Allikmets and Lapin 1967). Thus, in addition to the suppressant effects of tricyclics on amygdaloid function, the effects observed after amygdalectomy suggest that tricyclics may also possess an intrinsic stimulating component. Further support for an intrinsic stimulant effect of tricyclics in the absence of the amygdala is provided by the finding that the antireserpine effects of imipramine were much more marked after bilateral amygdalectomy (Allikmets and Lapin 1967). In contrast Górka *et al.* (1979) reported that unilateral amygdalectomy abolished the immobility-reducing effects of imipramine in the 'behavioural despair' test in rats (Porsolt 1981). This latter finding is difficult to interpret, however, because bilateral amygdalectomy is usually necessary to induce behavioural effects (see below).

Effects of intralimbic injection of antidepressants during behavioural tests for antidepressant activity

Learned helplessness in the rat

Sherman and Petty (1982*a*) have shown that chronic systemic administration of various antidepressants, but not other psychotropic compounds, decreases the deficit in passive avoidance learning which is observed when animals have previously been exposed to unavoidable electric shocks ('learned helplessness'). The same authors (Sherman and Petty 1980) have also shown that similar effects can be observed after local injections of imipramine or desipramine into the hippocampus but not other limbic areas. Furthermore, in animals chronically treated with imipramine before being exposed to inescapable shock, those showing reduced 'helplessness' had higher total levels of intrahippocampal imipramine/desipramine with a higher proportion of desipramine than those whose 'helplessness' remained unaffected (Sherman and Allers 1980). The mitigating effects of antidepressants on 'learned helplessness' appear to depend on a GABAergic mechanism because only intrahippocampal injections of GABA produce effects similar to those observed with tricyclic antidepressants (Sherman and Petty 1980). Furthermore, hippocampal release of GABA is diminished in 'helpless' animals, an effect which is mitigated by chronic treatment with imipramine, whereas chronic treatment with imipramine in naïve animals increases hippocampal GABA release (Sherman and Petty 1982*b*). Although tricyclic antidepressants are inactive in the same test when injected directly into the septum, septal serotonin (5-HT) appears, nonetheless, to be implicated in the effects of tricyclics on 'learned helplessness'. The decrease in septal 5-HT release observed in 'helpless' animals is abolished by tricyclic treatment, and 5-HT, injected directly into the septum, has mitigating effects on 'learned helplessness' similar to those produced by hippocampal injections of tricyclics or GABA.

These results, together with the observation that desipramine or 5-HT injected into the frontal cortex also mitigate 'learned helplessness' (Sherman and Petty 1980), suggest that tricyclics activate a cortico–hippocampal–septal circuit depending respectively on cortical serotonergic, hippocampal GABAergic, and septal serotonergic neurotransmission. A finding discordant with this scheme, however, is the lack of effect of intraseptal injections of desipramine.

Muricide behaviour in rats

Numerous authors have shown that antidepressants decrease, in a fairly specific manner, spontaneous mouse-killing behaviour in the rat (for a review see Vogel 1975). Similar decreases in muricide behaviour are seen after bilateral centro-medial amygdalectomy (Karli *et al.* 1969). Injections of imipramine or d-amphetamine into the centro-medial amygdala, but not in the basolateral amygdala and the septum, reduce mouse killing whereas similar injections of chlordiazepoxide or chlorpromazine are ineffective. Intra-amygdaloid injections of atropine, noradrenaline (NA), or 5-HT also reduce mouse killing whereas mouse killing is increased by injections of eserine (Vogel 1975).

Mouse killing behaviour can also be induced by bilateral olfactory bulbectomy (Vergnes and Karli 1963). Similarly to spontaneous mouse killing, bulbectomy-induced mouse killing is specifically inhibited by various antidepressant therapies (for a review see Ueki 1982). Local administration of tricyclic anti-depressants, methamphetamine, or NA into the centro-medial amygdala but not into the basolateral amygdala, septum, or hippocampus inhibits bulbectomy-induced mouse killing whereas nialamide, chlorpromazine, or atropine are only weakly active and diphenhydramine, chlordiazepoxide, acetylcholine, and 5-HT are without effect (Watanabe *et al.* 1979). Several lines of evidence suggest that bulbectomy-induced mouse killing depends primarily on activity within the central noradrenergic system. It is inhibited more potently by desipramine, a specific NA-uptake inhibitor, than by imipramine which is more potent in inhibiting the uptake of 5-HT. Furthermore mouse killing is inhibited by stimulation of the locus ceruleus, an area rich in noradrenergic cells, whereas lesions in the dorsal noradrenergic bundle enhance mouse killing. On the other hand stimulation of the raphe nucleus, rich in serotonergic cells, or treatment with the 5-HT precursor L-5-HTP have little effect. Phenoxybenzamine, an α-noradrenergic receptor blocker, is highly potent in inhibiting mouse killing, whereas the β-noradrenergic receptor blocker, sotalol, is ineffective suggesting that mouse killing depends primarily on α-noradrenergic neurotransmission.

Passive avoidance deficit in bulbectomized rats

Bilateral bulbectomy causes deficits in the acquisition of a passive avoidance task which can be corrected by chronic but not acute treatment with several antidepressants including mianserin, trazodone, and viloxazine – newer compounds with profiles of pharmacological activity different from those of classical antidepressants (Van Riezen *et al.* 1977; Cairncross *et al.* 1978; Broekkamp *et al.* 1982). D-amphetamine also decreases the passive avoidance deficit whereas the deficit is increased by the MAOI tranylcypromine (Van Riezen *et al.* 1977). The passive avoidance deficit is also decreased by acute systemic injections of 5-HT- and GABA-mimetics but is not affected by apomorphine or atropine and is even enhanced by chlordiazepoxide and chlor-promazine. Drug injection of amitriptyline or imipramine into the centro-medial but not the basolateral amygdala decreases the passive avoidance deficit. These effects appear to depend on the 5-HT system because they are blocked by systemic injection of the 5-HT blocker metergoline. Further evidence implicating a role for 5-HT is the fact that the passive avoidance deficit is blocked by local injections of 5-HT, but not NA or GABA, although GABA-mimetics are active after systemic administration. In contrast atropine is active after central but not after systemic administration (Garrigou *et al.* 1981; Broekkamp *et al.* 1982; Lloyd *et al.* 1982).

Amygdaloid kindling

Repeated low-intensity electrical stimulation of various brain regions, initially without overt effects eventually results in seizures at each stimulation ('kindling') (Babington 1981). Tricyclic antidepressants and electroconvulsive shock after acute but particularly after chronic treatment inhibit convulsions originating from the amygdala at doses lower than those which inhibit cortically- or septally-induced convulsions.

MAOI are less effective and some atypical compounds, such as iprindole and zimelidine, are without effect (Babington 1975; Stach *et al.* 1980). Minor tranquillizers also decrease kindled amygdaloid convulsions but, in contrast to tricyclics, have similar effects on both amygdaloid- and cortically-induced convulsions. Amygdaloid kindling is enhanced by pretreatment with reserpine and tetrabenazine, suggesting that kindling is controlled by monoaminergic inhibitory processes (Wilkison and Halpern 1979; Stach *et al.* 1980).

ELECTROPHYSIOLOGICAL ASPECTS

Intralimbic evoked potentials

The effects of antidepressants on intralimbic evoked potentials have been studied by numerous authors using a wide variety of different experimental conditions (for a review see Gogolak 1980). The results, obtained mostly with tricyclics, indicate that in general the potentials evoked by stimulation of the septum with recording from the amygdala or from the ventral or dorsal hippocampus, or those evoked in the ventral hippocampus by stimulation of the amygdala, are not modified by tricyclics. Potentials evoked in the septum by stimulation of the dorsal hippocampus or in the dorsal hippocampus by stimulation of the amygdala are increased in amplitude. Only intra-amygdaloid evoked potentials are decreased by tricyclic antidepressants (Guerrero-Figueroa and Gallant 1967). Interpretation of these findings is difficult, however, because it is not possible to conclude, even when only intralimbic potentials are taken into consideration, that the changes observed are due to a direct action of the compounds studied on limbic structures. The results obtained are the summation of the effects of excitability in several structures. This difficulty is illustrated by the findings of a Japanese group (Oishi *et al.* 1977; Shibata *et al.* 1979) who showed that the amplitude of the bulbo–amygdaloid evoked potential in the rat is diminished by prior stimulation of the locus ceruleus, by injection of NA into the ventricles or the centro-medial amygdala, or by systemic injection of desipramine. Desipramine, however, decreased the inhibitory effect of locus ceruleus stimulation whereas it increased the inhibitory effect of injection of NA into the ventricles or into the amygdala. These differences can probably be explained by the ability of desipramine to block the uptake of NA and thereby to potentiate the inhibitory action of NA at postsynaptic receptor sites in the amygdala, and at the same time to inhibit the discharge of locus ceruleus noradrenergic cells by increasing NA levels at their somato–dendritic inhibitory noradrenergic receptors.

Limbic afterdischarges

Numerous authors (for a review see Gogolak 1980), using diverse experimental conditions and many different species, have studied the changes in the threshold, duration, or amplitude of afterdischarges in various brain regions after stimulation of the amygdala, septum, and the dorsal or ventral hippocampus. The antidepressants which have been most studied are the tricyclics. In general it appears that tricyclics decrease the duration or the amplitude and/or increase the threshold for afterdischarges originating from the amygdala or septum. They appear to have an opposite effect on amygdaloid and septal afterdischarges when administered in low doses. Tricyclics have no effect on hippocampal afterdischarges. In contrast to tricyclics, MAOI drugs and mianserin increase amygdaloid afterdischarges (Ossowska and Wolfarth 1980) and MAOIs decrease hippocampal afterdischarges (Gogolak 1980).

BIOCHEMICAL ASPECTS

Neurotransmitters

In contrast to the wealth of data existing for neuroleptics, very few publications have dealt with differential effects of antidepressants on the metabolism of the various putative neurotransmitters in different brain regions. Miyauchi's group (Miyauchi *et al.* 1981, 1982) have studied the changes in NA metabolism in different brain structures after acute or chronic administration of amitriptyline or desipramine in rats which were either naïve or had been exposed to forced swimming ('behavioural despair'). As an index of NA turnover they studied the levels of MOPEG-SO$_4$. In naïve animals, MOPEG-SO$_4$ levels in the hippocampus were significantly decreased after amitriptyline and increased after desipramine whereas desipramine but not amitriptyline decreased the MOPEG-SO$_4$ level in the septum. In animals exposed to forced swimming, MOPEG-SO$_4$ levels were increased in all brain regions studied. Acute or chronic treatment with amitriptyline or desipramine decreased these levels in the septum, whereas the MOPEG-SO$_4$ levels in the hippocampus were further increased by desipramine but not by amitriptyline. The complexity of these findings renders their interpretation particularly difficult, a difficulty which is not diminished by the fact that changes in MOPEG-SO$_4$ levels could reflect either an increase or a decrease in the interaction of NA with its postsynaptic receptor (Reigle *et al.* 1982).

Even less information exists concerning possible regional differences in the effects of antidepressants on the metabolism of 5-HT. Lindberg *et al.* (1982), using a new specific blocker of 5-HT uptake, alaproclate, showed that this compound inhibited 5-HT uptake more potently in the hippocampus and hypo-thalamus than in the cerebral cortex or the striatum.

Tricyclic binding

Langer and his colleagues (for a review see Langer *et al.* 1982) have demon-strated the existence of specific high-affinity binding sites for [^{3}H]-imipramine in human and animal brain and in blood platelets. These endogenous binding sites seem to be associated with the neuronal uptake mechanism of 5-HT. Furthermore, the density of these binding sites, but not their affinity constant, differs according to the brain structure both in normal men (Langer *et al.* 1981) and in rats (Palkovits *et al.* 1981). In men, as in rats, the region with the greatest density of [^{3}H]-imipramine binding sites is the hypothalamus. In the human limbic system, the limbic cortex (area 34) contains a higher density than the amygdala whereas in rats the principal binding sites classified in decreasing order of density are the laterobasal amygdala, the centromedial amygdala, the ventral hippocampus, the dorsal hippocampus, and the septum. The high density of [^{3}H]-imipramine-binding sites in the amygdala – as high as that in the hypothalamus – suggests the importance of the amygdala for the pharmaco-logical effects of antidepressants. The distribution of specific [^{3}H]-desipramine-binding sites, on the other hand, is much more homogeneous between different brain regions apart from the cerebellum. In particular, the distribution is identical between the septum and hippocampus (Biegon and Samuel 1979). It must be pointed out, however, that the Scatchard plots of the binding sites for desipramine are not linear, suggesting a heterogeneous set of binding sites whereas Scatchard analysis of specific [^{3}H]-imipramine binding in the brain regions studied by Langer *et al.* (1981) gave straight lines, indicating a single population of non-interacting binding sites in each case.

DISCUSSION

Before attempting any general conclusions concerning the action of anti-depressants on the limbic system we should like to draw attention to some of the problems encountered when trying to characterize the effects of drugs on a single brain system, interpreting them in terms of brain function as a whole and determining their relevance to the therapeutic effects of drugs in pathological conditions.

First, it is evident that antidepressants, or any other class of psychotropic drug, do not act on a single anatomical system. Indeed, antidepressants have been shown to act at different parts of the synaptic cycle of several neuro-transmitters each belonging to separate but interacting histochemical systems with each system being distributed in a heterogeneous fashion throughout several anatomical structures within the brain. It is therefore difficult to attribute a given pharmacological or clinical effect of an antidepressant to changes observed within a particular structure unless it can be shown that the effect in question occurs in a relatively specific manner in the structure under consideration and not elsewhere.

Second, despite the large number of test models in neuropharmacology for the detection of antidepressants, only two of them (learned helplessness and muricide behaviour) have been used for studying the effects of intralimbic injections of antidepressants and lesions to limbic structures. Only one publication has investigated the effects of antidepressants on the reserpine syndrome after lesions in a single limbic structure. No study has been performed to investigate the effects of antidepressants after selective lesions of the different afferent pathways or cell types within the limbic system.

Third, all the results concerning the effects of antidepressants on the limbic system have been obtained after administration of antidepressants to animals which were either 'normal' or were manipulated in a manner which cannot unequivocally be identified as resembling 'real' depression. Furthermore, at present there is no clear biochemical (Langer and Karobath 1980), electrophysiological (Itil and Soldatos 1980), or radiochemical (Dagani 1981) information about limbic function in depressed patients with or without antidepressant treatment. Virtually all we know is that there are regional differences in normal man in the density of high-affinity [³H]-imipramine-binding sites.

Bearing these reservations in mind, what can we conclude concerning the role of the limbic system in the mode of action of antidepressant drugs? Most of the results described above agree in suggesting that the amygdala, in particular the centro-medial amygdala, might represent an important site of action for tricyclic antidepressant drugs. This is suggested first by the fact that local injections of tricyclics directly into the centro-medial amygdala have effects identical to those observed after systemic injections, but not after injections into other parts of the brain (Vogel 1975; Watanabe *et al.* 1979; Lloyd *et al.* 1982). Further suggestive evidence for this hypothesis is the fact that the amygdala is implicated in many of the animal test models in neuropharmacology for the detection of antidepressants. In some models (e.g. amygdaloid kindling), the amygdala is directly involved, whereas in others (e.g. spontaneous and bulbectomy-induced muricide behaviours) a role for the amygdala is suggested indirectly by the fact that bilateral amygdalectomy produces effects similar to those observed after systemic or local injections of antidepressants. Participation of the amygdala in the action of tricyclics is also suggested by the finding that unilateral amygdalectomy abolishes the effects of imipramine in the 'behavioural despair' test (Gorka *et al.* 1979) and that bilateral amydalectomy enhances the effects of imipramine in the reserpine test (Allikmets and Lapin, 1967) even though these later two findings cannot be clearly interpreted.

The available evidence suggests moreover that tricyclic antidepressants act by inhibiting amygdaloid function. They block kindled amygdaloid seizures (Babington 1981) and also inhibit emotional hyperreactivity (Penaloza-Rojas *et al.* 1961) and attention behaviour (Allikmets and Lapin 1967) induced by electrical amygdaloid stimulation. These behavioural data are confirmed by the results of electrophysiological experiments which show that both intra-amygdaloid evoked potentials and amygdaloid afterdischarges are inhibited by systemic administration of tricyclic antidepressants (Gogolak 1980). One result discordant with the hypothesis of an inhibitory action of antidepressants on the amygdala is the finding that intra-amygdala injection of imipramine or desipramine did not affect 'learned helplessness' in the rat (Sherman and Petty 1980). This negative finding cannot, however, be readily interpreted because the injections were only unilateral and the precise site of injection within the amygdala was not specified. In general, to obtain consistent effects with tricyclics, it seems crucial that lesions, injections, or stimulation of the amygdala are bilateral and preferably in the centro-medial amygdala. For example the corrective effect of local injections of tricyclic antidepressants on bulbectomy-induced passive avoidance deficits or muricide behaviour only occurred after bilateral infusion into the centro-medial amygdala (Lloyd *et al.* 1982; Watanabe *et al.* 1979).

As indicated by their pharmacological and biochemical effects in experiments with 'learned helplessness' (Sherman and Petty 1980, 1982*a, b*), it is evident that the tricyclics also act on the hippocampus and the septum. The positive correlation between the mitigating effects of imipramine on 'learned helplessness' and the increases in the levels of imipramine and in particular desipramine in the hippocampus indicates the importance of the hippocampus for the pharmacological and probably clinical effects of tricyclic antidepressants. Nonetheless, in view of the paucity of studies devoted to effects of tricyclics on the hippocampus and septum, it is not possible to determine the mode of action of these compounds in the two structures. The mode of action of the MAOI drugs is even more difficult to define. As with tricyclic antidepressants, direct injection of MAOIs into the amygdala inhibits bulbectomy-induced muricide behaviour (Ueki 1982) whereas electrophysiological findings suggest that, in contrast to tricyclics, MAOIs increase the excitability of the amygdala (Gogolak 1980). In further contrast to the tricyclics, MAOIs appear to decrease hippocampal excitability because they attenuate afterdischarges originating from the hippocampus (Gogolak 1980). There are, however, too few results available with MAOIs for valid conclusions to be drawn, as is also true for the various compounds (e.g. iprindole, mianserin, viloxazine) classified as atypical antidepressants.

CONCLUSION

Despite the limited findings available, it would appear that tricyclic antidepressants inhibit activity in the amygdala, particularly in the centro-medial region. While continually bearing in mind the tentative nature of any such speculation, a heuristic hypothesis can be proposed to the effect that tricyclics achieve their therapeutic effect in part by decreasing a pathological hyperactivity or abnormality in the amygdala which might be related to the clinical manifestations of depression (Horovitz 1967). In this perspective it can be noted that some of the symptoms of depression – exaggerated negative affective reaction to environmental events, loss of appetite, decrease in libido – in some ways present a mirror image to certain symptoms known to accompany bilateral amygdalectomy in man and in animals – indifference to environmental stimuli, hyperphagia, and hypersexuality (Karli 1976). In this view antidepressant treatment might be functionally equivalent to a partial and reversible amygdalectomy.

ACKNOWLEDGEMENT

We thank Françoise Latour for typing the text.

REFERENCES

Allikmets, L. H. and Delgado, J. M. R. (1968). Injection of antidepressants in the amygdala of awake monkeys. *Arch. int. Pharmacodyn.* **175**, 170–8.

—— and Lapin, I. P. (1967). Influence of lesions of the amygdaloid complex on behaviour and on effects of antidepressants in rats. *Int. J. Neuropharmacol.* **6**, 99–108.

Babington, R. G. (1975). Antidepressives and the kindling effect. In *Industrial pharmacology, Vol. II: Antidepressants* (ed. S. Fielding and H. Lal), pp. 113–24. Futura, New York.

—— (1981). Neurophysiological techniques and antidepressive activity. In *Antidepressants: neurochemical, behavioral and clinical perspectives* (ed. S. J. Enna, J. B. Malick, and E. Richelson), pp. 157–73. Raven Press, New York.

Biegon, A. and Samuel, D. (1979). Binding of a labeled antidepressant to rat brain tissue. *Biochem. Pharmacol.* **28**, 3361–6.

Broekkamp, C. L., Garrigou, D., and Lloyd, K. G. (1982). The importance of the amygdala in the effect of antidepressants on olfactory bulbectomised rats. In *Typical and atypical antidepressants: molecular mechanisms* (ed. E. Costa and G. Racagni), pp. 371–5. Raven Press, New York.

Cairncross, K. D., Cox, B., Forster, C., and Wren, A. F. (1978). A new model for the detection of antidepressant drugs: olfactory bulbectomy in the rat compared with existing models. *J. Pharmacol. Meth.* **1**, 131–43.

Dagani, R. (1981). Radiochemicals key to new diagnostic tool. *Chem. Engng News,* 9 Nov, 30–7.

Garrigou, D., Broekkamp, C. L., and Lloyd, K. G. (1981). Involvement of the amygdala in the effect of antidepressants on the passive avoidance deficit in bulbectomised rats. *Psychopharmacology* **74**, 66–70.

Gogolak, G. (1980). Neurophysiological properties (in animals). In *Psychotropic agents: Part I: Antipsychotics and antidepressants.* (ed. F. Hoffmeister and G. Stille), pp. 415–35. Springer-Verlag, Berlin.

Gorka, Z., Ossowska, K., and Stach, R. (1979). The effect of unilateral amygdala lesion on the imipramine action in behavioural despair in rats. *J. Pharm. Pharmacol.* **31**, 647–8.

Guerrero-Figueroa, R. and Gallant, D. M. (1967). Effects of pinoxepin and imipramine on the mesencephalic reticular formation and amygdaloid complex in the cat: neurophysiological and clinical correlations in human subjects. *Curr. Ther. Res.* **9**, 387–403.

Horovitz, Z. P. (1967). The amygdala and depression. In *Antidepressant drugs* (ed. S. Garattini and M. N. G. Dukes), pp. 121–9. Excerpta Medica, Amsterdam.

Itil, T. M. and Soldatos, C. (1980). Clinical neurophysiological properties of antidepressants. In *Psychotropic agents: Part I: Antipsychotics and antidepressants* (ed. F. Hoffmeister and G. Stille), pp. 437–69. Springer-Verlag, Berlin.

Karli, P. (1976). Les bases neurophysiologiques des processus de motivation. *J. Physiol., Paris* **72**, 503–16.

——, Vergnes, M., and Didiergeorges, F. (1969). Rat–mouse interspecific aggressive behaviour and its manipulation by brain ablation and by brain stimulation. In *Aggressive behaviour* (ed. S. Garattini and E. B. Sigg), pp. 47–55. Excerpta Medica, Amsterdam.

Langer, G. and Karobath, M. (1980). Biochemical effects of antidepressants in men. In *Psychotropic agents: Part I: Antipsychotics and antidepressants* (ed. F. Hoffmeister and G. Stille), pp. 491–504. Springer-Verlag, Berlin.

Langer, S. Z., Briley, M., Raisman, R., Sette, M., and Pimoule, C. (1982). Specific tricyclic antidepressant binding sites and their relationship with the neuronal uptake of different monoamines. In *New vistas in depression* (ed. S. Z. Langer, R. Takahashi, T. Segawa, and M. Briley), pp. 21–8. Pergamon Press, Oxford.

——, Javoy-Agid, F., Raisman, R., Briley, M., and Agid, Y. (1981). Distribution of specific high-affinity binding sites for [^{3}H]-imipramine in human brain. *J. Neurochem.* **37**, 267–71.

Lindberg, U. H., Hall, H., and Ögren, S. O. (1982). Alaproclate: inhibition de la recapture neuronale de la 5-hydroxytryptamine dans différentes régions cérébrales. Sélectivité hippocampale indicative d'un traitement médicamenteux de la démence sénile. *Proc. 18th Rencontres Int. Chim. Ther.,* Rennes, p. 81.

Lloyd, K. G., Garrigou, D., and Broekkamp, C. L. (1982). The action of monoaminergic, cholinergic and gabaergic compounds in the olfactory bulbectomised rat model of depression. In *New vistas in depression* (ed. S. Z. Langer, R. Takahashi, T. Segawa, and M. Briley), pp. 179–86. Pergamon Press, Oxford.

Miyauchi, T., Kitada, Y., and Satoh, S. (1981). Effects of acutely and chronically administered antidepressants on the brain regional 3-methoxy-4-hydroxyphenylethyleneglycol sulfate in the forced swimming rat. *Life Sci.* **29**, 1921–8.

——, ——, and —— (1982). Effect of acute and chronic treatment with tricyclic antidepressants on 3-methoxy-4-hydroxyphenylethyleneglycol (MHPG-SO$_4$) contents in various regions of rat brain. *Prog. Neuropsychopharmacol. Biol. Psychiat.* **6**, 137–42.

Oishi, R., Shibata, S., and Ueki, S. (1977). Inhibitory noradrenergic mechanism in the medial amygdaloid nucleus. *Jap. J. Pharmacol.* **27**, suppl. 39P.

Ossowska, K. and Wolfarth, S. (1980). The effect of imipramine and mianserin on the behavior and EEG after-discharges induced by single electric stimulation of the rabbit amygdala. *Pol. J. Pharmacol. Pharm.* **32**, 513–22.

Palkovits, M., Raisman, R., Briley, M., and Langer, S. Z. (1981). Regional distribution of [^{3}H]-imipramine binding in rat brain. *Brain Res.* **210**, 493–8.

Penaloza-Rojas, J. H., Bach-y-Rita, G., Rubio-Chevannier, H. F., and Hernandes-Peon, R. (1961). Effects of imipramine upon hypothalamic and amygdaloid excitability. *Exp. Neurol.* **4**, 205–13.

Porsolt, R. D. (1981). Behavioral despair. In *Antidepressants: neurochemical, behavioral and clinical perspectives* (ed. S. J. Enna, J. B. Malick, and E. Richelson), pp. 121–39. Raven Press, New York.

Reigle, T. G., Wilhoit, C. S., and Moore, M. J. (1982). Analgesia and increases in limbic and cortical MOPEG-SO$_4$ produced by periaqueductal gray injections of morphine. *J. Pharm. Pharmacol.* **34**, 496–500.

Sherman, A. D. and Allers, G. L. (1980). Relationship between regional distribution of imipramine and its effect on learned helplessness in the rat. *Neuropharmacology* **19**, 159–62.

—— and Petty, F. (1980). Neurochemical basis of the action of antidepressants on learned helplessness. *Behav. Neurol. Biol.* **30**, 119–34.

—— and —— (1982*a*). Specificity of the learned helplessness animal model of depression. *Pharmacol. Biochem. Behav.* **16**, 449–54.

—— and —— (1982*b*). Additivity of neurochemical changes in learned helplessness and imipramine. *Behav. Neurol. Biol.* **35**, 344–53.

Shibata, S., Oishi, R., and Ueki, S. (1979). Effect of desipramine on noradrenergic inhibition of amygdaloid evoked potential. *Jap. J. Pharmacol.* **29**, 489–92.

Stach, R., Lazarova, M. B., and Kacz, D. (1980). The effects of antidepressant drugs on the seizures kindled from the rabbit amygdala. *Pol. J. Pharmacol. Pharm.* **32**, 505–12.

Ueki, S. (1982). Mouse-killing behavior (muricide) in the rat and the effect of anti-depressants. In *New vistas in depression* (ed. S. Z. Langer, R. Takahashi, T. Segawa, and M. Briley), pp. 187–94. Pergamon Press, Oxford.

Van Riezen, H., Schnieden, H., and Wren, A. F. (1977). Olfactory bulb ablation in the rat: behavioural changes and their reversal by antidepressant drugs. *Br. J. Pharmacol.* **60**, 521–8.

Vergnes, M. and Karli, P. (1963). Déclenchement du comportement d'agression inter-spécifique rat-souris par ablation bilatérale des bulbes olfactifs. Action de l'hydroxy-zine sur cette agressivité provoquée. *C.R. Soc. Biol.* **157**, 1061–3.

Vogel, J. R. (1975). Antidepressants and mouse killing (muricide) behavior. In *Industrial pharmacology: Vol. II: Antidepressants* (ed. S. Fielding and H. Lal), pp. 99–112. Futura, New York.

Watanabe, S., Inoue, M., and Ueki, S. (1979). Effects of psychotropic drugs injected into the limbic structures on mouse-killing behavior in the rat with olfactory bulb ablations. *Jap. J. Pharmacol.* **29**, 493–6.

Wilkinson, D. M. and Halpern, L. M. (1979). The role of biogenic amines in amygdalar kindling: I. Local amygdala after-discharges. *J. Pharmacol. exp. Ther.* **211**, 151–8.

14

Neuropeptides and limbic system function

LOUIS STINUS, ANN E. KELLEY, AND MARIA WINNOCK

INTRODUCTION

There is no doubt that in recent years, the existence and multiplicity of neuro-peptides in the central nervous system (CNS) has represented a fundamental challenge to neuroscientists. For researchers concerned with the neurochemical substrates of behaviour as well as the brain mechanisms underlying drug responses, these discoveries have indeed changed our perspective. Almost 20 years ago, with the discovery of the brain's widespread monoamine systems, a similar challenge was faced; it became clear that the biogenic amines were of crucial importance in behavioural function and consequently a great amount of brain-behaviour research was focused on this domain. This endeavour has been fruitful, although much remains to be learned concerning the functions of dopamine, serotonin, and noradrenaline. Now we face considerably more complexity; the limbic system is generously endowed with neuropeptides, and many regions rich in monoaminergic innervation contain high levels of several neuropeptides. In this chapter, rather than comprehensively reviewing this rapidly developing field, we will attempt to present general concepts and repre-sentative examples of peptide research which may be pertinent to behavioural mechanisms. The first section will underscore the general significance of peptides in the limbic system; in the second section, we will give examples of our own work on the opioid peptides and substance P.

PEPTIDES AND LIMBIC FUNCTION: GENERAL CONCEPTS

The limbic system

First we would like to define, very briefly, what we mean by 'limbic system'. This concept and definition has undergone considerable evolution since the days when Broca (1878) described 'le grand lobe limbique'. This evolution can be viewed as having four conceptual stages, which largely developed through both anatomical and behavioural findings. The first stage encompasses the classic limbic system proposed by Papez (1937) and MacLean (1949). These structures, as depicted in Fig. 14.1(A), included such areas as hippocampus, septum, amygdala, mediodorsal thalamus, and hypothalamus and, along with their connecting pathways (e.g. fornix, stria terminalis), were strongly impli-cated in emotion and affective behaviour. The second stage expands the defini-tion of the limbic system to include further subcortical limbic circuits. Based on the work of Nauta and colleagues (Nauta 1958; Nauta and Domesick 1981;

(A)

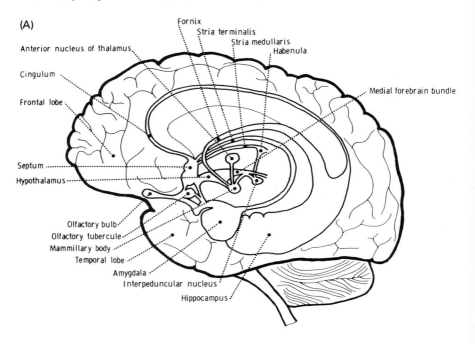

(B)

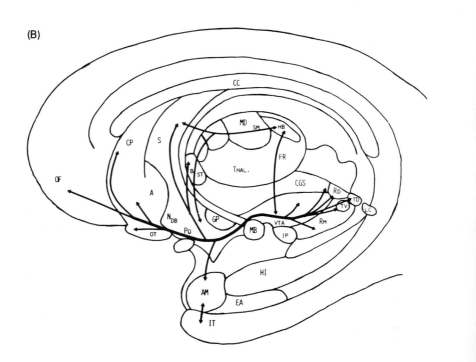

Chapter 1, this volume), this is known as the 'limbic forebrain–midbrain continuum' (Fig. 14.1(B)). In essence, these findings have shown that limbic structures and their corresponding forebrain circuits are intimately connected with the mesencephalon; thus medial midbrain areas such as the ventral tegmental area and raphe nuclei are reciprocally connected with telencephalic limbic structures via the medial forebrain bundle and stria–medullaris-fasciculus retroflexus. A large majority of ascending components are comprised of the monoaminergic systems.

The third development can be viewed as the gradual inclusion of certain neocortical areas in the limbic system. Thus, the neuropsychological literature over the past 30 years has provided much support for the involvement of prefrontal, inferotemporal, and cingulate cortices in limbic function; further, the intimate anatomical connections of these cortical areas with the subcortical limbic system are well known. Finally, the fourth fundamental modification to be considered is the association between forebrain limbic structures (particularly amygdala and hypothalamus) with brainstem nuclei concerned with visceral motor and sensory functions, notably the parabrachial nucleus, peri-peduncular nucleus, and the nucleus of the solitary tract (see Ricardo and Koh 1978; Nauta and Domesick 1981; Chapter 1, this volume).

How can we integrate these four notions into a definition of limbic function? At the risk of greatly generalizing, we can say that, within the network of limbic circuitry, there exists a superimposition of the animal's internal and external sensory states, as well as the imprint of past experience. This superimposition is fundamental in setting the affective and motivational set of the animal, in mediating appropriate and organized behavioural responses, and ultimately regulating biological adaptation and survival.

Peptide distribution in limbic structures

The striking abundance of neuropeptides within the limbic system has been emphasized repeatedly in anatomical distribution studies. Indeed, the discovery and early work on peptidergic neurones grew out of study on a convergence centre of the limbic system: the hypothalamus. The magnocellular neurones (supraoptic and paraventricular nuclei) of the hypothalamus were the first

---

Fig. 14.1. (A) Schematic diagram of structures and pathways comprising the so-called 'classic' limbic system proposed initially by Papez (1937) and MacLean (1949). (B) Diagrammatic representation of the limbic–midbrain continuum (based on Nauta and Domesick 1981). The limbic forebrain is reciprocally connected with the 'limbic midbrain area', which includes the ventral tegmental area (VTA) and the raphe nuclei (Rd, Rm). Abbreviations: OF, orbital frontal cortex; CP, caudate–putamen; S, septum; OT, olfactory tubercle; A, nucleus accumbens; CC, corpus callosum; Ndb, nucleus of the diagonal band; Po, preoptic area; MFB, medial forebrain bundle; FR, fasiculus retroflexus; SM, stria medullaris; HB, habenula; Bst, bed nucleus of stria terminalis; GP, globus pallidus; Am, amygdala; IT, infero–temporal cortex; HI, hippocampus; MB, mammillary body; MD, mediodorsal nucleus of thalamus; IP, interpeduncular nucleus; CGS, central gray substance; Tv, ventral tegmental nucleus of Gudden; Td, dorsal tegmental nucleus of Gudden; LC, locus ceruleus; EA, entorhinal area.

peptidergic neurones described, containing the pituitary hormones vasopressin and oxytocin (Scharrer 1976), and soon after further substances previously characterized as hormones were localized in hypothalamic neurones, such as thyrotropin-releasing factor (TRF), luteinizing hormone-releasing hormone (LHRH), and somatostatin (Guillemin 1978). We now know, thanks to immunohistochemical techniques, that the hypothalamus contains nearly every peptide thus far described. Further, other surrounding limbic structures contain high levels of neuropeptides, examples of which are the opioid peptides, somatostatin, neurotensin, cholecystokinin (CCK), and substance P (SP). Further cases in point can be provided: the amygdala, a region which plays a fundamental role in motivation, is extraordinarily rich in neuropeptides; the central nucleus contains particularly high amounts (Roberts *et al.* 1982). Interestingly, the central nucleus of the amygdala has connections with the nucleus of the solitary tract and parabrachial nucleus, lower brainstem nuclei concerned with autonomic function. These nuclei are rich in enkephalin and opiate receptors (Haber and Elde 1982; Wamsley *et al.* 1982), substance P (Cuello and Kanazawa 1978), somatostatin (Finley *et al.* 1981), neurotensin (Uhl *et al.* 1979), as well as numerous other peptides described in immunohistochemical maps. The striatal complex (caudate–putamen and nucleus accumbens), a central 'interface' between limbic processes and motor output (see Kelley *et al.* 1982), has an abundant distribution of peptides, notably enkephalin, CCK, SP, and neurotensin. Finally, although in general the cerebral cortex does not contain very high amounts of endogenous peptides, it is interesting to note that in certain cases neuropeptides are preferentially distributed to limbic cortical areas; for example, SP is found almost exclusively in frontal cortex (Cuello and Kanazawa 1978), CCK is highest in cingulate, pyriform, and entorhinal areas (Beinfeld *et al.* 1981), opiate receptor density is reported to be higher in frontal and cingulate regions (Herkenham and Pert 1980; Wamsley *et al.* 1982), and a recent study reported that neurones containing angiotensin are localized in frontal and cingulate cortices (Brownfield *et al.* 1982). In summary, anatomical descriptions of peptide distribution indicate that they may play an important role in limbic-system processing, and consequently, in behaviour subserved by the limbic system.

Molecular mechanisms: implications for limbic-system function

Presently, the vast majority of research on neuropeptides is concerned with descriptions of regional and cellular distribution and with molecular and biochemical mechanisms. Indeed, particularly in reviews of peptide neurobiology, there is a noticeable lack of worthwhile theories and functional models within the realm of behaviour. This situation is not surprising considering the present state of knowledge; it would be somewhat premature to propose such models. However, it is clear that anatomical descriptions of the peptide systems can provide some general clues about function; similarly, molecular mechanisms of peptides may well tell us something about their role in behaviour.

Although information available on the synthesis, storage, release, and break-down of peptides is by no means complete, it is evident that many of the associated mechanisms are distinctly different from those of more 'classical' neurotransmitters. As far as it is known, peptides or peptide precursors must be synthesized at the ribosomes within the neurone cell body, and transported down the axon to the terminal; unlike amines, peptides cannot be synthesized within the nerve terminal. In addition, once released, peptides appear to be inactivated by extracellular peptidases; there is no known re-uptake mechanism (similar to that of conventional transmitters). It has been suggested that this relatively slow and inefficient mechanism may be compensated for by a long duration of action (Hokfelt *et al.* 1980). Thus, numerous examples from the molecular biology and neuronal actions of peptides have led to the widespread use of the term 'neuromodulator', that is, substances which affect the excitability of cells, transmitter release, receptor sensitivity, or perhaps even metabolic processes and trophic functions (Florey 1967; Bloom 1973; Skirboll *et al.* 1981; see Fig. 14.2). For example, somatostatin and TRF applied to motoneurones produce long-acting changes in membrane excitability (Nicoll 1978). CCK increases the activity of dopaminergic (DA) neurones, particularly when CCK and DA coexist within the same neurones (Skirboll *et al.* 1981; the problem of coexistence is discussed below). Recent studies indicate that SP, in addition to having a neuromodulator/neurotransmitter role, may facilitate trophic functions and development; a single subcutaneous treatment of neonatal rat pups with SP reduces the damage to noradrenaline neurones caused by 6-hydroxydopamine (Jonsson and Hallman 1982). Further, SP neurones are reported to appear at a very early ontogenetical stage (Inagaki *et al.* 1982), suggesting that SP may facilitate, for example, the development of the brain's monoaminergic systems. It is likely that peptides act through very diverse mechanisms, some of which (particularly those involving long-term action) may have important behavioural consequences.

Presently, one of the most challenging problems in peptide neurobiology is the problem of 'coexistence' – the existence of classical neurotransmitters with neuropeptides within the same neurone (Hokfelt *et al.* 1980; Johansson *et al.* 1981). There is mounting evidence that this is a fairly widespread phenomenon in the CNS, and it represents a complexity which is geometric leaps above previous notions of function based on the one neurone/one transmitter hypothesis (Dale's principle). However, some interesting attempts at speculation have been made: the simultaneous release of a peptide along with a transmitter, for example cholecystokinin with dopamine, could provide a local modulatory mechanism, regulating the post synaptic response to dopamine (Skirboll *et al.* 1981). Thus, the peptide may play a significant role in mediating the 'gain' at which synaptic communication operates (Barbaccia *et al.* 1982). The implications of coexistence for behavioural research are overwhelming and undoubtedly very important, but as yet remain completely elusive.

Finally, the size and complex structure of peptides suggest intriguing possi-bilities about function. When peptides were first discovered within neurones, it

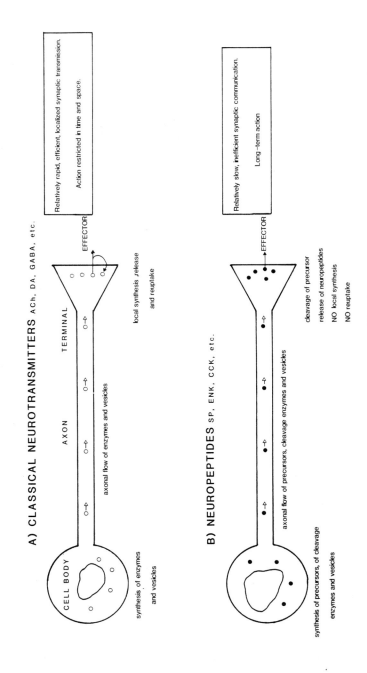

Fig. 14.2. Comparison of some of the neuronal events which may underlie (A) 'classical' neurotransmission, and (B) neurotransmission involving peptides. The relatively slower time course of such events in peptidergic neurones may be associated with long-term action, and hence neuropeptides are often termed 'Neuromodulators'. (Based on Hokfelt *et al.* (1980).)

was with considerable difficulty that they were considered as neurotransmitter candidates; one could not imagine such large molecules as having a synaptic function. However, as it is now well known that many neuronal peptides fulfil some of the criteria for neurotransmitters or neuromodulators, we can only speculate about the enormous coding flexibility and capacity for carrying information inherent within such complex molecules.

Functional considerations

Thus far, as noted above, there is a paucity of knowledge concerning the role of peptides in the control of behaviour. However, the multiplicity of peptides within the limbic system, as well as their apparent 'neuromodulatory' mechanisms, suggest that peptide action may lie within two general functional realms schematically depicted in Fig. 14.3. The first is concerned with what MacLean (1969) called 'primary affects', which he described as 'feelings that are informative of basic bodily needs'. MacLean elaborates further, 'Descriptive words for the primary affects associated with many of these basic needs come readily to mind, e.g. hunger, thirst, suffocation, fatigue, pain.' Examples of associated functions would be the intake of food and water, avoidance of pain, internal temperature and fluid regulation, and autonomic responses to emotion. There is convincing evidence that peptides are involved in these basic regulatory functions; dense distribution in the hypothalamus, the central nucleus of the amygdala, and the lower brainstem autonomic structures supports numerous experimental examples: the role of opioid peptides in pain (Gintzler 1980), cholecystokinin and opioid peptides in food intake and satiety (Sanger 1981; Morley 1982), bombesin in temperature regulation (Brown *et al.* 1977), angiotensin in fluid and osmotic regulation (Epstein 1978), to name just a few examples.

The second general domain concerns the function of peptides in forebrain and midbrain limbic structures. Peptidergic neurones within these regions may mediate emotional and adaptive behavioural responses, and may be involved in the processing of incoming sensory information (for example, in limbic cortical areas and hippocampus), as well as triggering highly integrated motor behaviour (for example, in nucleus accumbens and striatum). Particularly interesting in this regard would be the interaction between monoaminergic systems and neuropeptides. For example, the A10 mesolimbic dopamine system, originating in the midbrain (ventral tegmental area) and innervating widespread limbic regions, is crucial for the animal's motivational state and ability to behave flexibly; this system appears to be under the control of several neuropeptides both at the cell-body and terminal level, among which are neurotensin (Kalivas *et al.* 1982), CCK (Hokfelt *et al.* 1980), SP, and opioid peptides (see below).

Finally, it should be noted that, while it is heuristic to make speculations concerning functional considerations, it is impossible to think of any one peptide as having a unitary function; one peptide's function throughout the CNS may be quite diverse and would depend of course on the inputs, outputs, and local processing within a particular structure.

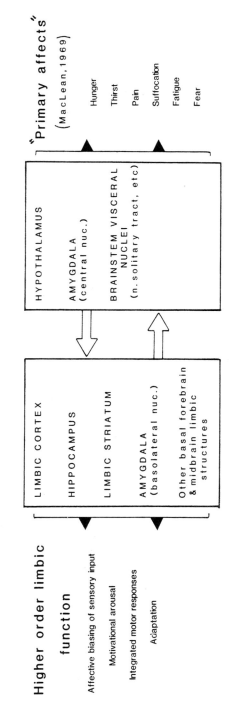

Fig. 14.3. Many limbic regions are rich in neuropeptides, and the overall function of these peptides may be hypothetically categorized into two general domains, described in text. Filled arrows indicate structures more likely to be associated with a given domain; open arrows signify the close anatomical and functional interactions between such domains.

STUDIES OF PEPTIDE FUNCTION IN THE LIMBIC SYSTEM: SUBSTANCE P AND OPIOID PEPTIDES

Introduction

We would now like to present two examples of interactions of peptides with the mesolimbic dopaminergic A10 neurones. First, with opioid peptides we show that these interactions are site-dependent, that is, they involve different mechanisms at the DA cell bodies and terminals. Second, we present evidence that in addition to their acute effects, peptides can induce long-lasting effects. It is further proposed that different mechanisms underlie the acute and long-term action of neuropeptides.

Interaction of opioid peptides with DA-A10 neurones in the ventral tegmental area and nucleus accumbens

In all of our intracerebral infusion experiments, rats are stereotaxically implanted with stainless-steel guide cannulae; in the studies described here, bilateral cannulae were placed either in the ventral tegmental area (VTA) or nucleus accumbens (N. accumbens). Such placements are schematically shown in Fig. 14.4. A complete description of infusion and behavioural methods has

INTERACTION BETWEEN OPIATE NEURONES AND MESOLIMBIC DA-10 NEURONES EITHER AT THE LEVEL OF DA CELL BODIES (A) IN VENTRAL TEGMENTAL AREA (VTA) OR AT THE LEVEL OF DA TERMINALS (B) IN THE NUCLEUS ACCUMBENS (N.ACC). LOCAL INFUSION OF OPIATES WAS PERFORMED ON AWAKE FREELY-MOVING RATS.

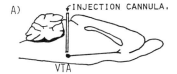
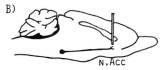

Fig. 14.4. Schematic representation of cannula placements in the rat brain.

been published previously (Kelley *et al.* 1980). We have shown that infusion of morphine or opioid peptides into the VTA elicits a behavioural activation, of which the amplitude is a function of dose and the duration a function of the substance injected (Joyce and Iversen 1979; Kelley *et al.* 1980; Stinus *et al.* 1980). An example of the locomotor response to VTA infusion of D-ala-met-enkephalin (a long-lasting enkephalin analogue) is shown in Fig. 14.5.

There is good evidence that this effect is dependent on the activity of DA-A10 neurones. As Fig. 14.6 (left panel) indicates, if a specific 6-hydroxydopamine (6-OHDA) lesion of the DA-A10 neurones is made, the response to any of the morphinomimetics infused into the VTA is completely abolished (Kelley *et al.* 1980; Stinus *et al.* 1980). While the action of the opioid peptides on dopaminergic cell bodies is most likely indirect (perhaps acting through a GABAergic interneurone), the behavioural response is mediated by stimulation

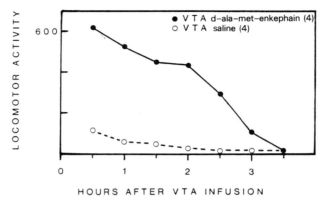

Fig. 14.5. Infusion of a long-acting analogue of enkephalin into the VTA induces a pronounced behavioural activation. (From Kelley *et al.* (1980).)

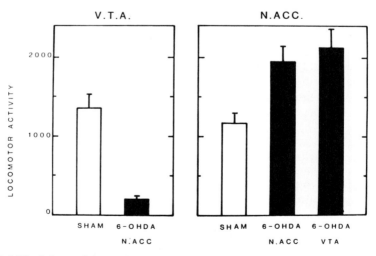

Fig. 14.6. The left panel shows that the behavioural response to opioid peptide infusion into VTA is completely blocked by a specific lesion of the DA-A10 neurones (bilateral 6-OHDA lesion of nucleus accumbens). The right panel shows that, in contrast, 6-OHDA lesion of the DA-A10 system (either at the level of DA terminals or cell bodies) enhances the behavioural response to nucleus accumbens infusion of opioid peptides. (From Stinus *et al.* (1980); and Stinus and Winnock, in preparation.)

of the A10 neurones. These results suggest that endogenous opioid peptide-containing neurones in the VTA may modulate DA neuronal activity (see also Johnson *et al.* 1980). There is further evidence that these opioid neurones may be a substrate for the euphoriant properties of morphine, since morphine is directly self-administered into the VTA (Bozarth and Wise 1981) and since local opiate stimulation in the VTA is shown to be rewarding in the place-preference paradigm (Phillips and Le Piane 1980).

We have compared these experiments to a related set of experiments on the nucleus accumbens (Stinus, 1982). Bilateral infusion of enkephalin or endorphin into nucleus accumbens also produces enhanced locomotor activity, similar in amplitude and duration to that obtained by opiate stimulation of the VTA. Although there is evidence that there is a large population of opiate receptors localized on DA terminals in striatum and n. accumbens (Schwartz 1979), it has been questioned whether the behavioural stimulant effect of opiate infusion into the n. accumbens is dependent on the DA-A10 neurones, since this effect is not blocked by DA antagonists (Pert and Sivit 1977). We have indeed found that, in contrast to our findings in the VTA, the behavioural response to endorphin infusion into n. accumbens is not at all blocked by specific (6-OHDA) destruction of DA-A10 neurones but is in fact greatly enhanced. The contrast between the effect of removal of the DA-A10 system on endorphin infusion into either the VTA or n. accumbens is shown in Fig. 14.6. Thus, in one case, we have the expression of the peptide effect *dependent* on the DA neurones and, in the other, the effect is *independent* of, or postsynaptic to, the DA neurones. Nevertheless, even in the second case, although the effect is not mediated through the DA neurones, the fact that the response is enhanced indicates that the DA lesion somehow altered postsynaptic sensitivity to opioid peptides. This finding is supported by our recent result that chronic treatment with a low dose of reserpine (1 mg kg^{-1} every 2 days for 10 days) induces the same augmented behavioural response to n. accumbens infusion of endorphin (Stinus, unpublished). In summary, the chronic absence of DA input induces a modification of opiate receptors in a site within the limbic system.

As was emphasized earlier, these results clearly demonstrate that, even within the limbic system, a peptide may have multiple actions, and the functions in which it is involved depend on the local neuronal circuitry in a given structure.

Interactions between substance P and dopaminergic A10 neurones in the ventral tegmental area

Evidence has accumulated that there is an important interaction between the neuropeptide substance P and the monoaminergic neurones; for example, SP is found within regions of monoaminergic cell bodies: the locus ceruleus, raphe nuclei, substantia nigra, and ventral tegmental area. Most of our work, in collaboration with Dr Susan Iversen, has focused on the VTA. We have shown previously that local infusion of SP into the VTA induces a marked behavioural activation (Stinus *et al.* 1978). These behavioural effects of SP are mediated through the activation of the DA-A10 system, since they are blocked by either prior administration of a neuroleptic into the nucleus accumbens, or by 6-OHDA induced destruction of the DA-A10 neurones (Kelley *et al.* 1979).

We have also demonstrated an effect of local infusion in the VTA which is relevant to our discussion of long-term action of neuropeptides. When the activity of SP-infused rats returns to baseline levels (after about 1 hour), if, at this moment, control and SP-treated rats are injected with a low dose (0.5 mg kg^{-1}) of the psychostimulant amphetamine, the SP rats show a greatly potentiated locomotor response to the drug (see Fig. 14.7; Stinus *et al.* 1978). 24 hours after this SP infusion, the reactivity of both groups to amphetamine is

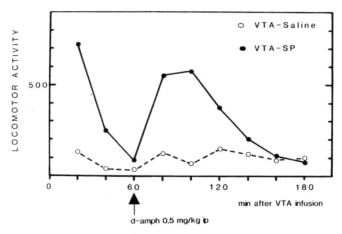

ENHANCEMENT OF d–AMPHETAMINE –INDUCED BEHAVIOURAL ACTIVATION

ONE HOUR AFTER VTA INFUSION OF SUBSTANCE P

Fig. 14.7. D-amphetamine administered i.p. one hour after VTA infusion of SP or saline results in a markedly potentiated locomotor response in the rats which had received SP. (From Stinus *et al.* (1978).)

similar. Since low doses of amphetamine act fairly selectively on the DA-A10 system, these results suggest that, in addition to the acute effect, the neuropeptide induces a longer-lasting change in the DA neurones, which considerably outlasts the half-life of the peptide itself. Although the baseline activity levels of SP-treated rats are similar to controls at the time of amphetamine injection, the pharmacological manipulation reveals this phenomenon.

While these results suggest a role for endogenous SP modulating the DA neurones, one might question the physiological relevance of the exogenous application of SP into the VTA. However, some recent studies done by our colleagues provide some very interesting speculations concerning the physiological function of SP in the VTA.

First, it has been shown that stressful situations such as isolation and electric footshocks selectively activate the mesolimbic and mesocortical dopaminergic neurones (Thierry *et al.* 1976; Herve *et al.* 1979; Blanc *et al.* 1980; Herman *et al.* 1982). It is possible, in light of a recent result reported by Lisoprawski *et al.* (1981), that this activation is mediated by the VTA SP innervation which originates at least in part in the habenula (Emson *et al.* 1977). Lisoprawski *et al.* found significantly decreased levels of SP in the VTA of rats submitted to a 20-min electric footshock session, while SP levels remained unchanged in the substantia nigra and prefrontal cortex. This effect was interpreted as an enhanced release of this peptide from SP nerve terminals within the VTA; that is, the enhanced utilization was not compensated for by the relatively slow SP synthesis and axonal flow. A related finding has been reported for the substantia nigra, in which chronic treatment with cataleptogenic neuroleptics (haloperidol, pimozide) significantly decreased endogeneous SP levels; this was also interpreted as enhanced release and subsequent neuronal depletion of SP (Hong *et al.* 1978; Hanson *et al.* 1981).

The increase in 3,4-dihydroxyphenyl acetic acid (DOPAC) in mesocortico-limbic areas induced by footshock stress can be detected during a short period following completion of the footshock session (J. P. Herman, personal communication). One hour after the session, the activity of DA-A10 neurones returns to normal levels as revealed by the DOPAC/DA ratio in frontal cortex and nucleus accumbens. However, we have recently found that, in rats subjected to chronic stress (one daily footshock session over 10 days), 24 hours after the last session, d-amphetamine-induced behavioural activation is greatly enhanced compared to non-shocked controls (Herman and Stinus, unpublished). It appears, therefore, that the chronic stress may induce an increase in the reactivity of DA-A10 neurones to amphetamine, even though this is not reflected in the DOPAC/DA ratio. Interestingly, we had previously shown (Iversen *et al.* 1978) that a similar enhancement of the amphetamine response can be observed after chronic application of exogenous SP in the VTA (five local infusions, one every two days). 24 hours after the last SP infusion, although the spontaneous activity of these rats is similar to that of controls, the response to amphetamine is greatly augmented. The effects of chronic stress and chronic SP-VTA treatment are compared in Fig. 14.8. In summary, acute stress may release SP from nerve terminals which would directly activate the DA mesocorticolimbic neurones, similar to the acute effect of local SP application. In contrast, the long-term effects observed in these two situations (potentiated d-amphetamine response) are due to the chronic release of SP which consequently induces a modification of the regulatory mechanisms which control the activity of the DA-A10 neurones. It is possible that further behavioural analysis of this phenomenon will also reveal abnormalities in different forms of adaptive behaviour, perhaps providing a model for neuronal regulatory mechanisms in psychopathology.

CONCLUSIONS

The attempt to integrate the ever-expanding research on neuropeptides with concepts of limbic system function and behaviour is a great challenge for those working or interested in this domain. In this chapter, we have tried to

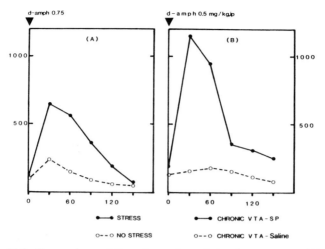

LOCOMOTOR RESPONSE TO d-AMPHETAMINE ,MEASURED 24 hrs AFTER CHRONIC ELECTRIC
FOOTSHOCK STRESS (A) OR CHRONIC INFUSION OF S P INTO THE VENTRAL TEGMENTAL AREA (B)

Fig. 14.8. Comparison of the enhanced amphetamine response: (A) in rats which had been chronically stressed (1 daily footshock session over ten days; session consisted of 1.5-mA shock for 2 seconds once every minute for 20 min) (Herman and Stinus, unpublished); (B) in rats which had received chronic infusion of SP into the VTA (one bilateral infusion every two days for 10 days). (From Iversen *et al.* (1978).)

emphasize how the consideration of anatomical maps and neurochemical mechanisms is an important basis for thinking about function. Many of our ideas are speculative, but we feel that the construction of hypothetical frameworks and conceptual bases is sometimes lacking in research on peptides and behaviour and should be developed further if this field is to advance. A number of important issues face us. What is the functional significance of coexistence of neuropeptides with classical transmitters? What is the significance of multiple forms of peptides, as well as multiple receptors? We are presently hampered by the lack of good pharmacological tools, such as peptide agonists and antagonists. As mentioned before, there is mounting evidence that peptide–monoamine interactions are fundamental in the control of behaviour, and this will undoubtedly be very promising for future research. Finally, the implications for biological psychiatry are paramount (Post *et al.* 1982; Diamond and Borison 1982). Clearly, limbic system dysfunction is involved in many mental disorders, and the added knowledge of how peptides function in these regions will enhance our perspective of the neural substrates of mental illness. Most importantly, further elucidation of neuronal mechanisms and function may open possibilities for the development of new and better therapeutic agents, interacting with peptide action in the brain. In particular, such new agents may be able to interact with long-term modulatory mechanisms without interfering with the basic functions underlying classical neurotransmission.

REFERENCES

Barbaccia, M. L., Chuang, D. M., and Costa, E. (1982). Is insulin a neuromodulator? In *Regulatory peptides* (ed. E. Costa and M. Trabucchi), Advances in Biochemical Psychopharmacology, Vol. 33, pp. 511–18. Raven Press, New York.

Beinfeld, M. C., Meyer, D. K., Eskay, R. L., Jensen, R. T., and Brownstein, M. J. (1981). The distribution of cholecystokinin immuno-reactivity in the central nervous system of the rat as determined by radio-immunoassay. *Brain Res.* 212, 51–7.

Blanc, G., Herve, D., Simon, H., Lisoprawski, A., Glowinski, J., and Tassin, J. P. (1980). Reduced activity and higher reactivity to stress of the mesocortico frontal dopaminergic neurons after long-term isolation in rats. *Nature* 284, 265–7.

Bloom, F. E. (1973). Dynamic synaptic communication: finding the vocabulary. *Brain Res.* 62, 299–305.

Bozarth, M. A. and Wise, R. A. (1981). Intracranial self-administration of morphine into the ventral tegmental area in rats. *Life Sci.* 28, 551–5.

Broca, P. (1878). Anatomie comparée des circonvolutions cérébrales. Le grand lobe limbique et la scissure dans la série des mammifères. *Rev. d'Anthropol. Ser. 2* 1, 285–498.

Brown, M., Rivier, J., and Vale, W. (1977). Bombesin: potent effects on thermoregulation in the rat. *Science* 196, 998–1000.

Brownfield, M. S., Reid, I. A., Ganten, P., and Ganong, W. F. (1982). Differential distribution of immunoreactive angiotensin and angiotensin-converting enzyme in rat brain. *Neuroscience* 7, 1759–70.

Cuello, A. C. and Kanazawa, I. (1978). The distribution of substance P immunoreactive fibers in the rat central nervous system. *J. comp. Neurol.* 178, 129–56.

Diamond, B. I. and Borison, R. L. (1982). Regulatory peptides in animal paradigms of neuropsychiatric illness. In *Regulatory peptides* (ed. E. Costa and M. Trabucchi), Advances in Biochemical Psychopharmacology, Vol. 33, pp. 541–8. Raven Press, New York.

Emson, P. C., Cuello, A. C., Paxinos, G., Jessell, T., and Iversen, L. L. (1977). The origin of substance P and acetylcholine projections to the ventral tegmental area and interpeduncular nucleus in the rat. *Acta physiol. scand. Suppl.* 542, 43–6.

Epstein, A. W. (1978). The neuroendocrinology of thirst and salt appetite. In *Frontiers in neuroendocrinology* (ed. W. F. Ganong and L. Martini), Vol. 5, pp. 101–34. Raven Press, New York.

Finley, J. C. W., Maderdrut, J. L., Roger, L. J., and Petrusz, P. (1981). The immunocytochemical localization of somatostatin-containing neurons in the rat central nervous system. *Neuroscience* 6, 2173–92.

Florey, E. (1967). Neurotransmitters and modulators in the animal kingdom. *Fed. Proc.* 26, 116–78.

Gintzler, A. R. (1980). Endorphin-mediated increases in pain threshold during pregnancy. *Science* 210, 193–5.

Guillemin, R. (1978). Peptides in the brain: the new endocrinology of the neuron. *Science* 202, 390–402.

Haber, S. and Elde, R. (1982). The distribution of enkephalin immuno-reaction fibers and terminals in the monkey central nervous system: an immunohistochemical study. *Neuroscience* 7, 1019–96.

Hanson, G. R., Alphs, L., Wolf, W., Levine, R., and Lovenberg, W. (1981). Haloperidol-induced reduction of nigral substance P-like immunoreactivity: a probe for the interactions between dopamine and substance P neuronal systems. *J. Pharm. exp. Ther.* 218, 568–74.

Herkenham, M. and Pert, C. B. (1980). In vitro autoradiography of opiate receptors in rat brain suggests loci of ''opiatergic'' pathways. *Proc. Nat. Acad. Sci. (USA)* 77, 5532–6.

Herman, J. P., Guillonneau, D., Dantzer, R., Scatton, B., Semerdjian-Rouquier, L., and Le Moal, M. (1982). Differential effects of inescapable footshocks and of stimuli previously paired with inescapable footshocks on dopamine turnover in cortical and limbic areas of the rat. *Life Sci.* **30**, 2207–14.

Herve, D., Tassin, J. P., Barthelemy, C., Blanc, G., Lavielle, S., and Glowinski, J. (1979). Difference in the reactivity of the mesocortical dopaminergic neurons to stress in the Balb/C and C97 BI/6 Mice. *Life Sci.* **25**, 1659–64.

Hokfelt, T., Johansson, O., Ljungdahl, A., Lundberg, J. M., and Schultzberg, M. (1980). Peptidergic neurons. *Nature* **284**, 515–21.

Hong, J. S., Yang, H. Y. T., and Costa, E. (1978). Substance P content of substantia nigra after chronic treatment with antischizophrenic drugs. *Neuropharmacology* **17**, 83–5.

Inagaki, S., Sakanaka, M., Shiosaka, S., Senba, E., Takatsuki, K., Takagi, H., Kawai, Y., Minagawa, H., and Tohyama, M. (1982). Ontogeny of substance P-containing neuron system of the rat: immuno histochemical analysis. 1. Forebrain and upper brainstem. *Neuroscience* **7**, 251–78.

Iversen, S. D., Joyce, E. M., Kelley, A. E., and Stinus, L. (1978). Behavioural consequences of the interaction between morphine, substance P and non-striatal dopamine neurons. In *Neuropsychopharmacology* (ed. C. Dumont), Advances in Pharmacology and Therapeutics, Vol. 5, pp. 263–72. Pergamon, Oxford.

Johansson, O., Hokfelt, T., Pernow, B., Jeffcoate, S. L., White, N., Steinbusch, H. W. M., Verhofstad, A. A. J., Emson, P. C., and Spindel, E. (1981). Immunohistochemical support for three putative transmitters in one neuron. Coexistence of 5-hydroxytryptamine, substance P and thyrotropin releasing hormone-like immunoreactivity in medullary neurons projecting to spinal cord. *Neuroscience* **6**, 1857–82.

Johnson, R. P., Sar, M., and Stumpf, W. E. (1980). A topographic localization of enkephalin on the dopamine neurons of the rat substantia nigra and ventral tegmental area demonstrated by combined histofluorescence-immunocytochemistry. *Brain Res.* **194**, 566–71.

Jonsson, G. and Hallman, H. (1982). Substance P counteracts neurotoxin damage on norepinephrine neurons in rat brain during ontogeny. *Science* **215**, 75–7.

Joyce, E. M. and Iversen, S. D. (1970). The effect of morphine applied locally to mesencephalic dopamine cell bodies on spontaneous activity in the rat. *Neurosci. Lett.* **14**, 207–12.

Kalivas, P. W., Nemeroff, C. B., and Prange, A. J. (1982). Neuroanatomical site specific modulation of spontaneous motor activity by neurotensin. *Eur. J. Pharmacol.* **78**, 471–4.

Kelley, A. E., Domesick, V. B., and Nauta, W. J. H. (1982). The amygdalostriatal projection in the rat. An anatomical study by anterograde and retrograde tracing methods. *Neuroscience* **7**, 615–30.

——, Stinus, L., and Iversen, S. D. (1979). Behavioural activation induced in the rat by substance P infusion into ventral tegmental area: implication of dopaminergic A10 neurones. *Neurosci. Lett.* **11**, 335–9.

——, ——, and —— (1980). Interactions between d-ala-met-enkephalin A10 dopaminergic neurones and spontaneous behaviour in the rat. *Behav. Brain Res.* **1**, 3–24.

Lisoprawski, A., Blanc, G., and Glowinski, J. (1981). Activation by stress of the habenulo-interpeduncular substance P neurons in the rat. *Neurosci. Lett.* **25**, 47–51.

MacLean, P. D. (1949). Psychosomatic disease and the "visceral brain". *Psychosom. Med.* **11**, 338–53.

—— (1969). The hypothalamus and emotional behaviour. In *The hypothalamus* (ed. E. Anderson and W. J. H. Nauta), pp. 659–72. Charles C. Thomas, Springfield, Illinois.

Morley, J. E. (1982). Minireview: the ascent of cholecystokinin (CCK) – from gut to brain. *Life Sci.* **30**, 479–94.

Nauta, W. J. H. (1958). Hippocampal projections and related neural pathways to the midbrain in the cat. *Brain* **81**, 319–40.

—— and Domesick, V. B. (1981). Neural associations of the limbic system. In *Neural substrates of behavior* (ed. A. Beckman). Spectrum, New York.

Nicoll, R. A. (1978). Neurophysiological studies. *Neurosci. Res. Prog. Bull.* **16**, 272–85.

Papez, J. (1937). A proposed mechanism of emotion. *Arch. Neurol. Psychiat.* **38**, 725–43.

Pert, C. B. and Sivit, C. (1977). Neuroanatomical focus for morphine and enkephalin-induced hypermotility. *Nature* **265**, 645–7.

Phillips, A. G. and Le Piane, F. G. (1980). Reinforcing effects of morphine micro-injection into the ventral tegmental area. *Pharmacol. Biochem. Behav.* **12**, 965–8.

Post, R. M., Gold, P., Rubinow, D. R., Ballenger, J. C., Bunney, W. E., and Goodwin, F. K. (1982). Peptides in the cerebrospinal fluid of neuropsychiatric patients: an approach to central nervous system peptide function. *Life Sci.* **31**, 1–15.

Ricardo, J. A. and Koh, E. T. (1978). Anatomical evidence of direct projections from the nucleus of the solitary tract to the hypothalamus amygdala and other forebrain structures in the rat. *Brain Res.* **153**, 1–26.

Roberts, G. W., Woodhams, P. L., Polak, J. M., and Crow, J. J. (1982). Distribution of neuropeptides in the limbic system of the rat: the amygdaloid complex. *Neuroscience* **7**, 99–133.

Sanger, D. J. (1981). Endorphinergic mechanisms in the control of food and water intake. *Appetite: J. Intake Res.* **2**, 193–208.

Scharrer, B. (1976). Peptides in neurobiology: historical introduction. In *Peptides in neurobiology* (ed. H. Gainer), pp. 1–8. Plenum, New York.

Schwartz, J. C. (1979). Opiate receptors on catecholaminergic neurones in brain. *Trends Neurosci.* **2**, 137–9.

Skirboll, L. R., Grace, A. A., Hommer, D. W., Rehfeld, J., Goldstein, M., Hokfelt, T., and Bunney, B. S. (1981). Peptide-monoamine coexistence: studies of the actions of cholecystokinin-like peptides on the electrical activity of midbrain dopamine neurons. *Neuroscience* **6**, 2111–24.

Stinus, L. (1982). Interaction entre systèmes peptidergiques et neurones dopaminergiques mésolimbiques: analyse comportementale. *Actualités de Chimie Thérapeutique*, Vol. 9, pp. 275–84. Technique et Documentation, Paris.

——, Kelley, A. E., and Iversen, S. D. (1978). Increased spontaneous activity following substance P infusion into A10 dopaminergic areas. *Nature* **276**, 616–18.

——, Koob, G. F., Ling, N., Bloom, F. E., and Le Moal, M. (1980). Locomotor activation induced by infusion of endorphins into the ventral tegmental area: evidence for opiate-dopamine interactions. *Proc. Nat. Acad. Sci. (USA)* **77**, 2323–7.

Thierry, A. M., Tassin, J. P., Blanc, G., and Glowinski, J. (1976). Selective activation of the mesocortical DA system by stress. *Nature* **263**, 242–4.

Uhl, G. R., Goodman, R. R., and Snyder, S. H. (1979). Neurotensin-containing cell bodies, Fibers and nerve terminals in the brain stem of the rat. Immunohisto-chemical mapping. *Brain Res.* **167**, 77–91.

Wamsley, J. K., Zarbin, M. A., Young, W. S., III, and Kuhar, M. J. (1982). Distribution of opiate receptors in monkey brain: an autoradiographic study. *Neuroscience* **7**, 595–614.

15

Peptide circuitry of the limbic system

G. W. ROBERTS, J. M. POLAK, AND T. J. CROW

The concept of nerves storing and releasing peptides within the central nervous system arose from studies on the neurohypophyseal hormones, oxytocin and vasopressin, three decades ago (Scharrer 1951). These two peptide hormones are released from nerves (originating in the hypothalamus) in the posterior pituitary into the systemic circulation. For a long time these 'peptidergic' neurosecretory nerves were believed to be a special case, unique to the hypothalamus; however, in the last decade an ever-increasing number of peptides (see Table 15.1) have been located within the mammalian central nervous system (Buck *et al.* 1981; Cooper *et al.* 1981; Emson 1979; Emson and Lindval 1979; Emson *et al.* 1982*a, b*; Geola *et al.* 1981; Hokfelt *et al.* 1980*a*) and evidence is accumulating that many of these possess neurotransmitter/modulator properties (Emson 1979; Haskins *et al.* 1982; Iversen *et al.* 1980; Kelly and Dodd 1981; Nicol *et al.* 1980; Phillis and Kirkpatrick 1980; Pittman and Siggins 1981).

Immunological techniques (immunocytochemistry and radioimmunoassay, which render visible and measurable the minute quantities of peptide present in the brain tissue) have demonstrated that peptides, far from being a local curiosity confined to the hypothalamus, are widely distributed in the brain and form complex networks which link different brain structures together. Studies to date point to the fact that peptides appear as important as 'classical neurotransmitters' (acetylcholine, noradrenaline, etc.) in terms of cell-body groups, complexity of fibre pathways, and a bewildering array of physiological and behavioural effects (Emson 1979; Fuxe *et al.* 1980; Gibbs *et al.* 1973; Haskins *et al.* 1982; Haubrich *et al.* 1982; Hokfelt *et al.* 1980*a*; Kelly *et al.* 1979; Ljungdahl *et al.* 1978; Magistretti *et al.* 1981; Onali *et al.* 1981; Samson *et al.* 1981). If this is so, profound changes will be necessary in our current paradigms of brain function, undoubtedly leading to the abdication of many cherished hypotheses concerning the role of transmitters in normal and abnormal behaviour.

SOME GENERAL OBSERVATIONS ON THE TOPOGRAPHY OF BRAIN PEPTIDES

The peptides present in the brain can be divided into (at least) two major groups on the basis of the localization of their cell bodies. One group has cell bodies confined to the hypothalamic area (Table 15.2). The other has cell bodies present in the hypothalamus and also in brainstem (where they may be co-localized in catecholaminergic neurones) and forebrain structures, particularly

TABLE 15.1. *Peptide neurotransmitter candidates present in the CNS*

ACTH	Met-enkephalin (ME)	Neuropepide y (NPY)
Angiotensin	Leu-enkephalin	Neurotensin (NT)
Bombesin	Glucagon	Oxytocin
Bradykinin	Growth hormone	PHI
Calcotinin	Insulin	PYY
Carnosine	Kyotorphine	Secretin
Cholecystokinin octapeptide (CCK)	Lipotropin	Somatostatin (SOM)
Dynorphin	LHRH	Substance P (SP)
Head activator	αMSH	TRH
β-endorphin	Motilin	Vasopressin
		Vasoactive intestinal polypeptide (VIP)

TABLE 15.2. *Peptide groups based on the distribution of immunoreactive cell bodies*

(A) Hypothalamic peptides	(B) Extra-hypothalamic peptides
ACTH	Cholecystokinin octapeptide
Angiotensin	Dynorphin
Bombesin	Met-enkephalin
Bradykinin	Leu-enkephalin
Calcitonin	Motilin
β-Endorphin	Neuropeptide Y
Glucagon	Neurotensin
Growth hormone	Secretin
Insulin	Somatostatin
Lipotropin	Substance P
LHRH	THR
αMSH	Vasoactive intestinal peptide
Oxytocin	
Vasopressin	

(A) Cell bodies confined to hypothalamus.
(B) Cell bodies in hypothalamus and other brain areas.

the limbic system (Table 15.2; Allen *et al.* 1983; Emson 1979; Emson and Lindval 1979; Hokfelt *et al.* 1980*a, b*).

Hypothalamic peptides follow highly divergent trajectories in which peptide-containing fibres from the hypothalamus reach out to target areas in distant brain regions bearing no obvious direct functional relationship to each other (Fig. 15.1). These target regions are areas where large numbers of fibre pathways intersect – 'nodal way stations' – and probably represent the most important regions of hypothalamic peptide function. By selectively innervating these regions, neurones in the hypothalamus can exert a great influence on many channels of information at once at a comparably small metabolic cost to themselves. This type of neuronal arrangement could be expected to have a generalized effect on numerous systems at once – acting as a local hormone, e.g. raising or lowering the excitatory thresholds of large fibre systems such as the medial forebrain bundle or fornix or brain regions such as the pre-optic area.

The extra-hypothalamic peptides (Table 15.2) are organized into tighter, better defined systems, with numerous cell-body groups giving rise to many local projections and some fibre bundles which project and terminate topographically in distant brain areas (Allen *et al.* 1983; Crowley and Terry 1980; Greenwood *et al.* 1981; Ljungdahl *et al.* 1978; Roberts *et al.* 1980, 1982, 1984; Vincent and McGeer 1981; Williams *et al.* 1981; Woodhams *et al.* 1983). These peptides form circuits which are characterized by their sharp delimitation from adjacent structures and strong spatial relationships to specific functional systems. This arrangement implies a synaptic (transmitter-like) mode of transmission for their physiological actions.

An interesting (functionally significant?) gradient is found in the number of different peptides present in cell bodies within the various brain regions. The

COMMON PROJECTION PATTERNS OF HYPOTHALAMIC PEPTIDES

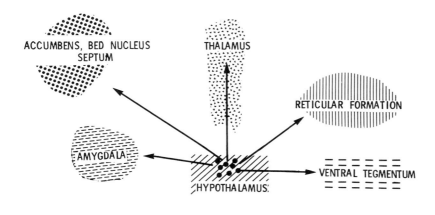

ACCUMBENS, BED NUCLEUS
SEPTUM

THALAMUS

RETICULAR FORMATION

AMYGDALA

VENTRAL TEGMENTUM

HYPOTHALAMUS

CELL BODIES •

Fig. 15.1. Schematic diagram of the areas receiving projections from the peptide-containing neurones (Table 15.2 (A)) in the hypothalamus.

cortex contains five different peptides present in cell bodies (VIP, CCK, SOM, NT, NPY; for abbreviations see Table 15.1). Other peptides are found in the cortex but they are present in fibres which have a subcortical origin. The hippocampus contains five different peptides in cell bodies (VIP, CCK, SOM, NPY, NT (Adrian *et al.* 1983; Allen *et al.* 1983; Roberts *et al.* 1984); the amygdala and bed nucleus of the stria terminalis contain many more (all of those listed in Table 15.2(b)). Areas of brainstem contain larger numbers of different peptide-containing neurones and, at the opposite end of the gradient to the cortex, the hypothalamus contains 35 different peptides in neurones (i.e. all those listed in Table 15.1). The reasons for this sort of organization – fewer peptides found in the most recently evolved structures–are of course unknown, but it is interesting to note that this gradient parallels the ease with which functions are assigned to brain areas, from hypothalamus (at least 35 peptides present in cell bodies – highly specific functions localized in discrete regions) to cortex (five types of peptide cell bodies – Lashley's 'polyfunctional homogenate').

PEPTIDES IN THE LIMBIC SYSTEM
Radioimmunoassay studies show that the limbic system contains some of the highest peptide levels in the central nervous system (e.g. SOM in the amygdala, CCK in the hippocampus, VIP in the bed nucleus of the stria terminalis; Buck *et al.* 1981; Cooper *et al.* 1981; Crowley and Terry 1980; Emson *et al.* 1979, 1982*a, b*; Gale *et al.* 1978; Palkovits *et al.* 1981). Immunocytochemical studies have shown that peptide-containing cell bodies are numerous in limbic structures and that fibre networks are widespread (Fig. 15.2; Allen *et al.* 1983;

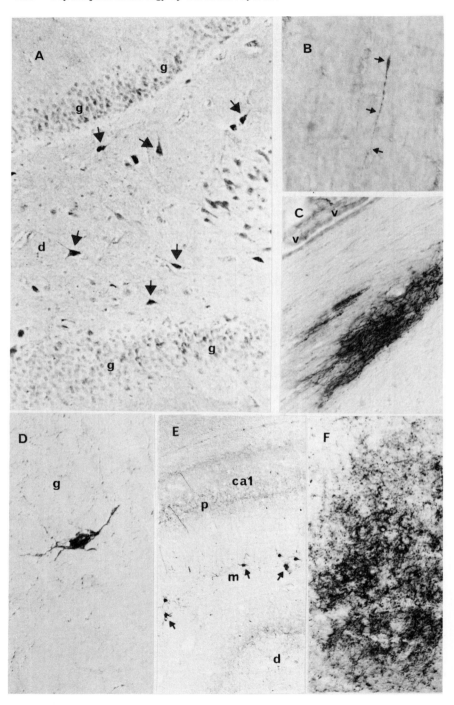

Bennet-Clarke *et al.* 1980; Del Fiacco *et al.* 1982; Finley *et al.* 1981; Ljungdahl *et al.* 1978; Vanderhagen *et al.* 1980; Walmsley *et al.* 1981). Recent detailed studies (Crowley and Terry 1980; Greenwood *et al.* 1981; Handelmann *et al.* 1981; Palkovits *et al.* 1981; Roberts *et al.* 1981*a, b*; 1982, 1984; Sakanaka *et al.* 1981; Vincent *et al.* 1981; Woodhams *et al.* 1983) have also begun to uncover the full extent of the peptide circuitry present in the limbic system and it is to this that we now turn. Most of this work has been done on rat and to a lesser extent monkeys. Preliminary studies indicate that peptides in the human brain have similar distributions (Adrian *et al.* 1983; Roberts *et al.* 1983; Sorensen 1982).

PEPTIDE CIRCUITRY OF THE LIMBIC SYSTEM

Limbic cortex

Appreciable quantities of peptide are found within the cortex. The distribution is heterogeneous, with limbic cortex containing more peptide than other cortical regions (Adrian *et al.* 1983; Buck *et al.* 1981; Cooper *et al.* 1981; Geola *et al.* 1981). To date only five peptides (VIP, CCK, NT, SOM, NPY) (Adrian *et al.* 1983; Allen *et al.* 1983; Emson and Lindval 1979; Inagaki *et al.* 1983) have been found within cell bodies in cortical regions. These peptides appear to be involved predominantly in local circuitry (Fig. 15.3). VIP- and CCK-positive neurones are present in cortical layers II–IV, have their long axes perpendicular to the pial surface, and project up to layer I and down to layers V and VI. NPY- and SOM-containing neurones have similar distributions; in addition they are frequently found in the deeper cortical layers (V–VI), where many have their long axes orientated parallel to the cortical surface and long fibres running over the surface of corpus callosum. Peptide-containing fibres appear to be present in a number of pathways which link cortical and subcortical structures, VIP, CCK, and SOM cell bodies in the entorhinal cortex give rise to fibres which join the perforant pathway to enter the hippocampus (Roberts *et al.* 1984). VIP, CCK, and SOM fibres originating in the temporal cortex cross the external capsule and may innervate lateral/basolateral amygdaloid regions (Roberts *et al.* 1982). CCK- and NT-containing cells in the temporal areas give rise to fibres which project to the caudate nucleus and olfactory nucleus/nucleus of the diagonal band of Broca respectively (Inagaki *et al.* 1983; Meyer *et al.* 1982*b*). Significant numbers of peptide-containing

---

Fig. 15.2. (A) Somatostatin-like immunoreactive neurones (arrowed) in the dentate gyrus (d) of the human hippocampus. × 300. (B) neurotensin-immunoreactive fibre (arrowed) within the human fornix. × 30. (C) Substance-P immunoreactive fibres within the stria terminalis (coronal section), adjacent to ventricle (v) of the rat. × 150. (D) CCK-containing basket-type cell close to granule layer (g) of the dentate gyrus in the hippocampus of the rat. × 338. (E) VIP-immunoreactive neurones and fibres in the dorsal CA1 region (coronal section) of the rat hippocampus. Note innervation of the pyramidal layer (p) and presence of neurones in striatum moleculare (m), d-dentate gyrus. × 188. (F) Dense plexus of methionine–enkephalin-immunoreactive fibres in the bed nucleus of the stria terminalis. × 135.

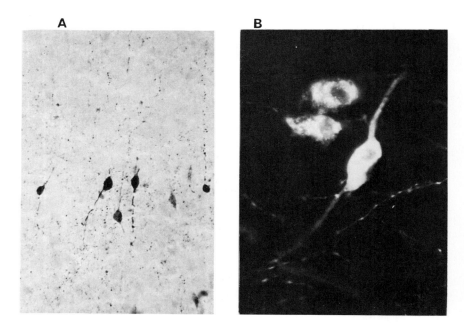

Fig. 15.3. (A) VIP-containing neurones in the cingulate cortex of the rat. × 300. (B) Neuropeptide-Y immunoreactive neurones in human temporal cortex. × 360.

fibres (VIP, CCK, SOM, NPY) are found in the induseum griseum running in a rostral–caudal direction. The same peptides (particularly NPY) are also found within fibres in the corpus callosum (Adrian *et al.* 1983; Allen *et al.* 1983; Meyer *et al.* 1982*a*; Sorensen 1982). These latter observations suggest that peptides play an important part in both inter- (commissural) and intracortical pathways and so may represent important elements of associational systems.

Hippocampus

Peptide-containing fibres appear to enter and leave the hippocampus in three ways – via

1. The fornix (the largest hippocampal efferent pathway);
2. The dorsal subiculum;
3. The ventral subiculum (Fig. 15.4) (Adrian *et al.* 1983; Allen *et al.* 1983; Greenwood *et al.* 1981; Roberts *et al.* 1981*b*, 1984; Vincent and McGeer 1981).

Findings with respect to each of these pathways are as follows

1. *Fornix.* At least seven peptides have been identified within fibres in this pathway (cf. stria terminalis). The distribution patterns of the peptides suggest that both afferent and efferent peptidergic projections could be contained within the fornix. The presence of VIP, CCK, NT, NPY, and SOM cell bodies within the hippocampus, more specifically within the subiculum (the common

PEPTIDERGIC CIRCUITRY OF THE HIPPOCAMPUS

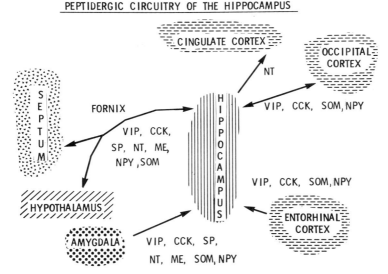

Fig. 15.4. Schematic summary diagram of the proposed peptidergic circuitry of the hippocampus.

source of many hippocampal efferents), and of their fibres within the alveus, dorsal fornix, and lateral fimbria suggests that each of these peptides may contribute hippocampal efferents to the fornix. However it must be emphasized that the peptidergic fibres form a small percentage of the fornix fibres and thus the peptidergic component represents a minor projection. The possibility that small numbers of VIP, CCK, NT, NPY, and SOM fibres also enter the hippocampus through the fornix cannot be discounted, although it has been shown that lesions of the fornix do not significantly alter hippocampal CCK (Handelmann *et al.* 1981) or SOM content. SP and ME appear to form the main peptidergic afferents via the fornix system (Roberts *et al.* 1984). These fibres arise in as yet unidentified diencephalic sites (e.g. for SP, possibly septal and supramamilliary regions) and project through the fornix to innervate the hippocampal formation.

2. *Dorsal subiculum.* Of the peptides examined, five are involved in this dorsal system – VIP, CCK, NT, NPY, and SOM. Each of these peptides has cell bodies and a fibre plexus within the dorsal subiculum. Although distributions overlap to a considerable extent, slight differences are apparent. The dorsal subiculum was the only hippocampal region containing NT cell bodies (Roberts *et al.* 1984) and is found to contain the highest hippocampal levels of NT (Emson *et al.* 1982). In addition to giving rise to alveus efferents (which possibly project via the fornix out of the hippocampus), the dorsal subiculum also gives rise to cortical projections. NT fibres originating from the dorsal subiculum turn dorsally and then rostrally, projecting as far as the frontal cortex along the cingulate bundle (Roberts *et al.* 1981*a*). NT fibres also emanate

from the subiculum and turn dorsally and then laterally to run along the surface of the corpus callosum ventrally as far as the piriform cortex and lateral amygdala. Whether VIP-, CCK-, and SOM-containing fibres run from the dorsal subiculum to the cortex (or vice versa) is uncertain as immunoreactive cell bodies are present in both areas.

3. *Ventral subiculum.* This region is the largest common target area of peptide-containing hippocampal afferents. Large numbers of VIP and CCK fibres arise from numerous VIP and CCK cell bodies in the entorhinal cortex. These fibres form projections which can be traced to the ventral hippocampal formation via the ventral subiculum. A similar though less dense pattern of SOM-containing fibres is also found. VIP, CCK, and SOM may represent peptidergic components of the perforant pathway. The perforant path is (qualitatively) the most important peptidergic afferent system. The VIP and CCK perforant projections terminate in different hippocampal regions. The VIP fibre projection turns dorsally and travels through the CA1, CA2 stratum oriens to terminate in the CA1 pyramidal/oriens layers and dentate gyrus. The CCK projection passes through the subiculum to terminate almost exclusively in the dentate gyrus. Peptides are also present in fibres which crossed from the caudal amygdala through the lateral hippocampal border and subiculum into the ventral hippocampal formation (Fig. 15.4). Similar amygdala–hippocampal connections have been shown using tritiated-leucine and horseradish-peroxidase tracing techniques (Price 1981). The course and terminations of VIP, CCK, and SOM fibres from the amygdala are similar to those described above for the perforant pathway. SP, NT, and ME fibres appear to enter the hippocampal formation from the amygdala, crossing the caudal amygdalo-hippocampal border, and form diffuse but widespread fibre projections. SP fibres travelled rostrally and dorsally through the CA1 molecular regions reaching as far as the dorsal CA1, 2, 3 pyramidal regions and the dorsal subiculum. SP fibres terminated in moderate densities in the ventral dentate gyrus and scattered fibres were found as far dorsally as the cingulate cortex. A ventral amygdalo–hippocampal SP pathway would explain why fornix transection results in only partial depletion of SP in the hippocampus (Vincent and McGeer 1981). ME and NT fibres show a similar distribution pattern to SP in the hippocampus.

Amygdaloid complex

This region has an even greater abundance of peptide-containing links with other structures (Buck *et al.* 1981; Emson 1979; Roberts *et al.* 1981*b*; Sakanaka *et al.* 1981; Woodhams *et al.* 1983). Peptide fibres enter and leave the amygdala in a number of ways – by the stria terminalis (the major pathway to the hypo-thalamus, pre-optic, and septal regions), the ventral–amygdalofugal pathway (a more diffuse hypothalamic connection), through fibres travelling to or from components of the basal ganglia (caudate–putamen and globus pallidus), to or from cortex, hippocampus, and via the medial forebrain bundle (Fig. 15.5).

1. *Stria terminalis.* Most peptide-containing fibres found within the stria appear to originate from cell bodies within the amygdaloid complex. Thus

Peptidergic circuitry of the amygdala

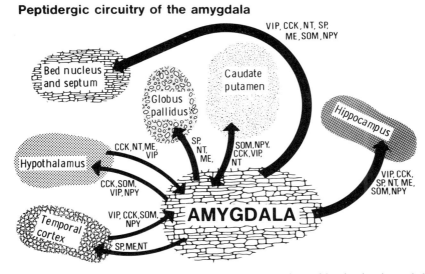

Fig. 15.5. Schematic summary diagram of the proposed peptidergic circuitry of the amygdala.

fibres from SP, NT, ME, NPY, and SOM cell bodies in the central nucleus are found ascending into and travelling within the stria terminalis. CCK fibres enter the stria from both the central and cortical amygdaloid nuclei. VIP- and NPY-containing fibres converge on the central nucleus (which has no VIP-containing cell bodies) from cortical, lateral, and basolateral nuclei and some of these continue into the strial terminalis. Thus the stria is the largest peptidergic efferent pathway from the amygdala (in addition it may also convey some incoming peptidergic fibres).

2. *Ventral amygdalofugal pathway.* Immunoreactivity to peptides is detectable in the fibres of this rather diffuse amygdalo–hypothalamic connection. VIP-containing fibres collect at the dorsal border of the medial nucleus and appear to enter or leave the amygdaloid complex by the ventral pathway at this point. SP fibres are also found in this region without any clear indication as to whether they are afferents or efferents. CCK-, NT-, and ME-positive fibres are found travelling in the ventral pathway towards the basolateral, cortical, and medial nuclei. It appears likely that these peptide fibres represent amygdaloid afferents. Some of these may originate in the ventral tegmentum and reach the amygdala by way of the median forebrain bundles.

3. *Connections with the basal ganglia.* Large numbers of NPY- and SOM-immunoreactive fibres link the lateral/basolateral and central nuclei and the caudate putamen; whether these form part of amygdaloid efferent or afferent systems or a mixture of both is unknown. CCK-, VIP-, and NT-positive fibres also run between amygdala and caudate putamen. In view of the absence of CCK- and NT-containing cell bodies in the caudate putamen it is assumed that CCK and NT fibres are travelling from the amygdaloid complex to the caudate–

putamen. For CCK, recent lesion studies support this view and indicate that virtually the entire CCK content of the rat caudate putamen originates from the amygdala and associated limbic cortex (Meyer *et al.* 1982*b*). SP-, ME-, and NT-containing fibres connect the central nucleus of the amygdala and the globus pallidus.

4. *Connections with the hippocampus and cortex.* Peptide-containing fibres (probably arising from cell bodies in the central nucleus and amygdalo–hippocampal area) cross the border between the caudal amygdala and hippocampus and provide a substantial innervation of the ventral hippocampal regions. SP-, NT-, and ME-containing fibres project from central/medial amygdaloid regions towards the piriform cortex.

Bed nucleus of the stria terminalis and septum

The bed nucleus of the stria terminalis and the septum are important termination areas for the efferent projections from both the hippocampus (fornix) and the amygdala (stria terminalis, ventral amygdalofugal pathway).

The amygdala projects topographically to the bed nucleus in two main divisions – lateral and medial. In this context it is noteworthy that the central amygdaloid nucleus, which projects to the lateral part of the bed nucleus and the lateral hypothalamus contains cell bodies immunoreactive for CCK, SP, NT, ME, SOM, and NPY (Allen *et al.* 1983; Crowley and Terry 1980; Palkovits *et al.* 1982; Roberts *et al.* 1980, 1982; Sakanaka *et al.* 1981; Woodhams *et al.* 1983), and a high density of fibres immunoreactive for these peptides is found in the lateral bed nucleus. CCK cell bodies are also numerous in the medial and cortical amygdaloid nuclei, and these areas are likely to contribute to the moderately dense fibre plexus seen in the medial bed nucleus. In addition, innervation of the medial bed nucleus may be derived from cells in the amygdalo–hippocampal area.

Very high levels of VIP immunoreactivity in the lateral bed nucleus represent amygdalofugal fibres, but the central nucleus does not contain VIP cell bodies and thus this innervation arises from the basolateral nucleus; the lateral nucleus, which has VIP-positive cells, has not been shown to project to the bed nucleus. The widespread SP-like immunoreactivity of the medial bed nucleus area may derive from cell bodies in the medial amygdala and that of the lateral bed nucleus from the cell bodies seen in the central nucleus, although the results of lesion studies have not been entirely clear (see Roberts *et al.* 1982 for discussion). Sakanaka *et al.* (1981) have shown prominent SP and SOM projections from the central and medial amygdala to the dorsal bed nucleus. Following central amygdaloid lesions they also showed small decreases of SP- and SOM-immunoreactivity in the lateral hypothalamus and a significant decrease of SOM in the lateroventral part of the anterior hypothalamic nucleus, but further studies with more discrete lesions and other tracer methods will be necessary to precisely depict the amygdalofugal stria terminalis projections for both SP and SOM. We found SOM cell bodies in the medial, basomedial, and cortical amygdaloid nuclei (Roberts *et al.* 1982) and thus these might be expected to project to a central zone of the bed nucleus between the lateral and medial divisions.

Terminals were indeed found in this area (Woodhams *et al.* 1983), although Crowley and Terry (1980) failed to confirm this projection by changes in immunoreactivity of the bed nucleus following basomedial amygdaloid lesions. In both the amygdala and bed nucleus, SOM immunoreactivity is rather diffusely spread and hence it is difficult to draw firm conclusions about the role of this peptide in the connectivity of the two regions.

Finally it should be noted that, in addition to areas of fibres and terminals, cell bodies immunoreactive for many peptides are also found in various parts of the bed nucleus. SP and SOM cells are clearly associated with fibres running into the stria terminalis. In the case of VIP, cells were seen in the lateral bed nucleus and the lesion study of Palkovits *et al.* (1981) suggested than an amygdalopetal projection exists to the lateral amygdala. Thus apart from local projections, it is likely that there are reciprocal peptidergic connections from the bed nucleus back to the amygdala for at least these three peptides and possibly others as well.

Peptide-containing fibres in the nucleus accumbens were continuous with those in the rostral bed nucleus and are probably part of the terminal field of the supracommissural division of the dorsal stria terminalis containing all six peptides (Woodhams *et al.* 1983). The presence of scattered VIP, ME, and SOM cell bodies in the accumbens suggests that there are local projections with the bed nucleus and there may even be reciprocal pathways via the stria terminalis back to the amygdala.

The septum is adjacent to the bed nucleus dorsally and medially, and in many cases the fibre plexi of the two areas are contiguous, indicating close functional links between the two structures. The lateral septum is well innervated by neuropeptides, and CCK, SP, NT, ME, and SOM cell bodies are seen here as well as fibres immunoreactive for these peptides. The VIP fibres appear continuous with the anterior component of the amygdaloid projection of the bed nucleus. Innervation of the medial septum was less extensive and only occasional VIP and SP cell bodies are seen, although NT and especially CCK fibres spread medially. SP, NT, and ME fibres run within the fornix and it is possible that these fibres project as far as the hippocampus. Indeed, Vincent and McGeer (1981) have shown that septal lesions decrease SP in the hippocampus. VIP, CCK, NT, and SOM fibres travel in a dorso-ventral direction in the midline in the medial septum and fornix, and the paucity of cell boides in this area suggests that these peptides could represent afferents from the hippocampus. All the peptides except VIP are found in fibres connecting the septum and the pre-optic area/hypothalamus, and also in the diagonal band of Broca, but the direction of these connections remains to be elucidated.

POSSIBLE INTERACTIONS WITH 'CLASSICAL' NEUROTRANSMITTERS

Monoaminergic afferents to the hippocampus arise in several brainstem areas (Kohler and Steinbusch 1982; Lindvall and Bjorklund 1978; Loy *et al.* 1980; Moore and Halaris 1975; Parent *et al.* 1981). The median raphe nucleus accounts for most of the serotonin innervation found within the hippocampal formation. The serotonin innervation arrives via the fornix and cingulum and terminates mostly in the stratum moleculare of CA1 and CA3 and in a restricted part of the hilus of the dentate gyrus beneath the granule cells. Inputs are also found to the strata radiatum and oriens though more densely in CA3 than in CA1. This serotonin innervation could affect primarily VIP and CCK (the SOM cell bodies being confined to the stratum oriens of CA1, CA2, and the hilus of the dentate gyrus). It has been suggested that the serotonin input is inhibitory in function. The noradrenaline innervation of the hippocampus is derived from the locus ceruleus via the dorsal noradrenaline bundle. The fibres enter the hippocampus via the retrospinal cortex and terminate specifically in the stratum moleculare of CA1 and CA3, the hilus of the dentate gyrus, and the stratum radiatum of the CA3. This input could also preferentially affect VIP- and CCK-containing systems. A large acetylcholine projection arising in the septum (Fonnum 1970; Lewis and Shute 1978) terminates in the stratum oriens and stratum radiatum and the granule-cell layer of the dentate gyrus. It is possible that the acetylcholine system could modulate somatostatin cell bodies and fibres in this region or vice versa. Recently it has been reported that dopamine could act as a transmitter within the dorsal hippocampus making interactions between the dopamine input and peptide in the dorsal subiculum (e.g. NT) possible (Ishikawa *et al.* 1982).

A dense dopaminergic innervation of the amygdala can be separated into two portions – one to the central and the other to the lateral amygdaloid nuclei (the medial parts of the amygdala contain little dopamine – Meibach and Katzman 1981). These innervations arise from different groups of dopamine-containing cell bodies in the brainstem (which can also be differentiated on the basis of their CCK content and their responses to this peptide – Skirboll *et al.* 1981). The innervation of the central nucleus is strategically placed to influence the main peptidergic output from the amygdala via the stria terminalis. However, the VIP pathway through the stria does not arise from the central nucleus and so may be unaffected by this branch of the dopaminergic input. Since there is also a dopaminergic input to the lateral and basolateral nulcei from which the VIP innervation is derived, it would be interesting to examine the physiological functions of each dopamine input and see how they contribute to amygdalas function. The highest levels of acetylcholinesterase activity in the amygdala are also found in the lateral nucleus (Lewis and Shute 1978). The central and lateral nuclei and the dorsal part of the basolateral nucleus receive a noradrenergic input from the locus ceruleus (Lindvall and Bjorklund 1978). This innervation could affect VIP-, CCK-, NT-, NPY-, ME-, and SP-containing neurones in these areas but would have little effect on the SP, CCK, SOM, and NPY neuronal cell bodies in the medial amygdala. The serotonergic innervation of the amygdala originates almost exclusively from the dorsal raphe nucleus and terminates in the central, medial, and cortical nuclei (Parent *et al.* 1981) and would therefore preferentially influence those areas and peptides (i.e. CCK, NPY, SOM, and SP in the medial nucleus) which are not affected by dopaminergic and noradrenergic inputs. GABA content (Fonnum and Storm-Mathisen 1978) and glutamic acid de-carboxylase activity are also reported as highest in the medial nuclear complex.

The bed nucleus has relatively low cholinergic activity (Lewis and Shute 1978) especially in contrast with the neighbouring nucleus accumbens. The ventral bed nucleus is reported to contain some cholinergic cells which project to the amygdala via the stria terminalis. Lesions to the stria terminalis do not affect levels of glutamate, asparate, taurine, or glycine in the bed nucleus or the central amygdaloid nucleus, and these amino acids are probably not important transmitters in the region. However, it seems likely that GABA is an important inhibitory transmitter not only of intrinsic amygdaloid and bed nucleus neurones but also in the major efferent pathway passing via the stria terminalis from the central nucleus to the rostral bed nucleus, especially the portion below the anterior commissure (Fonnum and Storm-Mathisen 1978). This distribution suggests some possible interactions with VIP, SP, and NT neurones in this part of the bed nucleus.

Monoamines provide an important input to the bed nucleus arriving via the median forebrain bundle (Lindvall and Bjorklund 1978). Whilst some of these fibres travel on through the stria terminalis to the amygdala, others terminate in the bed nucleus itself. Similarly, ascending dopamine fibres from the ventral tegmental area pass mainly through the ventral pathway to the amygdala, but part of this projection runs on dorsally via the bed nucleus and stria terminalis. Some of the mesolimbic dopamine neurones projecting to the bed nucleus may contain CCK (Hokfelt *et al.* 1980*b*), and Williams *et al.* (1981) have recently provided evidence which suggests that CCK may also be present in the projections of these mesolimbic cells to the anterior forebrain, including the bed nucleus. It thus seems likely that interactions between and even co-localization of CCK and dopamine occur in the bed nucleus (Hokfelt *et al.* 1980*b*).

CONCLUSION

From what is known about the functional neuroanatomy and neurophysiology of the limbic system it is thought that it mediates responsiveness to changing environmental stimuli (Gloor *et al.* 1981; Halgren 1981). Visual and auditory stimuli are processed and then transmitted to the higher neocortical areas of the temporal lobe where they are organized into the percepts which form the basis of 'modality specific engrams'. For these stimuli to be consciously experienced, the information elaborated by the temporal cortex has to be handed on to subcortical limbic structures – amygdala and hippocampus – the most likely function of these latter structures being the attachment of environmental significance to the engrams (Gloor *et al.* 1981; Halgren 1981).

Indeed it may be that such memories emerge into consciousness only if an affective component is attached to them. This output is then channelled through striatal (motor), hypothalamic, and brainstem regions enabling the organism to react to the changing environment. This view of limbic-system function re-emphasizes the close functional and anatomical relationships that bind limbic components together. It may be expected that the peptide circuitry within the limbic system has specific contributions to make in the integration of neocortical (Magistrett *et al.* 1981) and limbic processing (Haskins *et al.* 1982;

Kelley *et al.* 1979) – linking past and present experience and internalizing the resultant vectors in terms of affect and motivation, thus constantly redefining the environment in terms of biological significance to the organism.

REFERENCES

Adrian, T. E., Allan, J. N., Bloom, S. R., Ghatei, M. A., Rossor, M. N., Roberts, G. W., Tatemoto, J., Crow, T. J., and Polak, J. M. (1983). Neuropeptide Y: Distribution in human brain, *Nature* **306**, 584–6.

Allen, Y. S., Adrian, T., Allen, J., Tatemoto, K., Crow, T. J., Bloom, S. R., and Polak, J. M. (1983). Neuropeptide Y: Distribution in rat brain. *Science* **221**, 877–9.

Bennet-Clarke, C., Roviagnano, M., and Joseph, S. A. (1980). Distribution of somatostatin in the rat brain: Telencephalon and diencephalon. *Brain Res.* **188**, 473–86.

Buck, S. J., Deshmukh, P. P., Burks, T. F., and Yamamura, H. I. (1981). A survey of substance P. somatostatin and neutotensin levels in aging in the rat and human central nervous system. *Neutobiol Aging* **2**(4), 257–64.

Cooper, P. E., Fernstrom, M. H., Rorstad, O. P., Leeman, S. E., and Martin, J. B. (1981). The regional distribution of somatostatin, substance P and neurotensin in human brain. *Brain Res.* **218**, 219–32.

Crowley, W. R. and Terry, L. C. (1980). Biochemical mapping of somatostatinergic systems in rat brain: effects of periventricular hypothalamic and medial basal amygdaloid lesions on somatostatin-like immunoreactivity in discrete brain nuclei. *Brain Res.* **200**, 283–91.

Del Fiacco, M., Paximos, G., and Cuelo, A. C. (1982). Neostriatal enkephalin-immunoreactive neurones project to the globus pallidus. *Brain Res.* **231**, 1–17.

Emson, P. C. (1979). Peptides as neurotransmitter candidates in the mammalian CNS. In *Progress in Neurobiology* (ed. J. Phillis, J. Kirkaert), Vol. 13, pp. 61–116. Pergamon Press, London.

—— and Lindvall, O. (1979). Distribution of putative neurotransmitters in the neo-cortex. *Neuroscience* **4**, 1–30.

——, Fahrenkrug, Y., and Spokes, E. G. S. (1979). Vasoactive intestinal peptide distribution in normal human brain and in Huntington's disease. *Brain Res.* **173**, 174–8.

——, Goedert, M., Horsfield, P., Fioux, F., and St. Pierre, S. (1982*a*). The regional distribution and chromatographic characterisation of neurotensin-like immuno-reactivity in the rat central nervous system. *J. Neurochem.* **38**, 992–9.

——, Refeld, J. F., and Rossor, M. N. (1982*b*). Distribution of cholecystokinin-like peptides in the human brain. *J. Neurochem.* **38**, 1177–9.

Finley, J. C. W., Maderdrut, J. L., Roger, L. J., and Petrusz, P. (1981). The immuno-cytochemical localisation of somatostatin-containing neurons in the rat central nervous system. *Neuroscience* **6**, 2173–92.

Fonnum, F. and Storm-Mathisen, J. (1978). Localisation of GABA-ergic neurones in the CNS. In *Handbook of psychopharmacology* (ed. L. L. Iversen, S. D. Iversen, and S. H. Snyder), Vol. 9, pp. 357–401. Plenum, New York.

Fonnum, T. (1970). Topographical and subcellular localisation of choline acetyltrans-ferase in rat hippocampal region. *J. Neurochem.* **17**, 1029–37.

Fuxe, K., Andersson, K., Locatelli, V., Agnati, L. F., Hokfelt, T., Skirboll, L., and Mutt, V. (1980). Cholecystokinin peptides produce marked reduction of dopamine turnover in discrete areas in the rat brain following intraventricular injection. *Eur. J. Pharmacol.* **67**, 329–31.

Gate, J. S:, Bird, E. D., Spokes, E. G., Iversen, L. L., and Jessel, T. (1978). Human brain substance P: distribution in controls and Huntington's chorea. *J. Neurochem.* **20**, 663–4.

Geola, F. L., Yamada, T., Warwick, R. S., Tourtetotte, W. W., and Hershman, J. M. (1981). Regional distribution of somatostatin-like immunoreactivity in the human brain. *Brain Res.* **229**, 35–42.

Gibbs, J., Young, R. C., and Smith, G. P. (1973). Cholecystokinin decreases food intake in rats. *J. comp. Physiol. Psychol.* **84**, 488–95.

Gloor, P., Oliver, A., and Quesney, L. F. (1981). The role of the amygdala in the expression of psychic phenomena in temporal lobe seizures. In *The amygdaloid complex* (ed. Y. Ben-Ari), pp. 489–98. Elsevier/North Holland, Amsterdam.

Greenwood, R. S., Godar, S. E., Reaves, T. A., and Hayward, J. N. (1981). Cholecystokinin in Hippocampal pathways. *J. comp. Neurol.* **203**, 335–50.

Halgren, E. (1981). The amygdala contribution to emotion and memory: current studies in human. In *The amygdaloid complex* (ed. Y. Ben-Ari), pp. 395–408. Elsevier/North Holland, Amsterdam.

Handelmann, G. E., Meyer, D. K., Beinfeld, M. C., and Oertel, W. H. (1981). CCK-containing terminals in the hippocampus are derived from intrinsic neurons – an immunohistochemical and radioimmunological study. *Brain Res.* **224**, 180–4.

Haskins, J. T., Samson, W. K., and Moss, R. L. (1982). Evidence for vasoactive intestinal polypeptide (VIP) altering the firing rate of preoptic, septal and midbrain central gray neurons. *Regul. Pept.* **3**(2), 113–23.

Haubrich, D. R., Martin, G. E., Pflueger, A. B., and Williams, W. (1982). Neurotensin effects on brain dopaminergic systems. *Brain Res.* **231**, 216–21.

Hokfelt, T., Johansson, O. Ljungdahl, A., Lundberg, J. M., and Schultzberg, M. (1980*a*). Peptidergic neurones. *Nature* **284**, 515–21.

——, Rehfeld, J. F., Skirboll, L., Ivenmark, B., Goldstein, M., and Markey, K. (1980*b*). Evidence for co-existence of dopamine and CCK in meso-limbic neurones. *Nature* **285**, 476–8.

Inagaki, S., Shinoda, K., Kubota, Y., Shiosaka, S., Matsuzaki, T., and Tohyama, M. (1983). Evidence for the existence of a neurotensin-containing pathway from the endopiriform nucleus and adjacent prepiriform cortex to the anterior alfactory nucleus and nucleus of diagonal bond (Broca) of the rat. *Neuroscience* **8**, 487–95.

Ishikawa, K., Oh, T., and McGaugh, J. L. (1982). Evidence for dopamine as a transmitter in dorsal hippocampus. *Brain Res.* **232**, 222–6.

Iversen, L. L., Lee, C. M., Gilbert, R. F., Hunt, S., and Emson, P. C. (1980). Regulation of neuropeptide release. *Proc. R. Soc., Lond.* **B210**, 91–111.

Kelley, A. E., Stinus, L., and Iversen, S. D. (1979). Behavioural activation induced in the rat by substance P infusion into ventral tegmental area: implication of dopaminergic A10 neurones. *Neurosci. Lett.* **11**, 335–9.

Kelly, J. S. and Dodd, J. (1981). Cholecystokinin and gastrin as transmitters in the mammalian central nervous system in *Neurosecretion and brain peptides* (ed. J. B. Martin, S. Reichlin, and K. L. Bick), pp. 133–44. Raven Press, New York.

Kohler, C. and Steinbusch, H. (1982). Identification of serotonin and non-serotonin-containing neurons of the mid brain raphe projecting to the entorhinal area and the hippocampal formation. A combined immunohistochemical and fluorescent retrograde tracing study in the rat brain *Neuroscience* **7**, 951–75.

Lewis, P. R. and Shute, C . D. (1978). Cholinergic pathways in the CNS. In *Handbook of psychopharmacology* (ed. L. L. Iversen, S. D. Iversen, and S. H. Snyder), Vol. 9, pp. 315–55. Plenum, New York.

Lindvall, O. and Bjorklund, A. (1978). Organisation of catecholamine neurons in rat central nervous system. In *Handbook of psychopharmacology* (ed. L. L. Iversen, S. D. Iversen, and S. H. Snyder), Vol. 9, pp. 139–231. Plenum Press, New York.

Ljungdahl, A., Hokfelt, T., and Nilsson, G. (1978). Distribution of substance P-like immunoreactivity in the central nervous system of the rat. I. Cell bodies and nerve terminals. *Neuroscience* **3**, 861–943.

Loy, R., Koziell, D. A., Lindsey, J. D., and Moore, R. Y. (1980). Noradrenergic innervation of the adult rat hippocampal formation. *J. comp. Neurol.* **189**, 699–710.

Magistrett, P. S., Morrison, J. H., Shoemaker, W. J., Sapin, V., and Bloom, F. E. (1981). Vasoactive intestinal polypeptide induces glycogenolysis in mouse cortical slices: a possible regulatory mechanism for the local control of energy metabolism. *Proc. Nat. Acad. Sci. (USA)* **78**, 6535–9.

Meibach, R. C. and Katzman, R. (1981). Origin, course and termination of dopaminergic substantia nigra neurons projecting to the amygdaloid complex in the cat. *Neuroscience* **6**, 2159–71.

Meyer, D. K., Beinfeld, M. C., and Brownstein, M. J. (1982*a*). Corpus callosum lesions increase cholecystokinin concentrations in cortical areas with homeotopic connections. *Brain Res.* **240**, 151–3.

——, ——, Oertel, W. H., and Brownstein, M. J. (1982*b*). Origin of the cholecystokinin-containing fibres in the rat caudato–putamen. *Science* **215**, 187–8.

Moore, R. Y. and Halaris, A. E. (1975). Hippocampal innervation by serotonin neurons by the mid brain raphe in the rat. *J. comp. Neurol.* **164**, 171–84.

Nicol, R. A., Alger, B. E., and Jahr, C. E. (1980). Peptides as putative excitatory neurotransmitters:carnosine, enkephalin, substance P and TRH. *Proc. R. Soc., Lond.* **B210**, 133–49.

Onali, P., Schwartz, J. P., and Costa, E. (1981). Dopaminergic modulation of adenylate cyclase stimulation by vasoactive intestinal peptide in anterior pituitary. *Proc. Nat. Acad. Sci. (USA)* **78**, 6351–534.

Palkovits, M., Bessons, J., and Rotsztejn, W. (1981). Distribution of vasoactive intestinal polypeptide in intact stria terminals transected and cerebral cortex isolated rats. *Brain Res.* **213**, 455–9.

——, Tapia-Arancibia, L., Kordon, C., and Epelbaum, J. (1982). Somatostatin connections between the hypothalamus and the limbic system of the rat brain. *Brain Res.* **250**, 223–8.

Parent, A., Descarries, L., and Beaudet, A. (1981). Organisation of ascending serotonin systems in the adult rat brain. A radioautographic study after intraventricular administration of (H3) 5-hydroxytryptamine. *Neuroscience* **6**, 115–38.

Phillis, J. W. and Kirkpatrick, J. R. (1980). The actions of motilin, luteinizing hormone releasing hormone, cholecystokinin, somatostatin, vasoactive intestinal polypeptide and other peptides on rat cerebral cortical neurons. *Can. J. Physiol. Pharmacol.* **58**, 612–23.

Pittman, Q. J. and Siggins, G. R. (1981). Somatostatin hyperpolarises hippocampal pyramidal cells in vitro. *Brain Res.* **221**, 402–8.

Price, J. L. (1981). The efferent projections of the amygdaloid complex in the rat, cat and monkey. In *The amygdaloid complex* (ed. Y. Ben-Ari), pp. 121–32. Elsevier/North Holland, Amsterdam.

Roberts, G. W., Allen, Y. A., Crow, T. J., and Polak, J. M. (1983). Immunocytochemical localisation of peptides in the fornix of rat monkey and man. *Brain Res.* **263**, 151–5.

——, Crow, T. J., and Polak, J. M. (1981*a*). Neurotensin: first report of a cortical pathway. *Peptides* **2** (Suppl. 1), 37–43.

——, Polak, J. M., and Crow, T. J. (1981*b*). The peptidergic cirtuitry of the amygdaloid complex. In *The amygdaloid complex* (ed. Y. Ben-Ari), pp. 185–96. Elsevier/North Holland, Amsterdam.

——, Woodhams, P. L., Crow, T. J., and Polak, J. (1980). Loss of immunoreactive VIP in the bed nucleus following lesions of the stria terminalis. *Brain Res.* **195**, 471–6.

——, ——, Polak, J. M., and Crow, T. J. (1982). Neuropeptides in the limbic system of the rat: the amygdaloid complex. *Neuroscience* **7**, 99–131.

——, ——, ——, and —— (1984). The distribution of peptides in the limbic system of the rat: the hippocampal formation. *Neuroscience* [In press].

Sakanaka, M., Shiosaka, S., Takatsuki, K., Inagaki, S., Takagi, H., Semba, E., Kawai, Y., Matsuzaki, T., and Tohyama, M. (1981). Experimental immunohistochemical studies on the amygdalofugal peptidergic (substance P and somatostatin) fibres in the stria terminalis of the rat. *Brain Res.* **221**, 231–42.

Samson, W. K., Burton, K. P., Reeves, J. P., and McCann, S. M. (1981). Vasoactive intestinal peptide stimulates luteinizing hormone-releasing hormone release from median eminence synaptosomes. *Regul. Pept.* **2**, 253–64.

Scharrer, E. (1951). Neurosecretion X. A relationship between the paraphysis and the paraventricular nucleus in the garter snake. *Biol. Bull.* **101**, 106–13.

Skirboll, L. R., Grace, A. A., Hommer, D. W., Rehfeld, J., Goldstein, M., Hokfelt, T., and Bunney, S. (1981). Peptide–monoamine co-existence: studies of the actions of cholecystokinin-like peptide on the electrical activity of midbrain dopamine neurons. *Neuroscience* **6**, 2111–24.

Sorensen, K. V. (1982). Somatostatin: localisation and distribution in the cortex and the subcortical white matter of human brain. *Neuroscience* **7**, 122–33.

Vanderhagen, J. J., Lostra, F., De Mey, J., and Gilles, C. (1980). Immunocytochemical localisation of cholecystokinin and gastrin like peptides in the brain and hypophysis of the rat. *Proc. Nat. Acad. Sci. (USA)* **77** (2), 1190–4.

Vincent, S. R. and McGeer, E. G. (1981). A Substance P projection to the hippocampus. *Brain Res.* **215** (1–2), 349–51.

——, Kimura, H., and McGeer, E. G. (1981). Organization of substance-P fibres within the hippocampal formation demonstrated with a biotin–avidin immunoperoxidase technique. *J. comp. Neurol.* **199**, 113–23.

Walmsley, J. K., Young, W. S., and Kuhar, M. J. (1981). Immunohistochemical localisation of enkephalin in rat forebrain. *Brain Res.* **190**, 153–70.

Williams, R. G., Gayton, R. J., Zha, W., and Dockray, G. J. (1981). Changes in brain cholecystokinin octapeptide following lesions of the medial forebrain bundle. *Brain Res.* **213**, 227–30.

Woodhams, P. L., Roberts, G. W., Polak, J. M., and Crow, T. J. (1983). Distribution of neuropeptides in the limbic system of the rat: the bed nucleus of the stria terminalis, septum and pre-optic area. *Neuroscience* **8**, 677–703.

16

Alterations in neuropeptides in the limbic lobe in schizophrenia

I. N. FERRIER, T. J. CROW, G. W. ROBERTS, E. C. JOHNSTONE,
D. G. C. OWENS, Y. LEE, A. BARACESE-HAMILTON, G. McGREGOR,
D. O'SHAUGHNESSY, J. M. POLAK, AND S. R. BLOOM

INTRODUCTION

The belief that the primary disturbance in schizophrenia is chemical has long been plausible. Several neurohumeral hypotheses of schizophrenia have been advanced during the past 20 years. These have been tested in two main ways: first, by observing whether drugs with a known profile of neurochemical activity alter schizophrenic symptoms in a way that is predicted by the hypothesis and second, the measurement of appropriate biochemical indices in blood, urine, CSF, or post-mortem brains of schizophrenic patients. Many early hypotheses have been refuted in this way, but the dopamine overactivity hypothesis (Randrup and Munkvad 1972) is supported by some evidence. A consistent elevation in striatal dopamine receptors has been demonstrated in schizophrenic brains (Owen *et al.* 1978; Lee *et al.* 1978) and dopamine antagonists are therapeutically effective – an effect correlated with dopamine blockade (Crow *et al.* 1977; Johnstone *et al.* 1978).

However it is increasingly being recognized that neuroleptic drugs are not effective in treating the entire range of schizophrenic symptoms. Negative symptoms (loss of emotional responsitivity, poverty of speech, and loss of drive) are commonly seen in patients with chronic schizophrenia and such patients benefit much less from neuroleptic treatment (Letemendia and Harris 1967). It has also been demonstrated that these negative symptoms are associated with impaired cognitive and behavioural performance (Owens and Johnstone 1980) and also, perhaps, with abnormalities on CAT scanning (see Crow 1982 for recent references). These observations, taken together, have led to the hypothesis that two separate, but overlapping, syndromes of schizophrenia may be distinguishable – one characterized by drug-responsive positive symptoms which relate to dopamine overactivity and the other characterized by drug-unresponsive negative symptoms which may relate to structural changes in the brain (Crow 1980). The search for the pathophysiological basis of these syndromes is under way, particularly for the site of possible structural brain damage.

Extensive studies have revealed several associations between cerebral disease and schizophrenia. Davison and Bagley (1969) concluded that there was an association between cerebral tumours and injuries and schizophrenia, and more specifically, that this association was between lesions in the temporal lobe and schizophrenia. Several other observations suggest an association between schizophrenia and the temporal lobe. Slater *et al.* (1963) reported higher rates of schizophrenia-like psychoses among individuals whose EEG pattern showed abnormal electrical discharges from the temporal lobe than among epileptics without such a pattern, confirming Hill's (1948) and Pond's (1957) original observations. This increased incidence of schizophreniform psychoses in cases of temporal-lobe epilepsy (TLE) has recently been confirmed in a prospective study of Ounsted and Lindsay (1981) who also reported on the appearance of Schneiderian first-rank symptoms in these cases. Similarly Perez and Trimble (1980) have reported a very similar profile of psychopathology between schizo-phrenics and psychotic patients with TLE, particularly those with lesions of the left side. Malamud (1967) drew attention to the relationship between tumours of the limbic system and the development of schizophrenic symptoms.

Studies involving non-epileptic schizophrenics have also implicated the temporal lobe as a possible locus of dysfunction. Electroencephalographic evidence of temporal-lobe abnormalities in schizophrenic patients have been noted by Hill (1948), Abenson (1970), and Stevens *et al.* (1979). In addition, depth electrode studies by Heath (1954) revealed seizure-like discharges from the septal nuclei and hippocampus – two limbic structures associated with the temporal lobe.

The limbic system (temporal lobe, amygdala, and hippocampus) is often regarded as the area in which cognition, attention, and emotion are integrated. Retention deficits following amygdaloid and hippocampal lesions have been reported in primates (Kling 1972; Mahut *et al.* 1981) and following bilateral temporal-lobe surgery in man (Terzian and Ore 1955). In this latter study, features resembling negative symptoms were also observed. In view of the cognitive deficits of chronic schizophrenia (Crow and Stevens 1978; Owens and Johnstone 1980), all the evidence of associations between the temporal lobe and schizophrenia and the presence of dopamine terminals in the limbic system, a study of biochemical indices in the limbic lobes of post-mortem schizophrenics seemed appropriate.

A large number of neuronally-localized peptides are now known to exist in the brain (Hokfelt *et al.* 1980). These peptides have been found to be widely distributed within the central nervous system. In particular, cholecystokinin (CCK), somatostatin (SRIF), neurotensin (NT), vasoactive intestinal polypeptide (VIP), and substance P (SP) have been noted to have extensive and characteristic distribution patterns. Studies reported here (Chapter 15) and elsewhere (Roberts *et al.* 1981) have revealed complex but well-defined neuronal networks for these peptides in the limbic system of the rat. Numerous cell bodies are scattered throughout this area; however, for each peptide discrete localizations are discernible.

It is becoming recognized that the measurement of neuropeptides may provide information on the integrity of specific neuronal systems and that neuropeptides are thus useful markers in degenerative diseases (Rossor and

Emson 1982). For example, in Parkinson's disease, selective losses of met-enkephalin and CCK in the nigro-striatal system have been demonstrated (Tacquet *et al.* 1982; Studler *et al.* 1982) which appear to correlate well with the known loss of dopaminergic neurones in these areas in this condition. In Huntington's disease, marked reductions in SP, met-enkephalin, and CCK are found in the substantia nigra indicating damage to striatal efferents in this condition (Emson *et al.* 1980). In Alzheimer's disease, reductions in SRIF but not CCK or VIP have been demonstrated in several cortical areas including temporal cortex and hippocampus (Rossor and Emson 1982 for review). A report of reduced SP levels (Crystal and Davies 1982) in Alzheimer's disease probably applies to only the most severest cases. The reduction in SRIF in Alzheimer's disease correlates closely with a reduction in choline acetyltransferase (CAT). Thus the pattern of changes indicates relative preservation of certain populations of intrinsic cortical cells along with damage to the ascending cholinergic system. Further elucidation of this pattern will lead to a much greater understanding of the underlying pathophysiology of Alzheimer's disease.

Neuropeptide measurements on post-mortem material from schizophrenics have been few. Perry *et al.* (1981) reported normal levels of CCK and VIP in the entorhinal cortex of a small number of schizophrenic brains. Recently, Nemeroff *et al.* (1982) have made a preliminary report of reduced SRIF in frontal cortex and decreased NT in several cortical areas in schizophrenia. Two studies have recently reported on CSF levels of neuropeptides in schizophrenia. Widerlov *et al.* (1982) have reported subnormal CSF levels of NT in a subgroup of schizophrenics and their normalization after neuroleptic treatment. Wood *et al.* (1982) have found normal SRIF levels in CSF in schizophrenia. However there is considerable uncertainty about the significance of these findings since the relationship between neuropeptide levels in lumbar CSF and levels in discrete brain areas is far from clear.

THE PRESENT STUDY

We have recently measured the distribution of five neuropeptides (CCK, NT, SRIF, SP, and VIP) in brains from schizophrenics and appropriate controls. The schizophrenic patients had all been rated in life which made it possible to relate peptide changes to clinical state. A preliminary report of this work has appeared elsewhere (Roberts *et al.* 1982) and the results presented here are in outline only and with particular reference to changes in the limbic system.

The sample studied consisted of brains from 14 schizophrenic patients and 12 control subjects. The schizophrenics conformed to the criteria of Feighner *et al.* (1972) and mental state had been assessed during life on the Modified Inpatient Rating Scale derived by Krawiecka *et al.* (1977). From these ratings a score for positive symptoms (delusions, hallucinations, incoherence of speech, and incongruity of affect) and for negative symptoms (poverty of speech and flattening of affect) was given to each patient. The schizophrenics were divided into two subgroups on the basis of the presence or absence of negative symptoms. One group (designated type I; $n = 7$) had only positive symptoms whereas the other group (type II; $n = 7$) with positive symptom scores similar to the type I group had negative symptoms to a morbid degree on clinical examination in life. Control brains were obtained from patients with no history of psychiatric or neurological illness. The groups were matched for age, sex ratio, death-to-autopsy delay, and storage of sample time. Five of the schizophrenic patients had received no neuroleptic medication for at least one year prior to death.

Ten brain areas were examined: frontal, parietal, cingulate and temporal cortex, hippocampus, amygdala, globus pallidus, putamen, and lateral and dorso-medial thalamus. Tissue samples were extracted by boiling in 0.5 M acetic acid followed by homogenization. An aliquot of the stirred homogenate was neutralized prior to the CCK assay. CCK, SRIF, NT, SP, and VIP were measured by radioimmunoassay. Methods for each of these assays have been published elsewhere (Bloom and Long 1982; Lee *et al.* submitted.)

RESULTS

The distribution profiles of these neuropeptides in schizophrenic and control brains are shown in Fig. 16.1 (CCK), Fig. 16.2 (SRIF), Fig. 16.3 (NT), Fig. 16.4 (SP), and Fig. 16.5 (VIP). As can be seen, each peptide has a characteristic

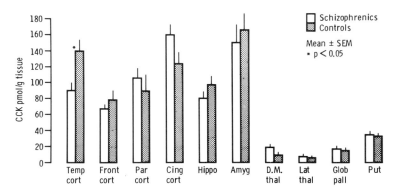

Fig. 16.1. CCK levels (pmol g^{-1} tissue) in 10 CNS areas in 14 schizophrenics and 12 controls.

distribution pattern within the CNS. It was found that there were no effects of delay between death and post-mortem on peptide levels confirming previous observations (e.g. Cooper *et al.* 1981; Emson *et al.* 1980) that these peptides are stable post-mortem.

In addition to this overall comparison between schizophrenic and control brains, the two schizophrenic subgroups (type I and type II) were compared with each other and to controls. Differences observed are summarized in Fig. 16.6 (CCKand SRIF) and Fig. 16.7 (VIP and SP). Statistical anaylsis was performed by analysis of covariance but a more complex multivariate analysis is underway.

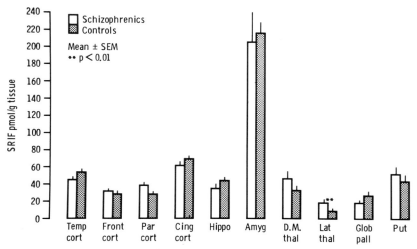

Fig. 16.2. SRIF levels (pmol g^{-1} tissue) in 10 CNS areas in 14 schizophrenics and 12 controls.

The important results were as follows

1. *CCK.* CCK was significantly reduced in the temporal cortex of the total schizophrenic group (Fig. 16.1) and in the hippocampus and amygdala of the type II schizophrenic group (Fig. 16.6).

2. *SRIF.* SRIF was elevated in the total schizophrenic group in the lateral thalamus (an area of very low peptide levels; Fig. 16.2). SRIF was reduced in the hippocampus of the type II schizophrenic group (Fig. 16.6).

3. *NT.* There was no difference in NT levels in the total schizophrenic group or either subgroup (Fig. 16.3).

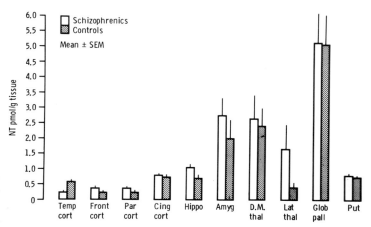

Fig. 16.3. NT levels (pmol g^{-1} tissue) in 10 CNS areas in 14 schizophrenics and 12 controls.

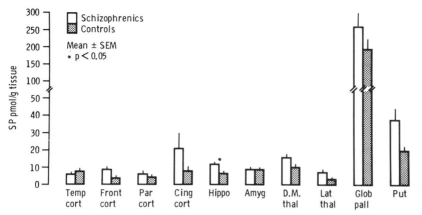

Fig. 16.4. SP levels (pmol g^{-1} tissue) in 10 CNS areas in 14 schizophrenics and 12 controls.

4. *SP.* SP was elevated in the hippocampus of the total schizophrenic group but there were no type I/type II differences (Fig. 16.7).

5. *VIP.* VIP was significantly elevated in the amygdala in type I schizophrenics (Fig. 16.7).

between those patients who had received neuroleptics up until the time of death and those who had been drug-free.

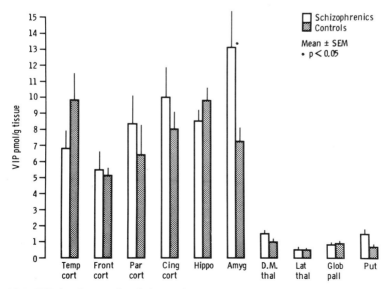

Fig. 16.5. VIP levels (pmol g^{-1} tissue) in 10 CNS areas in 14 schizophrenics and 12 controls.

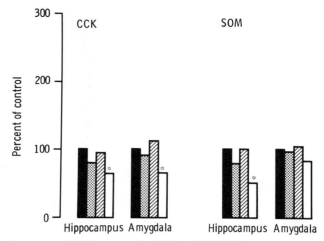

Fig. 16.6. Summary of differences in CCK and SRIF between total schizophrenic group (▒), type I (▨), and type II (□) schizophrenics as a percentage of controls (■) in amygdala and hippocampus. *$p < 0.05$.

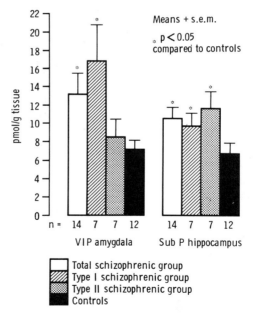

Fig. 16.7. Summary of differences in VIP and SP between the total schizophrenic group, type I and type II schizophrenics, and controls.

COMMENT

The differences between schizophrenics and controls and between the two subgroups of schizophrenics were almost completely confined to the limbic lobe. No differences in peptide concentrations were demonstrated in cortical, striatal, or thalamic areas (with the exceptions noted at points 1 and 2 above).

The pattern of change reported here is quite distinct from those reported for Alzheimer's disease, Parkinson's disease, or Huntington's disease. For example, widespread loss of cortical SRIF is noted in Alzheimer's disease but CCK levels are unchanged. While SRIF was reduced in hippocampus of schizophrenics with negative symptoms (type II schizophrenics), other cortical areas were unaffected and changes in CCK were noted. Neuropeptide changes in Parkinson's disease and Huntington's disease are limited to nigrostriatal areas; striatal levels of peptides were normal in the schizophrenic brain.

The findings of a reduction in CCK in temporal cortex in the schizophrenic patients contrasts with a previous report of Perry *et al.* (1981) of normal CCK in a small number of schizophrenic brains in entorhinal cortex – an area more medial than temporal cortex as sampled here. The temporal cortex is a possible source of efferents to the hippocampus and amygdala. However the deficits in these areas were confined to the group of patients with negative symptoms. This is of interest since this represents the first evidence of a biochemical abnormality associated with negative symptoms whose presence may identify a group of patients with poor outcome, intellectual impairment and neurological features (Owens and Johnstone 1980) and, perhaps, structural changes in the brain (Crow 1982). CCK has been recently shown to be co-localized with dopamine in some cells in the limbic lobe (Skirboll *et al.* 1981). This raises the possibility that the changes in CCK in the schizophrenic limbic lobe relate to changes in dopamine levels. However, the CCK content of the amygdala is highest in areas where there is sparse dopamine innervation and the hippocampus contains very little dopamine. Thus the changes in CCK reported here are unlikely to be related to dopaminergic overactivity.

VIP in the amygdala arises from both intrinsic and extrinsic sources. There are VIP-containing pathways connecting the amygdala with the temporal cortex, striatum, and hippocampus (Roberts *et al.* 1982). Since the VIP content of these areas was unaltered, it is likely that the VIP elevation found in the schizophrenic patients is due to changes in intrinsic amygdaloid VIP. It is conceivable that changes in VIP in patients with positive symptoms are a consequence of changes in the dopaminergic input to the lateral amygdala where large numbers of VIP-containing cell bodies are found in the rat (Roberts *et al.* 1982).

The finding of normal NT levels in schizophrenia contrasts with the findings of Nemeroff *et al.* (1982). The reasons for this difference are not clear at present. The findings of the present study cast doubt on the significance of the findings of reduced NT in CSF in a subgroup of schizophrenic patients (Widerlov *et al.* 1982) especially as this subgroup was not clinically distinct.

This study demonstrates differences in peptide content in schizophrenia occurring almost exclusively in the limbic lobe. The pattern of changes was different in two clinical subgroups of patients, where this difference in pattern depended on the presence or absence of negative symptoms. This may indicate different mechanisms underlying these two syndromes.

DISCUSSION

The precise significance of changes in neuropeptide levels in neuropsychiatric illness must await more detailed knowledge of their physiology. There is good evidence that many of the neuropeptides (including those studied here) are stored in vesicles, are released by depolarizing stimuli in a calcium-dependent fashion, and have electrophysiological effects on neurones (Emson 1979). These facts, together with the increasing evidence for co-localization of peptides with classical neurotransmitters, argue for their classification as neurotransmitters or neuromodulators. The behavioural effects of many peptides have been demonstrated when injected systemically, intraventicularly, or intracerebrally. However, specific roles for each peptide in discrete brain regions have not been established. Similarly, the effects or implications of a change in peptide concentration in a brain area are uncertain and, at present, we have no knowledge of whether a change in production, storage, or secretion is involved.

Nevertheless, specific patterns of changes in neuropeptides have been recognized in degenerative neuropsychiatric conditions (Rossor and Emson 1982). The pattern of changes in schizophrenia is not yet as clearly established. However, changes in several peptides were seen in the limbic lobe in schizophrenic brain in the present study and the changes were almost completely selective for this area. As pointed out in the introductory section, there is circumstantial evidence which points to an abnormality in limbic structures in schizophrenia. It is clear that much work must be done before any relationship between changes in peptides in the limbic lobe in schizophrenia and the psychosis of temporal-lobe epilepsy, EEG evidence of disturbance in the temporal lobe in schizophrenia, and the cognitive impairments of chronic schizophrenia is established.

Some associations between clinical state and limbic peptide changes were established in the present study. Changes in CCK in hippocampus and amygdala and SRIF in hippocampus were found in those patients with negative symptoms. There is evidence that these symptoms do not improve with neuroleptic medication and may be a marker of structural changes in schizophrenic brain. On the other hand VIP changes in the amygdala were found in patients who had only positive symptoms. These symptoms are neuroleptic-reversible and may relate to increases in dopamine receptors in schizophrenic brain. Thus the differential pattern of these peptide changes with these different symptom complexes may indicate that the mechanism of the disturbance underlying them is separate and distinct.

ACKNOWLEDGEMENTS

We thank the psychiatrists and pathologists for obtaining the tissue, especially Prof D. Hume Adams, Dr R. M. McCreadie, Dr A. Price, and Dr G. Slavin. We are grateful to Mr R. Lofthouse, Mr G. Locker, and the administrative, medical, and nursing staff of Shenley Hospital, Herts. for their assistance in collecting and storing this material. The support of the Wellcome Trust (INF) and of the MRC is gratefully acknowledged.

REFERENCES

Abenson, M. H. (1970). EEG's in chronic schizophrenia. *Br. J. Psychiat.* **115**, 421–5.
Bloom, S. R. and Long, W. L. (1982). *Radioimmunoassay of regulatory peptides.* Saunders, Philadelphia.
Cooper, P. E., Fernstrom, M. H., Rostad, O. P., Leeman, S. E., and Martin, J. B. (1981). The regional distribution of somatostatin, substance P and neurotensin in human brain. *Brain Res.* **219**, 219–32.
Crow, T. J. (1980). Molecular pathology of schizophrenia: more than one disease process? *Br. med. J.* **280**, 66–8.
—— (1982). Institutionalisation and the defects of schizophrenia. *Br. J. Psychiat.* **140**, 212–13.
——, Deakin, J. F. W., and Longden, A. (1977). The nucleus accumbens – possible site of antipsychotic action of neuroleptic drugs. *Psychol. Med.* **7**, 213–21.
—— and Stevens, M. (1978). Age disorientation in chronic schizophrenia: the nature of the cognitive deficit. *Br. J. Psychiat.* **133**, 137–42.
Crystal, H. A. and Davies, P. (1982). Cortical substance P-like immunoreactivity in cases of Alzheimer's disease and senile dementia of the Alzheimer's type. *J. Neurochem.* **38**, 1781–4.
Davison, K. and Bagley, C. R. (1969). Schizophrenia and the limbic system. In *Current problems in neuropsychiatry* (ed. R. N. Herington). Special publication, No. 4. *Br. J. Psychiat.*, Ashford.
Emson, P. C. (1979). Peptides as neurotransmitter candidates in the mammalian CNS. In *Progress in neurobiology* (ed. J. Phillis and J. Kirkaert), Vol. 13, pp. 61–116. Pergamon Press, London.
——, Rehfeld, J. F., Langevin, H., and Rossor, M. (1980). Reduction in CCK-like immunoreactivity in the basal ganglia in Huntington's disease. *Brain Res.* **198**, 497–500.
Feighner, J. P., Robins, E., Guze, S B., Woodruff, A., Winokur, G., and Munoz, R. (1972). Diagnostic criteria for use in psychiatric research. *Arch. gen. Psychiat.* **26**, 57–63.
Heath, R. G. (1954). *Studies in schizophrenia.* Harvard University Press. Cambridge, Massachusetts.
Hill, D. (1948). The relationship between epilepsy and schizophrenia. EEG studies. *Folia Psychiat. Neurol. Neurochir.* **51**, 95–111.
Hokfelt, T., Johansson, O., Ljungdahl, A., Lundberg, J. M., and Schultzberg, M. (1980). Peptidergic neurones. *Nature* **284**, 515–21.
Johnstone, E. C., Crow, T. J., Frith, C. D., Carney, M. W. P., and Price, J. S. (1978). Mechanism of the antipsychotic effect in the treatment of acute schizophrenia. *Lancet* **i**, 848–51.
Kling, A. (1972). Effects of amygdalectomy on social–affective behaviour in non-human primates. In *The Neurobiology of the amygdala* (ed. B. E. Eleftheriou), pp. 511–36. Plenum, New York.
Krawiecka, M., Goldberg, D., and Vaughan, M. (1977). A standardised psychiatric assessment for rating chronic psychotic patients. *Acta psychiat. scand.* **55**, 299–308.
Lee, Y. C., Roberts, G. W., Ferrier, I. N., Waddington, J. L., Crow, T. J., Ghatei, M. A., Bacarese-Hamilton, A. J., and Bloom, S. R. Failure of neuroleptic administration to affect the peptide content of the rat brain. *Life Sci.* (submitted).
Lee, T., Seeman, P., Tourtellotte, W. W., Farley, I. J., and Hornykiewicz, O. (1978). Binding of 3_H-neuroleptics and 3H-apomorphine in schizophrenic brain. *Nature* **274**, 897–900.
Letemendia, F. J. J. and Harris, A. D. (1967). Chlorpromazine and the untreated chronic schizophrenic: a long term trial. *Br. J. Psychiat.* **113**, 950.
Mahut, H., Moss, M., and Zola-Morgan, S. (1981). Retention deficits after combined amygdalo–hippocampal and selective hippocampal resections in the monkey. *Neuropsychologia* **19**, 201–25.

Malamud, N. (1967). Psychiatric disorder with intracranial tumours of the limbic system. *Arch. Neurol.* **17**, 113–23.

Nemeroff, C. B., Manberg, R. J., Youngblood, W. W., Prange, A. J., and Kizer, J. S. (1982). Alterations in regional brain levels of neurotensin thyrotrophin releasing hormone and somatostatin in schizophrenia and Huntington's chorea. Abstract 44.6. American Society for Neuroscience, Minneapolis, Minnesota.

Ounsted, C. and Lindsay, J. (1981). The long term outcome of temporal lobe epilepsy in childhood. In *Psychiatry and epilepsy*. (ed. E. H. Reynolds and M. R. Trimble), pp. 185–215. Churchill Livingstone, Edinburgh.

Owen, F., Cross, A. J., Crow, T. J., Longden, A., Poulter, M., and Riley, G. J. (1978). Increased dopamine-receptor sensitivity in schizophrenia. *Lancet* **ii**, 223–6.

Owens, D. G. C. and Johnstone, E. C. (1980). The disabilities of chronic schizophrenia – their nature and factors contributing to their development. *Br. J. Psychiat.* **136**, 384–95.

Perez, M. M. and Trimble, M. R. (1980). Epileptic psychoses – diagnostic comparison with process schizophrenia. *Br. J. Psychiat.* **137**, 245–9.

Perry, R. H., Dockray, G. J., Dimaline, R., Perry, E. K., Blessed, G., and Tomlinson, B. E. (1981). Neuropeptides in Alzheimer's Disease, depression and schizophrenia. *J. Neurol. Sci.* **51**, 465–72.

Pond, D. A. (1957). Psychiatric aspects of epilepsy. *J. Ind. med. Prof.* **3**, 1441–51.

Randrup, A. and Munkvad, I. (1972). Evidence indicating an association between schizophrenia and dopaminergic hyperactivity in the brain *Orthomol. Psychiat.* **1**, 2–7.

Roberts, G. W., Ferrier, I. N., Lee, Y. C., Adrian, T. E., O'Shaughnessey, D. J., Crow, T. J., Polak, J. M., and Bloom, S. R. (1982). Schizophrenia – peptidergic deficits in the limbic system. *Regul. Peptides* **3**, 81.

——, Polak, J. M., and Crow, T. J. (1981). The peptidergic circuitry of the amygdaloid complex. In *The amygdaloid complex* (ed. Y. Ben-Ari), pp. 185–96. Elsevier/North Holland, Amsterdam.

Rossor, M. N. and Emson, P. C. (1982). Neuropeptides in degenerative disease of the central nervous system. *Trends Neurosci.* **5**, 399–401.

Skirboll, L. R., Grace, A. A., Hommer, D. W., Rehfeld, J., Goldstein, M., Hokfelt, T., and Bunney, B. S. (1981). Peptide–monoamine co-existence: studies of the actions of cholecystokinin-like peptide on the electrical activity of midbrain dopamine neurons. *Neuroscience* **6**, 2111–24.

Slater, E., Beard, A. W., and Glithero, E. (1963). The schizophrenia-like psychoses of epilepsy. *Br. J. Psychiat.* **109**, 95–150.

Stevens, J. R., Bigelow, L., Denney, D., Lipkin, J., Livermore, A. H., Rauscher, F., and Wyatt, R. J. (1979). Telemetered EEG–EOG during psychotic behaviours of schizophrenica. *Arch. gen. Psychiat.* **26**, 251–263.

Studler, J. M., Javoy-Agid, F., Cesselin, F., Legrand, J. C., and Agid, Y. (1982). CCK-8 immunoreactivity distribution in human brain: selective decrease in the substantia nigra from Parkinsonian patients. *Brain Res.* **243**, 176–9.

Tacquet, H., Javoy-Agid, F., Cesselin, F., Hamon, M., Legrand, J. C., and Agid, Y. (1982). Microtopography of methionine–enkephalin, dopamine and noradrenaline in the ventral mesencephalon of human control and Parkinsonian brain. *Brain Res.* **235**, 303–14.

Terzian, H. and Ore, G. D. (1955). Syndrome of Kluver and Bucy: reproduced in man by removal of the temporal lobes. *Neurology, Minneapolis* **5**, 373–80.

Widerlow, E., Lindstrom, L. H., Beseve, G., Manberg, P. J., Nemeroff, C. B., Breese, G. R., Kizer, J. S., and Prange, A. J. (1982). Subnormal CSF levels of neurotensin in a subgroup of schizophrenic patients: normalization after neuroleptic treatment. *Am. J. Psychiat.* **139**, 1122–6.

Wood, P. L., Etienne, P., Lal, S., Gauthier, S., Cegal, S., and Nair, N. P. V. (1982). Reduced lumbar CSF somatostatin levels in Alzheimer's disease. *Life Sci.* **31**, 2073–9.

17

The limbic system and neurosecretion

D. A. POULAIN, C. J. LEBRUN, D. T. THEODOSIS, AND J. D. VINCENT

INTRODUCTION

Numerous studies have suggested a participation of the limbic system in the control of the activity of the magnocellular system of the hypothalamus. However, the magnocellular system is not a homogeneous structure, either anatomically or functionally. It is composed of four anatomically distinct nuclei, the paraventricular and supraoptic nuclei, and within each nucleus are found neurosecretory cells secreting either vasopressin or oxytocin. Vasopressin intervenes in the control of water balance and blood pressure, whilst oxytocin participates in the control of parturition and lactation, and it is therefore important to consider their physiology separately. The two hormones are involved in many neuroendocrine reflexes and, although we are well informed about some of the factors or stimuli implicated in their release, our knowledge of the afferent pathways in the central nervous system is still very poor. With respect to the limbic system, its influence on the magnocellular system has been known for a long time, but it has never been clearly established whether the limbic system is an integral part of the main afferent pathways, or whether it has to be considered as a structure operating in parallel. In relation to this question, we will here discuss a few observations limited to the septum, a major relay in the limbic system.

CONNECTIONS BETWEEN THE SEPTUM AND THE MAGNOCELLULAR SYSTEM

To determine whether a structure has a direct or an indirect influence on another structure, it is important to be precise as to whether or not direct neural connections exist between the two. The existence of reciprocal connections between the septum and the magnocellular nuclei lends some support to the hypothesis of a control of the activity of the latter by the former.

On the one hand, there are direct projections from septal neurones to both the supraoptic and paraventricular nuclei. This was first suggested by classical methods showing degeneration of terminals in the magnocellular nuclei after septal lesions (Powell and Rorie 1967; Tangapregassom *et al.* 1974; Zaborsky *et al.* 1975) and has been more directly demonstrated by recent techniques using anterograde and retrograde transport of tracers (Garris 1979; Tribollet and Dreifuss 1981). Electrophysiological studies corroborate these anatomical observations; an important proportion of septal neurones are activated antidromically upon stimulation of the supraoptic or paraventricular nuclei

(Poulain *et al.* 1981; Pittman *et al.* 1981). These projections are essentially ipsilateral.

On the other hand, immunohistochemical studies have revealed the existence of fibres immunoreactive to both vasopressin and oxytocin in the lateral septum (Buijs 1978; Weindl and Sofroniew 1976). These fibres, however, do not appear to arise directly from the magnocellular nuclei since no neurosecretory cell projecting to the neurohypophysis could be activated antidromically by electrical stimulation of the septum (Poulain *et al.* 1980; Pittman *et al.* 1981). It cannot be excluded that some neurosecretory cells project exclusively to other central structures, without projecting to the neurohypophysis, in which case they would have been overlooked by conventional electrophysiological studies in which neurosecretory cells are identified by antidromic stimulation from the neurohypophysis. It has been suggested recently that the immunoreactive septal fibres may originate from the bed nucleus of the stria terminalis (Hoornemann and Buijs 1982). Whatever the origin of these fibres, their presence in the lateral septum raises the question of their function, and suggests that the vasopressin- and oxytocin-releasing systems may exert a feedback on their control structures.

SEPTAL INFLUENCE ON VASOPRESSIN SECRETION

Vasopressin exerts a major role in water-balance regulation by its antidiuretic action on the kidney, and in blood-pressure regulation by its pressor effect on arterioles. The limbic system seems to participate in the control of vasopressin release – directly by acting on the vasopressin system itself and indirectly through its role in thirst and drinking (see Swanson and Mogenson 1981). For instance, destruction of the medial septum reduces both the basal levels of vasopressin and also the amount of vasopressin released in response to osmotic stimulation (Iovino *et al.* 1983). On the other hand, septal lesions produce polydipsia (Harvey and Hunt 1965).

It is now well established that the release of vasopressin depends on the electrical activity of the vasopressinergic neurones (for a review, see Poulain and Wakerley 1982). We have tried therefore to establish the importance of the septum in the regulation of vasopressin secretion by studying directly the electrical activity of vasopressinergic neurones after lesion or stimulation of the septal area. Massive bilateral septal lesions do not prevent vasopressinergic neurones from evolving their phasic pattern of action potentials, a form of electrical activity that is particular to vasopressin cells during hormone release. The phasic pattern does not therefore depend entirely on septal control. Nevertheless, single-pulse septal stimulation does cause a short inhibition lasting 60–80 ms after each shock, as seen on post-stimulus time histograms, and is capable of altering the phasic activity of vasopressinergic neurones (Poulain *et al.* 1980).

The various limbic structures may play other roles *vis-à-vis* vasopressin release. Single-pulse stimulation of the amygdala excites magnocellular neurones and, in particular, is capable of triggering prematurely a burst of action potentials in phasically firing neurones (Thomson 1982). These data point therefore to a modulatory role of the septum on vasopressin release, whereby limbic activity would modify the rate of hormone release by influencing the parameters of electrical activity of the vasopressinergic neurones.

CONTROL OF OXYTOCIN RELEASE

The release of oxytocin during parturition and lactation is well known. In recent years, considerable progress has been made toward understanding the pattern of activation of the oxytocin system during suckling and the milk-ejection reflex. In the rat, oxytocin release during suckling is pulsatile and occurs intermittently, every 5–15 min, even though the suckling stimulus from the pups is continuous. Oxytocin neurones in the hypothalamus usually exhibit a slow irregular pattern of electrical activity (1–3 spikes s^{-1}). However, in the few seconds preceding each milk ejection, the neurones display a striking high-frequency discharge of action potentials, up to 60–100 spikes s^{-1}, lasting 3 to 4 s (Fig. 17.1(A); for review, see Poulain and Wakerley 1982).

The limbic system appears implicated in the control of oxytocin release, since milk ejection can be inhibited by stress and emotional disturbances, or can be conditioned to visual or auditory signals. Furthermore, lesions of various areas of the limbic system disturb maternal behaviour and lactation (for a review, see Lebrun *et al.* 1983). The lactating rat has proved to be a very convenient model for studying septal influences on the milk-ejection reflex, because in this species the reflex persists even though the mother is anaesthetized, and it is therefore possible to eliminate behavioural components and to distinguish each milk ejection by an intramammary pressure recording after cannulation of one mammary gland. Massive bilateral destruction of the septum did not prevent the reflex from taking place, nor did it stop the reflex once it was set. Furthermore, massive bilateral destruction of the septum did not alter significantly the amplitude, time course, and rhythm of the milk ejections, nor did it affect the electrical activity of oxytocin neurones during suckling (Fig. 17.1(B)). We also investigated the long-term effect of maternal septal lesions on the growth of the litter. Lesions were made on the third day post-partum, and the body weights of the litters were measured daily. Contrary to other reports, we never observed any sign of the septal rage syndrome in the mothers, nor any cannibalism of the young. The growth of the litters was identical in litters from lesioned mothers and from intact rats (Lebrun *et al.* 1983).

Nonetheless, the septum may play an inhibitory role on the reflex. Single-pulse septal stimulation resulted in an inhibition of the electrical activity of oxytocinergic neurones in the supraoptic nucleus, as seen on post-stimulus time histograms (Poulain *et al.* 1980). Prolonged stimulation (for 20 min at 5 and 10 Hz) of the septum in anaesthetized animals caused a complete interruption of milk ejection during the time of stimulation and for 7–27 min after cessation of stimulation (Fig. 17.1(B)). At 1 Hz, the effect was less obvious, but the intervals between successive milk ejections during the period of stimulation were significantly increased. When septal stimulation was applied for 1 min just after a milk ejection had occurred, the following milk ejection was significantly

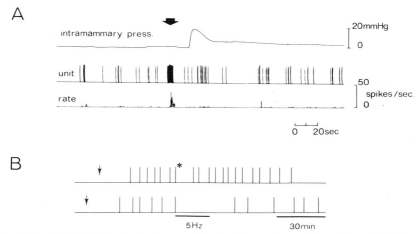

Fig. 17.1. (A) Milk ejection in the anaesthetized lactating rat. Recording of the intra-mammary pressure during suckling shows the brief increase in pressure observed at the time of a milk ejection. Below: the electrical activity of an oxytocin cell is shown on a polygraph recording (unit) and the concomittent ratemeter record (rate). Note the high-frequency discharge of action potentials occurring a few seconds before milk ejection (arrow). (B) Effects of bilateral lesions and stimulation in the septum on the milk-ejection reflex. Each horizontal line represents one experiment where each vertical line indicates one milk ejection, once the pups started suckling (arrow). Top: after the seventh milk ejection, bilateral lesions were made by passing a radiofrequency current. Note the absence of any change in the pattern of the milk ejections. Below: after the sixth milk ejection, a train of electrical stimuli (1 ms, 0.5 mA, 5 Hz) was applied bilaterally for 20 min. Note the inhibition of the reflex which reappeared 18 min after the end of stimulation.

delayed at 5–10 Hz, but not at 1 Hz (Lebrun *et al.* 1983). Septal stimulation therefore, when applied at a minimum frequency and for a minimum time, can exert an inhibitory action, which seems to bear more on the periodicity of the reflex than on the milk ejections themselves.

CONCLUSION

Among the neuroendocrine systems of the hypothalamus, oxytocin and vaso-pressin neurones offer unique advantages for the study of the afferent control of their activity. Representing final common pathways of the many neuro-endocrine reflexes by which they are set into activity, their location within the hypothalamus is clearly defined. Their activity can be easily manipulated by various stimuli and can be evaluated either by assaying hormone levels in blood or by recording the cells' electrical activity, which presents very characteristic patterns depending on the type of neurone.

Analysis of the influence of the septum on vasopressin or oxytocin secretion shows that, for both hormonal systems, the septum is not an essential structure, since the activity of the two types of neurone still persists after destruction of the septum. However, the septum, and probably other limbic structures, appear to operate in parallel with the main pathways, so that their activity modulates the effect of the main input to the neurones.

REFERENCES

Buijs, R. M. (1978). Intra and extrahypothalamic vasopressin and oxytocin pathways in the rat. *Cell Tissue Res.* **192**, 423–35.

Garris, D. R. (1979). Direct septo–hypothalamic projections in the rat. *Neurosci. Lett.* **13**, 83–90.

Harvey, J. A. and Hunt, H. F. (1965). Effect of septal lesions on thirst in the rat as indicated by water consumption and operant responding for water reward. *J. comp. Physiol. Psychol.* **59**, 49–56.

Hoornemann, E. M. D. and Buijs, R. M. (1982). Vasopressin fibers in the rat brain following suprachiasmatic nucleus lesioning. *Brain Res.* **243**, 235–41.

Iovino, M., Poenaru, S., and Annunziato, L. (1983). Basal and thirst-evoked vaso-pressin secretion in rats with electrolytic lesion of the medioventral septal area. *Brain Res.* **258**, 123–6.

Lebrun, C. J., Poulain, D. A., and Theodosis, D. T. (1983). The role of the septum in the control of the milk ejection reflex in the rat: effects of lesions and electrical stimulation. *J. Physiol., Lond.* **339**, 17–31.

Pittman, Q. J., Blume, H. W., and Renaud, L. P. (1981). Connections of the hypo-thalamic paraventricular nucleus with the neurohypophysis, median eminence, amygdala, lateral septum and midbrain periaqueductal gray: an electrophysiological study in the rat. *Brain Res.* **215**, 15–28.

Poulain, D. A., Ellendorff, F., and Vincent, J. D. (1980). Septal connections with identified oxytocin and vasopressin neurones in the supraoptic nucleus of the rat. An electrophysiological investigation. *Neuroscience* **5**, 379–87.

——, Lebrun, C. J., and Vincent, J. D. (1981). Electrophysiological evidence for con-nections between septal neurones and the supraoptic nucleus of the hypothalamus of the rat. *Exp. brain Res.* **42**, 260–8.

—— and Wakerley, J. B. (1982). Electrophysiology of hypothalamic magnocellular neurones secreting oxytocin and vasopressin. *Neuroscience* **7**, 773–808.

Powell, E. W. and Rorie, D. K. (1967). Septal projections to nuclei functioning in oxytocin release. *Am. J. Anat.* **120**, 605–10.

Swanson, L. W. and Mogenson, G. J. (1981). Neural mechanisms for the functional coupling of autonomic, endocrine and somatomotor responses in adaptive behavior. *Brain Res. Rev.* **3**, 1–34.

Tangapregassom, A. M., Tangapregassom, M. J., Soulairac, A., and Soulairac, M. L. (1974). Effets des lésions septales sur l'ultrastructure du noyau supraoptique. *Ann. Endocrinol., Paris* **35**, 149–52.

Thomson, A. M. (1982). Responses of supraoptic neurones to electrical stimulation of the medial amygdaloid nucleus. *Neuroscience* **7**, 2197–205.

Tribollet, E. and Dreifuss, J. J. (1981). Localization of neurones projecting to the hypothalamic paraventricular nucleus area of the rat: a horseradish peroxidase study. *Neuroscience* **6**, 1315–28.

Weindl, A. and Sofroniew, M. V. (1976). Demonstration of extrahypothalamic peptide secreting neurons. A morphologic contribution to the investigation of psychotropic effects of neurohormones. *Pharmacopsychologia* **9**, 226–34.

Zaborsky, L., Leranth, Cs., Makara, G. B., and Palkovits, M. (1975). Quantitative studies on the supraoptic nucleus in the rat. II – Afferent fiber connections. *Exp. brain Res.* **22**, 525–40.

Index